MAPPING NANOTECHNOLOGY INNOVATIONS AND KNOWLEDGE

Global and Longitudinal Patent and Literature Analysis

INTEGRATED SERIES IN INFORMATION SYSTEMS

E-BUSINESS MANAGEMENT: *Integration of Web Technologies with Business Models*/ edited by Michael J. Shaw
VIRTUAL CORPORATE UNIVERSITIES: *A Matrix of Knowledge and Learning for the New Digital Dawn*/ Walter R.J. Baets & Gert Van der Linden
SCALABLE ENTERPRISE SYSTEMS: *An Introduction to Recent Advances*/ edited by Vittal Prabhu, Soundar Kumara, Manjunath Kamath
LEGAL PROGRAMMING*: Legal Compliance for RFID and Software Agent Ecosystems in Retail Processes and Beyond*/ Brian Subirana and Malcolm Bain
LOGICAL DATA MODELING: *What It Is and How To Do It*/ Alan Chmura and J. Mark Heumann
DESIGNING AND EVALUATING E-MANAGEMENT DECISION TOOLS: *The Integration of Decision and Negotiation Models into Internet-Multimedia Technologies*/ Giampiero E.G. Beroggi
INFORMATION AND MANAGEMENT SYSTEMS FOR PRODUCT CUSTOMIZATION/ Thorsten Blecker et al
MEDICAL INFORMATICS: *Knowledge Management and Data Mining in Biomedicine*/ edited by Hsinchun Chen et al
KNOWLEDGE MANAGEMENT AND MANAGEMENT LEARNING: *Extending the Horizons of Knowledge-Based Management*/ edited by Walter Baets
INTELLIGENCE AND SECURITY INFORMATICS FOR INTERNATIONAL SECURITY: *Information Sharing and Data Mining*/ Hsinchun Chen
ENTERPRISE COLLABORATION: *On-Demand Information Exchange for Extended Enterprises*/ David Levermore & Cheng Hsu
SEMANTIC WEB AND EDUCATION/ Vladan Devedžić
INFORMATION SYSTEMS ACTION RESEARCH: *An Applied View of Emerging Concepts and Methods*/ edited by Ned Kock
ONTOLOGIES: *A Handbook of Principles, Concepts and Applications*/ edited by Raj Sharman, Rajiv Kishore, Ram Ramesh
METAGRAPHS AND THEIR APPLICATIONS/ Amit Basu and Robert W. Blanning
SERVICE ENTERPRISE INTEGRATION: *An Enterprise Engineering Perspective*/ Cheng Hsu
DIGITAL GOVERNMENT: *E-Government Research, Case Studies, and Implementation*/ edited by Chen et al
TERRORISM INFORMATICS: *Knowledge Management and Data Mining for Homeland Security*/ edited by Chen et al
VALUE-FOCUSED PROCESS ENGINEERING: *A Systems Approach*/ Dina Neiger, Leonid Churilov, Andrew Flitman

MAPPING NANOTECHNOLOGY INNOVATIONS AND KNOWLEDGE

Global and Longitudinal Patent and Literature Analysis

Hsinchun Chen
Mihail C. Roco

Springer

Authors
Hsinchun Chen
University of Arizona
Tucson, Arizona, USA
hchen@eller.arizona.edu

Mihail C. Roco
National Science Foundation
Arlington, Virginia, USA
mroco@nsf.gov

ISSN: 1571-0270
ISBN-13: 978-0-387-71619-0
e-ISBN-13: 978-0-387-71620-6
DOI: 10.1007/978-0-387-71619-0

Library of Congress Control Number: 2008936135

Printed on acid-free paper

springer.com

TABLE OF CONTENTS

PREFACE

Introduction

The description, planning and governance of nanotechnology development require data on knowledge creation and innovation in various areas of application, how these evolve in time and what is the international context. This book aims to selectively provide such information based on the analysis of databases for science and engineering articles (Thompson Citation Index) and patents (USPTO in the United States, EPO in Europe and JPO in Japan). After a survey of the investigative methods, comparative results per countries, technology fields and research organizations are presented for articles and patents in parts of the interval 1976 to 2006. Interesting features on the evolution of major research themes and connection between research awards and patents have been obtained via longitudinal investigation of the published articles and patent data, as well as connection between NSF funding in nanotechnology and patents awarded to their principal investigators. A web-based system has been developed for accessing and visualizing nanotechnology patents, articles and NSF awards.

Scope and Organization

The monograph aims to present its chapters in a manner understandable and useful to students, researchers, and nanotechnology professionals. The titles of the eleven chapters are listed below:

- Chapter 1. Nanotechnology: An Emerging Field
- Chapter 2. Knowledge Mapping: Foundation
- Chapter 3. Knowledge Mapping: Analysis Framework
- Chapter 4. Mapping Nanotechnology Innovations via USPTO Database: A Longitudinal Study, 1976-2002
- Chapter 5. Federal Funding and Nanotechnology Innovations: NSF Funding and USPTO Patent Analysis, 1991-2002
- Chapter 6. Topological Analysis of Patent Citation Networks: USPTO Nanotechnology Patents, 1976-2004
- Chapter 7. Federal Funding and Nanotechnology Innovations: NSF Funding and USPTO Patent Analysis, 2001-2004
- Chapter 8. Academic Literature Citation in Patents: A Longitudinal Study of USPTO Patents, 1976-2004
- Chapter 9. Worldwide Nanotechnology Development: A Comparative Study of USPTO, EPO, and JPO Patents, 1976-2004
- Chapter 10. Mapping Nanotechnology Knowledge via Literature

Database: A Longitudinal Study, 1976-2004
- Chapter 11. The Nano Mapper System: Accessing and Visualizing Nanotechnology Patents and Grants

Main Features in This Book

The nanotechnology market and potential (Chapter 1)

Nanotechnology is the control and restructuring of matter at the intermediate dimensions, ranging in size from one atom to approximately one hundred molecules (about 100 nanometers), where new phenomena enable new applications. Unlike information technology, which is better understood by the general pubic, nanotechnology can mean different things for different people. It has sparked the discovery/realization of new products and processes not believed possible as well as societal concerns about its transforming dimensions.

With industry input, it was estimated in 2000 (Roco and Bainbridge, 2001) that the market of final products incorporating nanotechnology will reach $1 trillion worldwide by 2015. Their estimation of 25% market growth rate per year stands in 2007. Several current estimations support that trend. Lux Research (2007) reached a comparable estimation of about $2.4 trillion by 2014 by summing the market chain contributions (primary nanostructures + intermediate nanoscale assemblies + final products incorporating nanotechnology).

Following the establishment of the U.S. National Nanotechnology Initiative in 2000, worldwide interest and investment in nanotechnology R&D have grown steadily. Today virtually every major industrial country has a dedicated nanotechnology initiative. The estimated worldwide government nanotechnology R&D spending in 2006 according to the NNI definition of nanotechnology totaled about $4.7 billion, with the United States, European Union, and Japan spending at about $1.35 billion, $1.15 billion, and $980 million, respectively (Roco, 2007). Worldwide nanotechnology R&D investments increased by more than tenfold from 1997 to 2006.

Worldwide nanotechnology patent publication trend at three main depositories: USPTO, EPO, and JPO (Chapters 4, 6 and 9)

We compare the number and contents of nanotechnology patents identified by keywords ("title-abstract" search) from the USPTO, EPO, and JPO between 1976 and 2004. We collected 5,363 nanotechnology patents at USPTO, 2,328 nanotechnology patents at EPO and 923 at JPO in this time interval. The numbers of nanotechnology patents published in USPTO and EPO have continued to increase exponentially since 1980, while those published in JPO stabilized after 1993.

We observe that the top 20 assignee countries and their rankings are very similar in USPTO and EPO, with the United States publishing the most patents in both databases. Although their ranks are reversed in USPTO and EPO, Japan and Germany have the most patent publications after the United States, followed by France and Republic of Korea. Canada and China (Taiwan) have much higher ranks and numbers of patents in the USPTO than in EPO.

Institutions and individuals located in the same region as a repository's patent office have a higher contribution to the nanotechnology patent publication in that repository, which illustrates the "home advantage" effect. In the USPTO, International Business Machines Corp. (IBM) produced the most nanotechnology patents, followed by The Regents of the University of California, The United States of America as represented by the Secretary of the Navy, Eastman Kodak Co., and Minnesota Mining and Manufacturing Co. (3M Co.). In the EPO, French cosmetic company L'Oreal held the most patents, followed by IBM, Rohm & Haas (an American special materials company), Eastman Kodak Co., and Samsung Electronics Co. Ltd. In the JPO, Nippon Electric Co. is the largest assignee institution, followed by Japan Science and Tech Corp, Agency of Industrial Science and Technology, Matsushita Electric Industrial Co. Ltd, and Tokyo Shibaura Electric Co.

Bibliographic analysis on USPTO and EPO patents shows that researchers with nanotechnology articles in the United States and Japan published larger numbers of patents than other countries and that their patents were more frequently cited by other patents. Nanotechnology patents covered physics research topics in all three repositories. However, USPTO showed the broadest coverage in biomedical and electronics areas.

A more in depth analysis was performed for USPTO. Based on knowledge mapping analysis of the USPTO patent documents between 1976 and 2002, we found that the United States, Japan, Germany, and France had the majority of nanotechnology patents. The United States was assigned about 70% of the U.S. nanotechnology-related patents in this interval. The United States dominated the citation network and the U.S. patents intensively interacted with patents of most other countries. Japan was the second largest patent citation center, followed by Germany, France, the United Kingdom, Canada, and Switzerland.

International Business Machines Corporation (IBM) was issued the largest number of patents, followed by the Xerox Corporation (Xerox). We observe that patents issued to Eastman Kodak, Du Pont, General Electric, and the Dow Chemical Company had an average patent age of over 10 years, while patents issued to the Regents of the University of California, NEC,

Micron Technology, and Advanced Micro Devices were of a much "younger" age, under five years. When considering both quantity and freshness of patents assigned, Micron Technology outperformed all other institutions. Based on technology field analysis we observe that nanotechnology-related research was dominated by the industries of chemical/catalysts/pharmaceuticals and electronics. IBM and Micron Technology were the institutional patent citation centers. Patents of these two companies were cited extensively by patents of other institutions.

Linkage between academic research funding and patent publication: NSF funding (Chapters 5, 7 and 8)

The National Science Foundation (NSF) funding of nanoscale science and engineering has been correlated to the innovations made by the NSF principal investigators, as reflected in the USPTO patent data. NSF funded 5,263 nanotechnology-related awards involving 38 Divisions and 245 Programs between 1991 and 2002. The top five NSF Programs funding the nanotechnology research during that time are: Electronics, Photonics, and Device Technologies, Condensed Matter Physics, Small Business Phase I, Polymers, and Major Research Instrumentation.

The key linkage between the NSF awards and USPTO patents is the set of nanotechnology patent inventors who are also principal investigators of NSF awards. They are referred to as *"PI-inventors."* We identified 307 PI-inventors by matching their names and their institutions in the award and patent data sets. These PI-inventors were associated with 760 nanotechnology-related patents and 628 NSF awards. Thomas J. Pinnavaia of Michigan State University (also Claytec, Inc.) topped the list with 30 patents, followed by George M. Whitesides of Harvard University (also with several companies), who had filed 24 nanotechnology-related patents. The remaining top six PI-inventors measured by number of nanotechnology-related patents were: Stuart M. Lindsay of Arizona State University (also Molecular Imaging Corporation), Mark E. Thompson of the University of Southern California, Sanford A. Asher of the University of Pittsburgh, and Charles R. Cantor of Boston University (also Genelabs Technologies, Inc).

Statistical analysis shows that the NSF-funded researchers and their patents have higher impact factors than other private and publicly funded reference groups. This may suggest the importance of fundamental research on nanotechnology development.

From 2001 to 2004, the NSF's PI-inventors and their patents had higher impact than the inventors and patents in other groups based on the number of citations. NSF group impact increases over time as compared to other groups, indicating the importance of fundamental research on innovation. In the shorter term (1-2 years), the difference between the impact level of inventor groups is small.

The link between journal publications and patents have been analyzed for the first time. Academic nanotechnology research provides a foundation for nanotechnology innovation reported in patents. About 60% of the nanotechnology-related patents identified by “full-text” keyword searching between 1976 and 2004 at the USPTO have an average of 18 academic citations. In our research, we evaluated the most cited academic journals, individual researchers, and research articles in the nanotechnology area over a 29-year period. The most influential articles were cited about 90 times on the average, while the most influential author was cited more than 700 times by other nanotechnology-related patents. Thirteen mainstream journals accounted for about 20% of all citations. *Science*, *Nature,* and *Proceedings of the National Academy of Sciences (PNAS)* have consistently been the top three most cited journals. There are also influential specialty journals, represented by *Biosystems* and *Origin of Life*, which have very few articles cited but have exceptionally high frequencies of cites. The number of academic citations per year from the ten most cited journals has increased by over 15 times in the time period 1990-1999 as compared to 1976-1989, and again over 3 times in the time period 2000-2004 as compared to 1990-1999. This is an indication of the increased influence of academic research on the nanotechnology-related patents

Worldwide nanotechnology literature publication trend using the Thomson SCI Database (Chapter 10)

We analyzed the nanotechnology papers published in the Thomson Science Citation Index (SCI) Expanded literature database to assess worldwide nanotechnology research status from 1976 to 2004. We identified 213,847 nanotechnology papers that were published in 4,175 journals from 1976 to 2004. These papers contain 120,687 unique first authors from 24,468 institutions in 156 countries/regions. Between 1976 and 2004, the United States authors generated the largest number of papers (61,068), followed by Japan (24,985), Germany (21,334), and China (20,389). Three of theses countries that were not ranked in the top 20 USPTO patenting list showed significant productivity in academic paper publications: China (#4 in published papers), Russian (#7), and India (#13). Most countries showed a similar publication growth pattern, with the exception of China and South Korea, who outpaced their competitors in recent years. China surpassed Japan in nanotechnology publication in 2003 and has been the second most prolific country since then. South Korea also showed rapid development after 2000, exceeding Italy, Russia, and England in number of papers and becoming the sixth most prolific country in nanotechnology publication in 2004. All top 20 productive institutions were universities and national research centers rather than private companies. “Chinese Academy of Sciences” and “Russian Academy of Sciences” were the most productive institutions. They showed rapid growth since about 1998, outpacing other competing institutions

The United States was found to be the largest citation center based on citation network analysis. Germany, Japan, France, China, and Russia were the secondary citation centers. These major citation centers have close citation relationships among them. China and Russia were new citation centers for published papers but they were not well represented in patent citation networks based on USPTO database.

Developing knowledge mapping systems for technology assessment: The Nano Mapper System (Chapters 2, 3 and 11)

The invisible college, which consists of a small group of highly productive and networked scientists and scholars, is believed to be responsible for the growth of scientific knowledge, particularly in emerging areas such as nanoscale science and engineering. By analyzing scholarly publications of these researchers using select content analysis, citation network analysis, and information visualization techniques, "knowledge mapping" helps reveal this interconnected invisible college of scholars and their ideas. Main online resources for such analyses include: abstracts and indexes, commercial full-text articles and digital libraries, free full-text articles and e-prints, citation indexing systems and services, electronic theses and dissertations, patents, and business and industry articles and reports. These resources can be used to identify important authors and inventors, publications and publication outlets, institutions, countries and regions, and subject and topic areas over time.

"Knowledge Mapping," based on content analysis, citation network analysis, and information visualization, has become an active area of research that helps reveal such an interconnected, invisible college or network of scholars and their seminal publications and ideas. We present our effort to build a knowledge mapping system, Nano Mapper (http://nanomapper.eller.arizona.edu), which integrates the analysis of nanotechnology patents and grants into a Web-based platform. The Nano Mapper system contains nanotechnology-related patents from the USPTO, EPO, and JPO, and grant documents from the National Science Foundation. It provides simple search functionalities and makes available a set of analysis and visualization tools that can be applied on different levels of analytical units at different time periods for the patent and grant data. Nano Mapper simplifies the patent/grant analysis processes and demonstrates the feasibility of building large-scale Web-based cross-database knowledge mapping systems to access and analyze innovations in various science and engineering disciplines.

Audience

The primary audience for the proposed monograph includes the following:

- **IT Academic Audience:** College professors, research scientists, graduate students, and select undergraduate juniors and seniors in computer science, information management, information science, and other related public policy disciplines who are interested in knowledge mapping methodology and its applications in various emerging technology fields.
- **Nanotechnology Academic Audience:** College professors, research scientists, graduate students, and select undergraduate juniors and seniors in nanotechnology-related fields who are interested in an overview of the nanotechnology knowledge creation, development, and transfer process and the global nanotechnology landscape.
- **Nanotechnology Industry Audience:** Executives, managers, analysts, and researchers in Fortune 500, mid-size, and start-up companies (e.g., pharmaceutical firms, information technology firms, and telecommunication companies) and research laboratories who are actively conducting various nanotechnology-related research and product development; and venture capitalists and industry analysts who are interested in identifying critical inventions and innovations that can lead to major commercial success in the nanotechnology industry.
- **Nanotechnology Government Audience:** Policy makers and analysts in federal governments of different countries who are interested in monitoring the competitive global nanotechnology landscape and designing strategic future research programs.

Hsinchun Chen,
University of Arizona, and
Mihail C. Roco,
National Science Foundation

AUTHOR BIOGRAPHIES

Dr. Hsinchun Chen is McClelland Professor of Management Information Systems (MIS) at the Eller College of Management of the University of Arizona. He received the Ph.D. degree in Information Systems from New York University in 1989, MBA in Finance from SUNY-Buffalo in 1985, and BS in Management Science from the National Chiao-Tung University in Taiwan. Dr. Chen is a Fellow of IEEE and AAAS. He received the IEEE Computer Society Technical Achievement Award in 2006. He is author of 20 books and more than 150 SCI journal articles covering knowledge management, digital library, homeland security, Web computing, and biomedical informatics in leading information technology publications. He serves on ten editorial boards, including: *Journal of the American Society for Information Science and Technology, ACM Transactions on Information Systems, ACM Journal on Educational Resources in Computing, IEEE Transactions on Systems, Man, and Cybernetics, International Journal of Digital Library,* and *Decision Support Systems*. He has served as a Scientific Advisor/Counselor of National Library of Medicine (USA), Academia Sinica (Taiwan), and National Library of China (China).

Dr. Chen founded The University of Arizona Artificial Intelligence Lab in 1990. The group is distinguished for its applied and high-impact AI research. Since 1990, Dr. Chen has received more than $20M in research funding from various government agencies and major corporations. He has been a PI of the NSF Digital Library Initiative Program, NSF Digital Government Program, and the NIH NLM's Biomedical Informatics Program. His group has developed advanced medical digital library and data and text mining techniques for gene pathway and disease informatics analysis and visualization since 1995. Dr. Chen's work also has been recognized by major US corporations and been awarded numerous industry awards for his contribution to IT education and research, including: AT&T Foundation Award in Science and Engineering and SAP Award in Research/Applications.

Dr. Chen has been heavily involved in fostering digital library, medical informatics, knowledge management, and intelligence informatics research and education in the US and internationally. He has been a PI for more than 30 NSF and NIH research grants since 1990. Dr. Chen is conference chair of ACM/IEEE Joint Conference on Digital Libraries (JCDL) 2004 and has served as the conference general chair or international program committee chair for the past six International Conferences of Asian Digital Libraries (ICADL), 1998-2005. He has been instrumental in fostering the ICADL activities in Asia. Dr. Chen is the founder and also conference co-chair of the IEEE International Conference on Intelligence and Security Informatics (ISI), 2003-2008. The ISI conference has become the premiere meeting for international, national, and homeland security IT research. Dr. Chen's COPLINK public safety information sharing system and Dark Web terrorism informatics research have been featured in *Associated Press, New York Times, USA Today, Washington Post, Chicago Tribune, BBC, PBS, ABC News, Fox News,* and *Newsweek*, among others.

Dr. Mihail C. Roco is the Senior Advisor for Nanotechnology at the National Science Foundation (NSF) and a key architect of the National Nanotechnology Initiative. Dr. Roco is the founding Chair of the U.S. National Science and Technology Council's subcommittee on Nanoscale science, Engineering and Technology (NSET), and leads the Nanotechnology Group of the International Risk Governance Council. He also coordinated the programs on academic liaison with industry (GOALI). Prior to joining National Science Foundation, he was Professor of Mechanical Engineering at the University of Kentucky (1981-1995), and held visiting professorships at the California Institute of Technology (1988-89), Johns Hopkins University (1993-1995), Tohoku University (1989), and Delft University of Technology (1997-98).

Dr. Roco was a researcher in multiphase systems, visualization techniques, computer simulations, nanoparticles, and nanosystems. Dr. Roco is credited with thirteen patents and contributed over two hundred articles and fifteen books, including most recently "Nanotechnology: Societal Implications - Maximizing Benefits to Humanity" (Springer, 2006) and "Managing Nano-Bio-Info-Cogno Innovations" (Springer, 2007). Dr. Roco coordinated the preparation of the U.S. National Science and Technology Council (NSTC) reports on "Nanotechnology Research Directions" (NSTC, 1999) and "National Nanotechnology Initiative" (NSTC, 2000). Under his stewardship the U.S. nanotechnology federal investment has increased from about $3 million to $1.3 billion in 2006.

Dr. Roco is a Correspondent Member of the Swiss Academy of Engineering Sciences, and a Fellow of ASME, AIChE, and Institute of Physics. Forbes magazine recognized him in 2003 as first among "Nanotechnology's Power Brokers" and Scientific American named him one of 2004's top 50 Technology Leaders. In 2005, he received the AIChE Forum award "for leadership and service to the national science and engineering community through initiating and bringing to fruition the National Nanotechnology Initiative." He is the Editor of several journals, including the *Journal of Nanoparticle Research.* He is recipient of the Carl Duisberg Award in Germany, Burgers Professorship Award in Netherlands and the University Research Professorship award in the U.S. Dr. Roco is the recipient of the National Materials Advancement Award from the Federation of Materials Societies as “The primary coordinator of the U.S. nanotechnology science, engineering and technology effort, and …widely recognized as the individual most responsible for support and investment in nanotechnology by government, industry, and academia worldwide.” He is a member of several honorary boards and was elected Engineer of the Year by the U.S. Society of Professional Engineers and NSF in 1999 and again in 2004.

ACKNOWLEDGEMENTS

We thank our research partners, collaborators, and selected organizations that have contributed to the projects reported in the book: United States Patent and Trademark Office, European Patent Office, Japanese Patent Office, Thomson Scientific, and Springer Publishing. We thank Dr. Maria Zemankova for her comments and discussion on various information analysis and visualization ideas. We also thank our publisher, Gary Folven, and the Information Systems series co-editor, Dr. Ramesh Sharda, for their encouragement and support during the course of the project. The research reported in this book also owes much to support from the Nano Mapper research team at the Artificial Intelligence Lab of the University of Arizona, in particular, Zan Huang, Xin Li, Ying Lin, Zhi-Kai Chen, Lijun Yan, and Fei Guo.

The first co-author acknowledges support from the following National Science Foundation (NSF) grants: "Worldwide Nanotechnology Development: A Comparative Study of Global Patents," NSF 0654232, January 2007-December 2007; "Mapping Nanotechnology Development Based on the ISI Literature-Citation Database," NSF 0549663, September 2005-August 2006; "NanoMap: Mapping Nanotechnology Development," NSF 0533749, August 2005-July 2007; "Intelligent Patent Analysis for Nanoscale Science and Engineering," NSF 0311652, May 2003-April 2004; and "NanoPort: Intelligent Web Searching for Nanoscale Science and Engineering," NSF 0204375, January 2002-December 2002. The second co-author was partially supported by the Directorate for Engineering, NSF.

Chapter 1

NANOTECHNOLOGY: AN EMERGING FIELD

CHAPTER OVERVIEW

Science and technology are at the core of human endeavor, and the process of creating new tools and products has been accelerated by reaching at the basic building blocks at the nanoscale. In today's world of laptop computers, cell phones, regenerative medicine, targeted drugs, fuel cells, environmentally friendly technologies and carbon sequestration, it is natural to imagine that technology can take us even further. Nanotechnology is an enabler and catalyst of current and future possibilities. It can help us realize a wide spectrum of applications not only in engineered materials, nanomanufacturing, electronics, and communication, but also in energy, environment, biomedicine, food and agricultural systems. This chapter provides a brief overview of this field beginning with a definition and brief history of nanotechnology development, along with new applications and a description of the emerging nanotechnology industry. It also offers a look at the competitive international nanotechnology research and development landscape, particularly research programs in the United States, Asia and Europe. Lastly, this chapter discusses the potential environmental and health impacts of nanotechnology and approaches to addressing these concerns.

1. DEFINITION

Nanotechnology is the control and restructuring of matter at the intermediate dimensions, between the sizes of one atom to about hundred molecules shoulder to shoulder (about 100 nm), where new phenomena enable new applications. Nanoscale science and engineering operates at the first level of organization of atoms and molecules for both living and anthropogenic systems. The nanoscale is a natural threshold between discrete behavior of single atoms or single molecules with given properties, on one side, and collective behavior of assemblies of atoms and molecules where the properties are a function of size, structure and composition, on the other side. It is where the fundamental properties and functions of all materials and systems are defined, and where novel properties can be exploited and changed. Such fundamental control promises a broad and revolutionary technology platform for industry, biomedicine, environmental engineering, safety and security, shared resources (food, water, energy, nanoinformatics), and countless other areas.

Unlike information technology, which is better understood by the general public, nanotechnology can mean different things for different people. It has sparked creative visualization of new products and processes not before possible, as well as economical manufacturing of existing products. It has also triggered societal concerns about its transforming dimensions.

The U.S. National Nanotechnology Initiative (NNI) (www.nano.gov; see overview in Roco, 2007) has defined nanotechnology as: "Encompassing the science, engineering, and technology related to the understanding and control of matter at the length scale of approximately one to 100 nanometers." Nanotechnology includes the research and development of materials, devices, and systems that have novel properties and functions due to their nanoscale structures or components. It goes beyond working with individual nanoscale objects and structures, and looks for their assembling. It is also about leveraging knowledge of the nanoscale to design and manufacture new products.

2. HISTORY

As long ago as 3.5 billion years, cells developed the ability to perform photosynthesis, mainly inside chloroplast. Chloroplast may have been one of the first nanoscale machines, a nanoscale biomachine in this case (*Scientific American*, 2002). In fact, all matter as we know it has nanostructure naturally; nanotechnology aims to harness nanoscale properties by controlling and manufacturing matter at that scale. The control and restructuring at the nanoscale is new.

In 1959, Richard Feynman gave the talk, "There's Plenty of Room at the Bottom," in which he invited listeners to consider a new research field, one of "manipulating and controlling things on a small scale." In 1974, Naorio Tangiguchi conceived the term "nanotechnology" to represent surface asperities of less than a micron after machining, a meaning slightly different from the broader term used now that includes using quantum phenomena and selfassembling. In 1981, the first scanning probes to detect nanostructures on surfaces were created at IBM Zurich. In 1991, Sumio Iijima of DEC Japan experimentally identified carbon nanotubes.

By 2000, the potential of understanding nanoscale properties of matter and the formulation of key research opportunities were becoming clearer at the confluence of many disparate disciplines (Roco, Williams and Alivisatos, 1999). That year, the Clinton administration announced the National Nanotechnology Initiative, which provided a big boost in nanotechnology funding and visibility, both in the United States and internationally.

Nanotechnology offers the potential for ground-breaking scientific research. It also inspires general fascination about futuristic possibilities of gaining ultimate control over matter, from creating selfassembling materials and systems to curing chronic diseases via nanomedicine.

Roco and Bainbridge (2001) estimated that, with industry input, the market for final products incorporating nanotechnology as a key functional component would reach \$1 trillion worldwide (Figure 1-1) by 2015. This figure, which corresponds to a rate of market increase of 25% per year, stands in 2007. The Lux Research (2007) estimation reached a comparable conclusion. By summing the three-step market chain contributions (primary nanostructures, intermediate nanoscale assemblies, and final products incorporating nanotechnology), they estimated the market at about \$2.4 trillion, or if divided by three, about \$0.8 trillion of final products by 2014.

These estimations have been based on direct communication with leading experts in large companies with related R&D programs in the United States, Japan and Europe, and on the international study completed between 1997 and 1999 (Siegel et al., 1999).

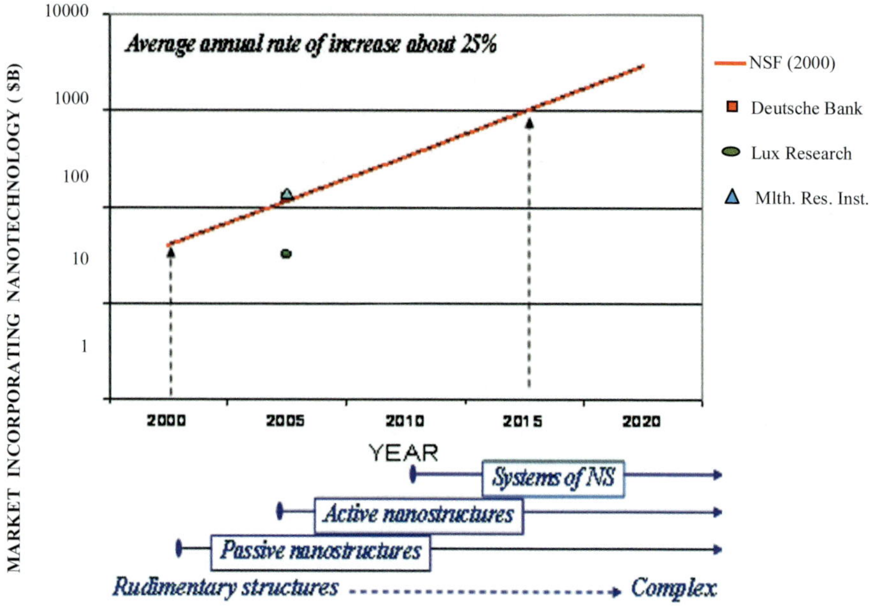

Figure 1-1. Estimation of the worldwide market incorporating nanotechnology (estimation made in 2000 at the National Science Foundation; Roco and Bainbridge, 2001).

3. INTERNATIONAL NANOTECHNOLOGY R&D INVESTMENT

Following the establishment of the U.S. National Nanotechnology Initiative in 2000, worldwide interest and investment in nanotechnology R&D have grown steadily. Today, virtually every major industrial country has a dedicated nanotechnology initiative. The estimated worldwide government nanotechnology R&D spending in 2007—based on the NNI definition of nanotechnology and excluding microtechnology, microstructured materials, etc.—totaled about \$6.3 billion (Table 1-1), slightly less than the industry R&D investment estimated at about \$7 billion in the same year. The United States, European Union and Japan spent about \$1.35 billion, \$1.5 billion, and \$980 million, respectively (Roco, 2007). Worldwide nanotechnology R&D investments have increased by more than fourteen-fold from 1997 (\$430 million; Siegel et al., 1999) to 2007.

In addition to federal government investments, regional, state, and local nanotechnology initiatives provide another vehicle for R&D funding and a vital avenue for commercialization and economic development. Lux

Research reported that U.S. state and local governments and organizations invested about $500 million in 2006 in nanotechnology research, facilities and business incubation programs (Lux Research Inc., 2007). In addition, local initiatives supported the development of a technically skilled workforce through the creation or promotion of education and training programs in undergraduate colleges, community colleges and high schools.

Private investment is also critical for nanotechnology development and commercialization. Measures of private investment include both corporate internal investment and venture capital activity. Of the $12.4 billion that Lux Research (2007) estimates was spent on nanotechnology R&D worldwide in 2006, $5.3 billion was by corporations: 38 percent ($2 billion) by North American companies, mostly in the United States; 42 percent ($2.2 billion) by Asian companies; 19 percent ($1.02 billion) by European companies; and less than 2 percent by firms in other regions. Venture capital investment totaled roughly $650 million in 2006: $581 million in North America, $36 million in Europe, $4 million in Asia, and $28 million in other regions. In 2006, the annual rate of increase of worldwide R&D investment was 19 percent for corporations, 10 percent for governments, and 3 percent for venture funds.

Table 1-1. Estimated government nanotechnology R&D expenditures, 1997–2007 (millions of dollars/year).

Region	**1997**	**1998**	**1999**	**2000**	**2001**	**2002**	**2003**	**2004**	**2005**	**2006**	**2007**
EU+	126	151	179	200	~225	~400	~650	~950	~1050	~1150	~1500
Japan	120	135	157	245	~465	~720	~800	~900	~950	~980	~980
USA*	116	190	255	270	464	697	862	989	1200	~1351	1354
Others	70	83	96	110	~380	~550	~800	~900	~1000	~1200	~2500
Total	432	559	687	825	1534	2367	3112	3739	4200	4681	6334
(% of 1997)	100%	129%	159%	191%	355%	547%	720%	866%	972%	1083%	1475%

Explanatory notes: National and EU funding are included. The EU+ includes countries in EU(15)/ EU(25)/ EU(27) - as EU has expanded in time - and Switzerland (CH). The rate of exchange is $1 U.S. = 1.1 Euro until 2002, $1 U.S.= 0.9 Euro in 2003, and $1 U.S.= 0.8 Euro in 2004–2007. The Japan rate of exchange is $1 U.S. = 120 yen until 2002, $1 U.S.= 110 yen in 2003, $1 U.S.= 105 yen in 2004–2007. "Others" includes Australia, Canada, China, Eastern Europe, Former Soviet Union, Israel, Korea, Singapore, Taiwan, and other countries with nanotechnology R&D. Estimates use the U.S. National Nanotechnology Initiative definition of nanotechnology (this definition does not include micro-electrical-mechanical systems (MEMS), microelectronics, or general research on materials; see Roco, Williams and Alivisatos, 2000; and Springer, formerly Kluwer, also on http://nano.gov). Estimates include the publicly reported government allocations for the respective financial years.

* A fiscal year (FY) begins in the United States in October and six months later, around April 1, in most other countries. This chart reports the actual budget recorded at the end of the respective fiscal years.

4. INTERNATIONAL PUBLICATIONS OUTCOMES

One metric often used to gauge scientific leadership is the number of peer-reviewed scientific articles. The total number of articles globally has increased at a faster rate after 2000 as compared to the previous decade (Zucker and Darby, 2005; Kostoff et al., 2007; Porter et al., 2007; and the study discussed in Chapter 8 of this book). The total number of nanotechnology-related papers reached more than 60,000 in 2006. The United States is dominant in terms of the number of nanotechnology research articles published, accounting for more than 15,000 articles in 2006, about 40 percent more than People's Republic of China and more than double that of any other country (Figure 1-2). The U.S. portion of nanotechnology articles remained approximately constant after 2000, while those from Japan and Europe slightly decreased. Other countries, most notably the People's Republic of China and South Korea, have increased their share, reflecting their larger R&D investment rates in this field.

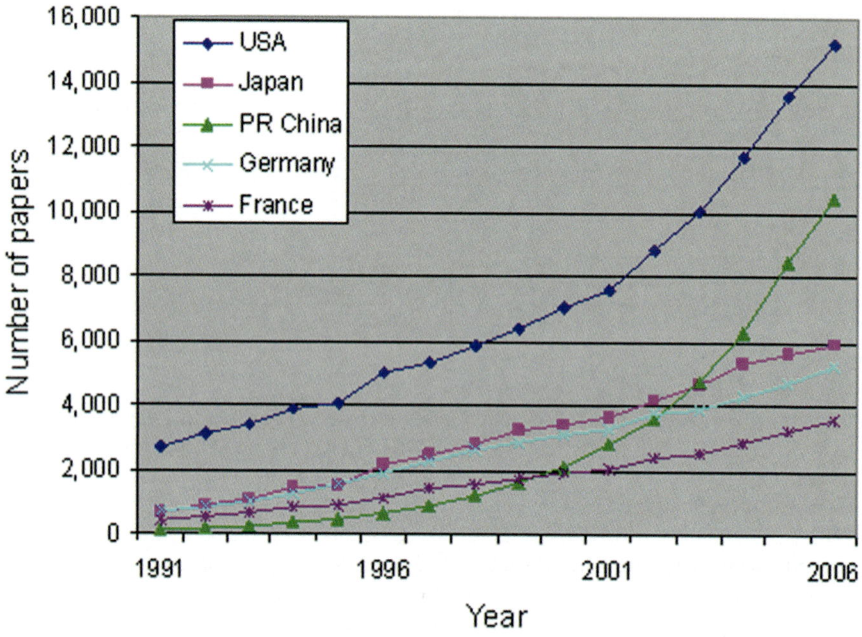

Figure 1-2. Nanotechnology papers identified by "title-abstract" keyword search in Thomson ISI's SCI database through Web of Science, 1991–2006 (in 2006: the United States has 15,212 papers, People's Republic of China has 10,474 papers, Japan has 5,905 papers, Germany has 5,250 papers and France has 3,596). The keyword search approach is explained in Chapter 4.

Using as a metric those papers that are frequently cited, such as those published in *Science*, *Nature* and *Proceedings of the National Academy of Sciences,* the contribution of U.S.-based authors is slightly higher than half after 2000 (Figure 1-3).

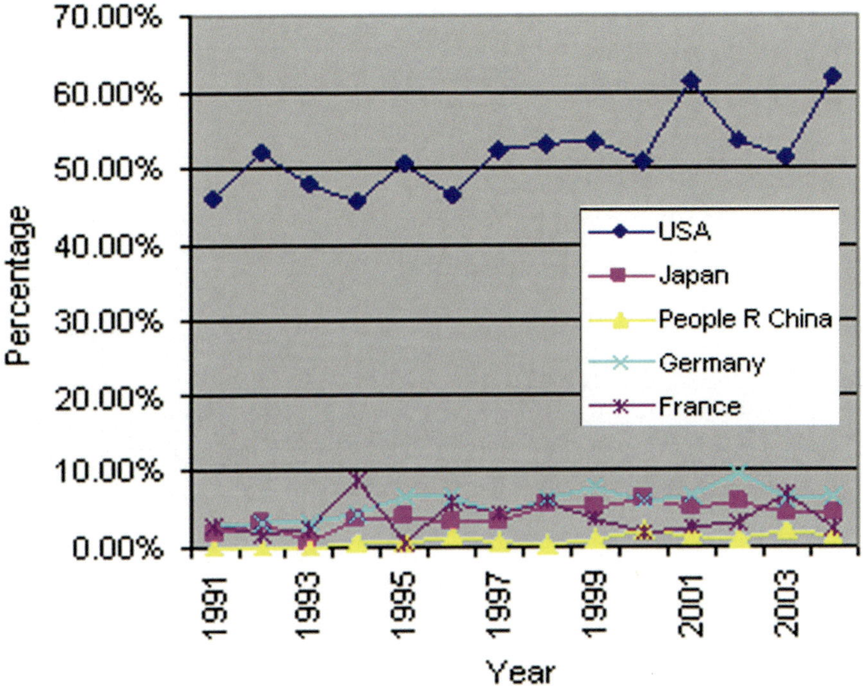

Figure 1-3. Contribution of different countries to nanotechnology-related papers published in *Science*, *Nature* and *Proceedings of the National Academy of Sciences*.

Another metric commonly used to gauge leadership in technology innovation, and one that is perhaps more indicative of commercialization potential, is the number of patents. A study by Huang et al. (2004) reveals the rapid, quasi-exponential growth of nanotechnology-related patents since 2000. Based on the analysis of nanotechnology patents in the U.S. Patent and Trademark Office (USPTO) database, U.S. entities accounted for over 60 percent of nanotechnology patents from 1976 to 2006. In addition, U.S. patents received the most citations by other patents, another indication of technology leadership. Overall, the ten countries receiving the highest number of nanotechnology-related patents in the interval 2000–2006 were the United States, Japan, Germany, Republic of Korea, and France, followed by China (Taiwan), Canada, Netherlands, the United Kingdom and Switzerland. The contribution of the most important regions is shown in Figure 1-4. More detailed analyses of international nanotechnology patents and scientific literature will be provided in the subsequent chapters.

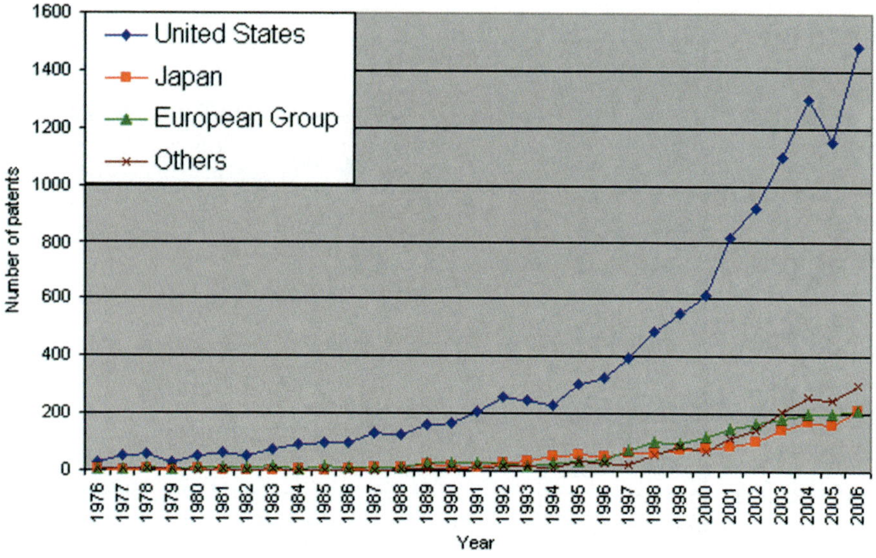

Figure 1-4. Nanotechnology patents identified by "title-claims" keyword search at the U.S. Patent and Trademark Office, 1976–2006 (in 2006: the United States has 1,488 patents, Japan has 212, the European group has 214, and Others have 298).

The keyword search approach is explained in Chapter 4. USPTO revised the definition of nanotechnology according to NNI documents, and this caused a discontinuity in the growth rate of the annual number of patents in 2005.

Due to resource constraints, many countries have adopted a strategy of making targeted investments, thereby positioning themselves to be leaders in a key industry or technology. The United States nanotechnology program is broad-based, with a large contribution to fundamental research in universities (65 percent of the U.S. government nanotechnology R&D funds go to universities) as well as supporting R&D in research laboratories (25 percent) and industry (10 percent, notably small businesses). Asia, Korea and Taiwan are focusing on nanoelectronics; Singapore has a particular focus on nanobiotechnology; and Japan's recognized strengths are in instrumentation and devices. In Europe, efforts exist at both the national level and at the European Union (EU) level. EU research is broad-based, while individual countries pursue targeted development. For example, Germany has developed a focus on nanoscale materials, mechanics, electronics and optical science.

5. TECHNICAL CHALLENGES AND POTENTIAL ADVANCES: 2007–2020

Nanotechnology has the potential to change our comprehension of nature and life, develop unprecedented manufacturing tools and medical procedures, and even affect societal and international relations (Roco, 2007). Nanotechnology holds the promise of increasing efficiency in traditional industries and bring radically new applications through emerging technologies.

The first set of nanotechnology grand challenges was established in 1999–2000. The NNI Subcommittee on Nanoscale Science, Engineering and Technology updated its strategic plan in 2004 and 2007. Now, let's imagine again what could be done in the future. Potential developments by 2015 are discussed below.

(1) ***At least half of the newly designed advanced materials and manufacturing processes are built using control at the nanoscale for at least one of their key components.*** Even if this control may be rudimentary as compared to the long-term potential of nanotechnology, this development will mark a milestone toward the new industrial revolution as outlined in 2000. Given the experience with information technology in the 1990s, an overall increase of societal productivity by at least 1 percent per year is estimated after 2015 because of these changes. Several examples are:

- Silicon transistors will reach dimensions smaller than 10 nanometers and will be integrated with molecular, nanotube or other kinds of nanoscale systems. Alternative technologies for replacing the electronic charge as information carrier with electron spin, phase, polarization, magnetic flux quanta, and/or dipole orientation are under consideration.
- Technologies will be developed for directed self assembly into hierarchically organized, device-oriented structures and creation of functional, nanoscale building blocks. New molecules will be designed to selfassemble in pre-designed patterns.
- Lighter composite nanostructured materials, fuels made less reactive and less polluting by additional of nanoparticles, and automated systems enabled by nanoelectronics will dominate automotive, aircraft and aerospace industries.
- Top-down miniaturization manufacturing is cxpected to integrate with bottom-up molecular assembling techniques using modular approaches.

- The use of nanoscale-designed catalysts will be expanded in "exact" chemical manufacturing to "cut" and assemble molecular assemblies, with minimal waste.
- Measurement and imaging applicable to large domains of biological and engineering interest are expected to reach atomic precision and achieve time resolution of chemical reactions. Visualization and numerical simulation of three-dimensional domains with nanometer resolution will be necessary for engineering applications.

(2) *Suffering from chronic illnesses is sharply reduced.* It is conceivable that by 2015, our ability to detect and treat tumors in their first year would be advanced such that suffering from cancer could be lessened and the deaths from cancer reduced.

- Today, based on results obtained during 2001–2007 in understanding the cell and in designing new instrumentation, researchers are working to treat cancer in its early stages and eventually eliminate it as a cause of death.
- Pharmaceutical synthesis, processing and delivery will be enhanced by nanoscale control, and about half of pharmaceuticals will use nanotechnology in a key component.
- Modeling the brain based on neuron-to-neuron interactions will be possible by using advances in nanoscale measurement and simulation.

(3) *Converging science and engineering from the nanoscale will establish a mainstream pattern for applying and integrating nanotechnology with biology, electronics, medicine, learning and other fields (Roco and Bainbridge, 2003).* This includes hybrid manufacturing, neuromorphic engineering, artificial organs, expanding life span, virtual reality applications, and enhancing learning and sensorial capacities. Science and engineering of nanobiosystems will become essential to human healthcare and biotechnology. The brain and nervous system functions are expected to be better understood and used for cognitive engineering and human-machine interface developments.

(4) *Product development will include life-cycle sustainability and biocompatibility.* Knowledge development in nanotechnology will lead to reliable safety rules for limiting unexpected environmental and health consequences of nanostructures. Measurement and control of contents of nanoparticles will be performed in air, soil and water using a national monitoring network. International agreements will address the nomenclature, standards, and risk governance of nanotechnology.

(5) ***Knowledge development and education will center on understanding nature from the nanoscale instead of the microscale.*** A new education paradigm will emerge, one not based only on disciplines but rather on the unity of nature. This new approach will seek to integrate education and research for K–16 (reversing the pyramid of learning that now begins with isolated disciplinary topics and ends with a unified view of matter at the nanoscale (Roco, 2003). Science and education paradigm changes will be at least as fundamental as those during the microscale science and engineering transition that originated in the 1950s, when microscale analysis and scientific analysis were stimulated by the space race and digital revolution. The new nanoscale science and engineering transition will change the foundation of analysis and the language of education, which will be stimulated by nanotechnology products. This new transition originated at the threshold of the third millennium.

(6) ***Nanotechnology businesses and organizations will restructure to integrate with other technologies, to enable distributed production, to support continuing education, and to form consortia of complementary activities.*** Traditional and emerging technologies will be equally affected. An important development will be creation of nanotechnology R&D platforms to serve various areas of applications with modular nanoscale components and the same investigative and productive tools. Two examples are the nanotechnology platform created at a newly built laboratory by General Electric and the discovery instrumentation platform developed at Sandia National Laboratories.

(7) ***The capabilities of nanotechnology for systematic control and manufacture at the nanoscale are envisioned to evolve in five overlapping generations of nanotechnology products,*** of which the first four maintain the coherence of a single technology (Roco, 2004b) (Figure 1-5). The *Fifth Generation* may be approximately established after 2020 and will be characterized by bifurcation of various nanosystem architectures, each of these combining with other emerging, converging technologies. The generations of nanotechnology products change based on the level of complexity, dynamic behavior and integration with other processes. A generation of new products is expected to include, at least partially as components, products from previous generations. Each generation of products is envisioned to begin with the creation of commercial prototypes using systematic control of the respective phenomena and manufacturing processing:

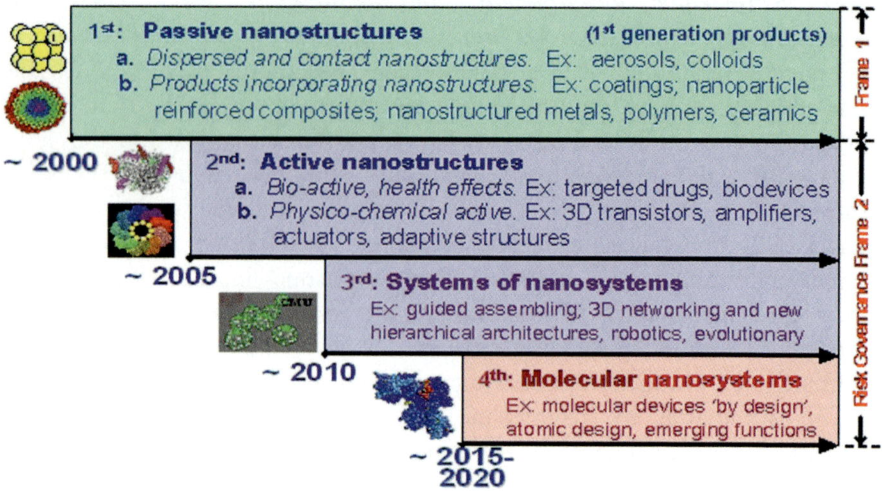

Figure 1-5. Timeline for beginning of industrial prototyping and nanotechnology commercialization: First four generations of nanoproducts in 2000-2020. (After about 2020 it is estimated a fifth generation will be characterized by bifurcation of various nanosystem architectures and their convergence with other emerging technologies.)

a. **First Generation of products (~2001–) is "passive nanostructures"** and is typically used to tailor macroscale properties and functions. The specific behavior is stable in time. Examples are nanostructured coatings, dispersion of nanoparticles, and bulk materials, such as nanostructured metals, polymers, and ceramics.

b. **Second Generation of products (~ 2005–) is "active nanostructures"** for mechanical, electronic, magnetic, photonic, biological, and other effects. These are typically integrated into microscale devices and systems. Examples are new transistors, components of nanoelectronics beyond CMOS (complementary metal oxide semiconductor), amplifiers, targeted drugs and chemicals, actuators, artificial "muscles," and adaptive structures.

c. **Third Generation (~ 2010–) is "systems of nanosystems."** Such systems would be designed using various syntheses and assembling techniques, such as bio-assembling, robotics with emergent behavior, and evolutionary approaches such as directed evolution. A key challenge is networking at the nanoscale to create hierarchical architectures. Research focus will shift toward heterogeneous nanostructures and supramolecular system engineering. This includes directed multiscale selfassembling, artificial tissues and sensorial systems, quantum interactions within nanoscale systems, processing of information using photons or electron spin, assemblies of nanoscale electromechanical systems (NEMS) and converging-technology (nano-bio-info-cogno) platforms integrated from the nanoscale.

d. **Fourth Generation (~ 2015–) will bring "heterogeneous molecular nanosystems,"** where each molecule in the nanosystem has a specific structure and plays a different role. Molecules will be used as devices and from their engineered structures and architectures will emerge fundamentally new functions. Designing new atomic and molecular assemblies is expected to increase in importance and to include macromolecules "by design" and nano-sized machines. It also includes directed and multiscale selfassembling, exploiting quantum control, nanosystem biology for healthcare, and human-machine interface at the tissue and nervous system level. Research will include topics such as: atomic manipulation for design of molecules and supramolecular systems; controlled interaction between light and matter with relevance to, among many applications, energy conversion; exploiting quantum control of mechanical-chemical molecular processes; and nanosystem biology for healthcare and agricultural systems.

(8) ***Advances in nanotechnology will directly contribute to several of society's crucial needs*:**

- **Energy conversion and storage** is a main objective of nanotechnology development. Exploratory projects in areas such as photovoltaic conversion and direct conversion of thermal to electrical energy, as well as for high density storage of energy in nanostructured materials, are under way.
- **Water filtration and desalinization** using nanotechnology has high promise, despite the fact that research investment was relatively low until the first 5-6 years after 2000.
- **Nano-informatics** refers to specific databases and the methods to use them. Nanoinformatics will be developed for characterization of nanocomponents and of materials and processes integrated at the nanoscale. Such databases will interface with those existing in fields such as bio-informatics and human and plant genomes.

6. NATIONAL NANOTECHNOLOGY INITIATIVE

The vision of the U.S. National Nanotechnology Initiative (NNI) is a future in which the ability to understand and control matter on the nanoscale leads to a revolution in technology and industry. NNI was launched in fiscal year 2001 under the Clinton administration and received continuing support under the Bush administration. The annual federal investment in nanotechnology R&D has more than quadrupled to almost $1.4 billion in

2007. The number of participating federal departments and agencies has grown from 6 to 26 (see http://www.nano.gov). The NNI goals include the following:

- Maintain a world-class research and development program aimed at realizing the full potential of nanotechnology;
- Facilitate transfer of new technologies into products for economic growth, jobs, and other public benefit;
- Develop educational resources, a skilled workforce, and the supporting infrastructure and tools to advance nanotechnology; and
- Support responsible development of nanotechnology.

Several cross-cutting areas of application are:

- Advanced materials and nanomanufacturing, as discussed above.
- Aerospace: High strength, low weight, multifunctional materials to improve performance and reduce operational costs for aircraft and spacecraft; faster, yet more compact electronics that enable fully automated, self-guided and unmanned air vehicles for reconnaissance and surveillance with nanoscale systems.
- Agriculture and food: Secure production, processing, and shipment of food products; improved agricultural efficiency; reduced agricultural waste through conversion of agricultural materials into valuable products.
- National security: Systems with the speed and capability to enable command, control, communications, surveillance, reconnaissance, and information dominance; automation and robotics with sufficient decision-making capabilities to minimize the exposure of combat personnel and first responders to harm.
- Energy: Tailored nanomaterials for solar energy conversion, transmission, and storage; thermoelectric converters; high-performance batteries and fuel cells; control of corrosion, friction, and wear; catalysts for energy-efficient manufacturing.
- Environmental improvement: Reduced pollution through the development of new "green" nanotechnologies; better environmental remediation through more efficient removal of containments, particularly ultra-fine particles.
- Information technologies: Nanoelectronic devices based on novel magnetic, spin, molecular, or quantum information approaches; nanotechnology-based systems to improve computer speed, expand mass storage, increase communication bandwidth, and enable brighter and sharper displays than what is available today.
- Medicine and health: Novel sensor arrays enabling inexpensive, rapid diagnosis; new composite structures for superior bone and tooth

implants; targeted treatment of diseases such as cancer or heart disease with reduced side effects.
- Transportation and civil infrastructure: Vehicles, bridges, and roadways which are easier to build and maintain through use of composite material structures.

Most of these applications have the potential to significantly improve technology, economy and quality life. Many nanotechnology applications also have immediate to mid-term commercialization potential. Despite the knowledge and economic promises of nanotechnology, however, significant environmental and health concerns about nanoparticles and nanosystems remain and need to be considered.

7. ENVIRONMENTAL, HEALTH AND SAFETY IMPLICATIONS

Technological progress can enhance the quality of life, but it also can raise societal concerns. Because nanotechnology operates at the very foundation of matter, at the first level of organization for both living and anthropogenic systems, the potential benefits are large—but so are the perceived risks, which must be addressed early (Roco, 2005b). Moreover, everyone must be involved—not just scientists from a few countries.

Nanotechnology promises to expand the limits of sustainable development through "green" manufacturing and environmental remediation. Yet, nanostructures and nanosystems can exhibit properties quite different from those of the corresponding bulk materials. Such properties include enhanced reactivity and greater ability to penetrate tissues and cell membranes. There are important health risks to consider which occur through various types of contact with nanoscale materials.

Research is needed to increase fundamental understanding of nanoscale material interactions at the molecular and cellular levels, as well as develop the necessary instrumentation and metrology. We need to better understand how nanoscale materials interact with the environment and human body, as well as the fate, transport, and transformation of nanoscale materials in the environment and their life cycle. In addition, we need to identify and characterize potential exposure, determine possible human health impact, and develop appropriate methods of controlling and mitigating exposure when working with nanoscale materials.

There are immediate and continuing issues related to environmental, health and safety (EHS) concerns. Even chemicals that are considered safe in bulk need to be tested as nanostructures for their effects. Several important health risks of reactive and nonbiocompatible nanoparticles in human tissues, such as the lung, the liver, the kidneys, and the nervous system, occur through the inhalation of nanoscale aerosols, contact with nanostructured surfaces, or the consumption of foods with nanosized colloids. Industry, specialized government agencies, civic organizations, and the public have rapidly recognized the near-term EHS aspects of nanotechnology. The following three immediate and continuing issues in particular must be addressed concurrently with the development of nanoscale products and the funding of R&D projects.

First, research laboratories and industrial units must understand the EHS consequences of nanotechnology and develop appropriate safety measures for both workers and consumers. Examples of immediate and continuing concerns include toxicity of nanoproducts; mechanisms and routes of exposure to nanoproducts in the air, water, and soil; effectiveness of personal protective equipment such as special clothing, respirators, and engineering controls; cell behavior in the presence of nanostructures; and preventing the release of synthetic nanoparticles into the environment. As knowledge about nanoproducts develops, existing EHS regulations must be strengthened to recognize the difference between chemicals in bulk and as nanomaterials.

Second, everyone must speak the same language. Researchers are from different disciplines and economic and industrial sectors. Internationally accepted nomenclatures, norms, standards, and regulations are necessary for the progress of science, engineering, technology, and new markets. Examples are the development of specific instrumentation, sensors, and standards for nanoparticles in air, water, and soil.

Third, methodologies for risk analysis and governance should be developed for management of new products and the technology created in the private sector and the government.

Long-term thinking is needed in parallel. Several issues that are related to responsible development of nanotechnology hinge on broader social and economic outcomes and require more time to be addressed. Government and civic organizations must work together to ensure an equitable and responsible strategy for addressing these issues.

Respect for human nature, dignity, and physical integrity is essential. Society should co-evolve harmoniously with nanoscale technology. Our rights to a good quality of life, long-term health and safety, and access to knowledge must be respected. The implications of the convergence of nanotechnology, biotechnology, information technology, cognitive science, and other technologies are important because nanoscale knowledge and tools enable and unify other transforming technologies at their foundations.

R&D nanotechnology investment should be balanced in such a way that the benefits and secondary consequences are fairly distributed throughout society. Examples include equitable access to advanced medicine and other nanotechnology services and products. NNI has this goal in mind. In addition, economic risks must be addressed: the loss of production because of polluted water and air, increased restrictions on the extraction of raw materials, limits on the use of polluted real estate, requirements to clean up existing pollution, and public opposition to industrial development. Early nanotechnology education and training should be focused on basic concepts, be interdisciplinary, and be relevant to preparing the workforce.

Environmental and human health protection and improvement. Principles of green chemistry and engineering can be used to establish preemptive, environmentally benign manufacturing methods rather than postmanufacturing cleanup. For example, some known approaches and criteria for sustainable development of technology, energy supply, and transportation include life-cycle analysis of products, materials flow analysis, clean-up techniques based on new principles, weather implications, and other global effects. In addition, industrial byproducts and natural nanomaterials must be controlled and mitigated. Examples of such nanomaterials include byproducts from combustion engines, furnaces, and welding, as well as natural nanoparticles in sandstorms and forest fires. If left untracked, nanoparticles could potentially make their way into air, water, and food and present a risk to human health. Government agencies must provide research and infrastructure facilities for safe food and clean water and air. Finally, we need a system that can monitor and label the nanostructures that may change from benign to toxic because of nanostructuring, such as surface materials that flake off nanoparticles.

Ethical, moral, and legal aspects. As new technology emerges, knowledge of socioeconomic implications must be developed and future scenarios should be anticipated to the extent possible. That knowledge should be developed through systematic research and disseminated through databases. This process should include two-way interaction between the public and various interested government and civic organizations.

Risk governance. Government organizations must adapt or correct their oversight activities to provide long-term risk governance in a complex societal system. Ideally, an adaptive systems approach is needed that looks ahead at global effects, such as the food chain and the path for technology replacement. Simple regulations applicable to single components of the societal system may not be effective over time. Ensuring continuous government funding for these activities will be another challenge. International collaboration is needed because the knowledge of nanoscience and the products of nanomanufacturing, as well as health and environmental implications, do not have borders.

The proposed framework for governance calls for several key functions: supporting the transformative impact; advancing responsible development that includes health, safety and ethical concerns; encouraging national and global partnerships; and commitment to long-term planning with effects on human development (Roco, 2008). Principles of good governance include participation of all those involved or affected by the new technologies, transparency, participant responsibility, and effective strategic planning. Introduction and management of nanotechnology must be done with respect for immediate concerns (such as addressing toxicity of new nanomaterials) and longer-term concerns (such as human development and concern for human integrity, dignity and welfare). Four levels of governance of converging technologies have been identified: (a) adapting existing regulations and organizations; (b) establishing new programs, regulations and organizations specifically to handle converging technologies; (c) national policies and institutional capacity building; and (d) international agreements and partnerships.

Nanotechnology research and development is expanding rapidly and it is important to maintain the balance between the promised benefits and the necessary measures to mitigate possible undesirable secondary effects. Because of market globalization, shared environment and common social development, the evaluation and mitigation measures must be correlated with the international efforts. Risk governance, beginning with risk identification, assessment and management and ending with regulatory and educational measures, is needed at the national and international levels for nanotechnology advancement. International organizations, such as the International Risk Governance Council (IRGC), may provide an independent and global framework for identification, assessment and mitigation of risk.

8. QUESTIONS FOR DISCUSSION

1. What are the most promising areas for immediate commercialization of nanotechnology?
2. What are the leading institutions and companies in nanotechnology development and what are their unique capabilities?
3. What are the unique nanotechnology capabilities in different regions and countries of the world?
4. What nanotechnology educational and training programs are available in the U.S. (see www.nano.gov and www.nsf.gov/nano)?

Chapter 2

KNOWLEDGE MAPPING: FOUNDATION

CHAPTER OVERVIEW

The “invisible college,” which consists of a small group of highly productive and networked scientists and scholars, is believed to be responsible for the growth of scientific knowledge. By analyzing the scholarly publications of these researchers using select content analysis, citation network analysis, and information visualization techniques, “knowledge mapping” can reveal this interconnected invisible college of scholars and their ideas. In this chapter, we discuss online resources that are often used for such analyses, including: abstracts and indexes, commercial full-text articles and digital libraries, free full-text articles and e-prints, citation indexing systems and services, electronic theses and dissertations, patents, and business and industry articles and reports. These resources can be used to identify important authors and inventors, publications and publication outlets, institutions, countries and regions, and subject and topic areas over time.

1. INVISIBLE COLLEGES AND KNOWLEDGE MAPPING

In Diane Crane's seminal book, *Invisible Colleges: Diffusion of Knowledge in Scientific Communities* (Crane, 1972), she suggests that it is the "invisible college," a small group of highly productive scientists and scholars, that is responsible for growth of scientific knowledge. Crane shows that many scientific disciplines go through similar stages of initiation, growth, expansion, maturation, and decline. The productive scientists and scholars form a network of collaborators in promoting and developing their fields of study. The presence of an invisible college or network of productive scientists linking separate groups of collaborators within a research area has been evident in many studies (Chen, 2003; Shiffrin & Börner, 2004).

"Knowledge Mapping" or "Science Mapping,", based on content analysis, citation network analysis, and information visualization, has become an active area of research that helps reveal such an inter-connected, invisible college or network of scholars and their seminal publications and ideas.

According to Chaomei Chen in his book, *Mapping Scientific Frontiers* (Chen, 2003), science mapping helps "depict the spatial relations between research fronts, which are areas of significant activity. Such maps can also simply be used as a convenient means of depicting the way in which research areas are distributed and conveying added meaning of their relationships… By using a series of chronically sequential maps, one can see how knowledge advances. Mapping scientific frontiers involves several disciplines, from the philosophy and sociology of science, to information science, scientometrics, and information visualization."

In a National Academy of Sciences colloquium entitled "Mapping Knowledge Domains" (Shiffrin & Börner, 2004), the term "mapping knowledge domains" (or knowledge mapping) was used to "describe a newly evolving interdisciplinary area of science aimed at the process of charting, mining, analyzing, sorting, enabling navigation of, and displaying knowledge."

Two forces are contributing to the rapid development and overwhelming interest in knowledge mapping (we will use the term to encompass science mapping in the rest of the book). First, the availability of online publications, from scientific abstracts and indexes (A&I), full-text articles, and online preprints to digital dissertations, multimedia (e.g., videos and audios) magazine and journal articles, and multilingual Web-accessible patent filings, has made it possible to more systematically examine the scientific output of members of the invisible colleges. Secondly, the recent advances in text mining, network analysis, and information visualization techniques have provided more scalable and accurate methods to understand and reveal the interconnections between scientific disciplines and scholars.

Excellent research has also been performed in adopting econometric approaches to improve our quantitative knowledge of the sources of economic growth (Pakes & Sokoloff, 1995; Jaffe & Trajtenberg, 2002). Some econometric researchers have used patent citations to investigate the diffusion of technological information across institutions and over time and space (Jaffe & Trajtenberg, 2002). Such works often involve economics-based theoretical modeling, econometric analysis, and parameter estimation and can help draw inferences for knowledge diffusion or spillover. Interested readers are referred to (Pakes & Sokoloff, 1995; Jaffe & Trajtenberg, 2002).

In this chapter, we provide an overview of online research resources that are increasingly available for knowledge mapping analysis and were used in the following chapters of this book. We then describe the units of analysis and representation issues of relevance to such resources.

2. ONLINE RESOURCES FOR KNOWLEDGE MAPPING

Various online resources are available for mapping scientific knowledge. They vary from formal to informal publications; from text-based to multimedia presentations; and from academic literature to industry-relevant international patents.

- *Abstracts & Indexes*: A&I contain abstract and index (bibliographic) information of a given article and are used to locate articles, proceedings, and occasionally books and book chapters in various subjects. Most abstracts and indexes are available electronically. Public and university libraries often subscribe to such databases and services. Only a very few biological or scientific databases are searchable for free on the Web, primarily databases generated by the National Library of Medicine (http://www.nlm.nih.gov/), such as MEDLINE (medicine) or TOXLINE (toxicology). There are A&I databases in almost every subject area, e.g., BIOSIS (biology), COMPENDEX (engineering and technology), ERIC (education), etc.
- *Commercial full-text journal articles and digital libraries:* Many commercial publishers have made their online content available on the Web. The most prominent service of such type is provided by the *Web of Science* (http://scientific.thomson.com/products/wos/), a product of Thomson Scientific. The *Web of Science* provides seamless access to current and retrospective information from approximately 8,700 research

journals in the world. More recently, many professional societies have made their articles available through various digital libraries. For example, the ACM Digital Library (http://portal.acm.org/dl.cfm) contains 54,000 online articles from 30 journals and 900 proceedings of the Association for Computing Machinery. The IEEE Computer Society Digital Library (http://www.computer.org/portal/site/csdl/index.jsp) provides online access to eighteen IEEE journals and 150 proceedings in computer science.

- *Free full-text articles and e-prints*: There is also a grassroots movement initiated by the academic community to provide free access to journals and books. For example, on the *Free Medical Journals* site (http://www.freemedicaljournals.com/), you can find many important academic journals made available online, free and in full-text. *HighWire Press* (http://highwire.stanford.edu/lists/freeart.dtl), a service affiliated with Stanford University, is believed to be the largest archive of free full-text science articles. As of December 20, 2006, it provides access to more than 1.5 million free full-text articles in many subject disciplines. In some scientific disciplines, *e-prints* (scientific or technical documents circulated electronically to facilitate peer exchange, including preprints and other scholarly papers) are strongly encouraged and accepted by the community. For example, the *arXiv.org* service (http://arxiv.org/), supported by Cornell University, provides open access to about 400,000 e-prints in Physics, Mathematics, Computer Science, and Quantitative Biology.
- *Citation indexing systems and services:* In addition to accessing bibliographic and full-text content of scientific articles, aggregated and individualized citation information is critical in the assessment of highly-cited, influential papers and authors. The *Science Citation Index* (http://scientific.thomson.com/products/sci/), a product of Thomson Scientific, provides access to bibliographic information, abstracts, and cited references in 3,700 of the world's scholarly science and technical journals covering more than 100 disciplines. A recent service provided by *Google Scholar* (http://scholar.google.com/intl/en/scholar/) also supports broad access to scholarly literature. A user can search across many disciplines and sources: peer-reviewed papers, theses, books, abstracts, and articles. The service features many advanced search functionalities, including ranking articles based on how often an article has been cited in other scholarly literature. *CiteSeer* (http://citeseer.ist.psu.edu/citeseer.html) is another example of an advanced search system (for computer and information science literature) that is built upon citation information. It was one of the first digital libraries to support automated citation indexing and citation linking.

- *Electronic Theses and Dissertations (ETD)*: In addition to formal literature published in journals, magazines, and conference proceedings, Ph.D. and Master's theses and dissertations constitute a significant part of scientific knowledge generated. University Microfilms (UMI) was founded in 1938 to collect, index, film, and republish doctoral dissertations in microfilm and print. Currently UMI's dissertation abstract database has archived over 2.3 million dissertations and Master's theses. Some two million of them are available in print, microfilm, and digital format, via its *ProQuest* system (http://il.proquest.com/brand/umi.shtml). More recently, the Networked Digital Library of Theses and Dissertations (NDLTD, http://www.ndltd.org/) was formed to promote the adoption, creation, use, dissemination, and preservation of electronic analogues to the traditional paper-based theses and dissertations. Via ETD, graduate students learn electronic publishing as they engage in their research and submit their own work, often in a rich multimedia format. Universities learn about digital libraries as they collect, catalog, archive, and make ETDs accessible to scholars worldwide.
- *Patents*: Patent publications have often been used in evaluating science and technology development status worldwide (Narin, 1994). While academic literature represents fundamental scientific knowledge advancement; patents reveal scientific and technological knowledge that has a strong potential for commercialization. There are several governmental or intergovernmental patent offices that control the granting of patents in the world. United States Patent and Trademark Office (USPTO, http://www.uspto.gov/), European Patent Office (EPO, http://www.european-patent-office.org/index.en.php), and Japan Patent Office (JPO, http://www.jpo.go.jp/) issue nearly 90 percent of the world's patents (Kowalski et al., 2003). USPTO handles over 6.5 million patents with 3,500 to 4,000 newly granted patents each week. EPO handles over 1.5 million patents with more than 1,000 newly granted patents each week. JPO handles over 1.7 million patents with 2,000 to 3,000 newly granted patents each week. All three patent offices provide search systems for Web-based access.
- *Business and industry articles and reports*: Critical science and technology knowledge eventually flows from academic literature and patents to various industries and companies. At the other end of the knowledge mapping resources are various business and industry articles and reports; some are reported in general-interest science and technology magazines and newspapers, while others can be purchased from industry-specific consulting firms. For example, timely, in-depth industry-specific or technology-specific reports are available at sites such as: Forrester

(http://www.forrester.com), IDC (http://www.idc.com), and Gartner (http://www.gartner.com), among others.

- *Web sites, forums, chat rooms, blogs, multimedia sites, social networking sites, and virtual worlds:* In addition to the abovementioned formal publications generated by scholars, students, and industry practitioners, the Web has enabled everyone to become an online publisher. Potentially interesting scientific, product, and marketing information has been generated and disseminated over the Web. However, the diversity and quality of such information vary significantly. It is often much more difficult to use such Web-based, self-produced information for technology assessment or knowledge mapping.

3. UNITS OF ANALYSIS AND REPRESENTATIONS

For knowledge mapping analysis, pre-processing of raw online resources is needed. Each article, patent, or report needs to be processed to identify key indicators for further analysis and comparison. Among the most common units of analysis for knowledge mapping are: authors or inventors, publications and publication outlets, institutions (companies or universities), countries or regions, subject and topic areas (broad categories or specific topics), and timeline (publication date).

- *Authors or inventors:* The most critical unit of analysis for knowledge mapping consists of the researchers, authors, and inventors who are the productive members in the invisible college. Extracting the author or inventor field from various knowledge sources is a non-trivial task. Although html, XML, and structured database representations have made automatic name identification easier (than in the paper-based format), author name extraction and identification is difficult in different cultural contexts (e.g., recognizing Chinese names), especially when a publication does not contain complete first and last names. For example, how many different researchers by the name of "W. Zhang" or "L. Liu" are there in the Chinese Academy of Sciences (one of the most productive and largest academic research institutions in the world)?
- *Publications and publication outlets:* Different academic publications have different levels of prestige; most are measured based on their Impact Factor (an aggregate, normalized number based on citation counts). For example, in 2005the Impact Factor of *Science* was 30.927; while the *Journal of Computational Biology* Impact Factor was 2.446. There are many other publications that do not even have an Impact Factor score. In order to determine the value and impact of a researcher's work, quality is more important than quantity. Quality is often determined

based on the prestige of a publication outlet. In addition, the number of citations is also a major determinant. A seminal or landmark paper can often help define a person's career or a particular field. For example, while many good academic articles are cited hundreds of times, Albert Einstein's seminal paper, "Can quantum-mechanical description of physical reality be considered complete?", that appeared in *Physical Review* in 1935 was cited 3,753 times (based on a search on *Google Scholar*). Based on analysis results reported by *ScienceWatch* (http://www.sciencewatch.com/), the most cited paper of the past two decades (1983-2002) was: Chomczynski, N. Sacchi, "Single-step method of RNA isolation by acid guanidinium thiocyanate phenol chloroform extraction," *Analytical Biochemistry,* 162(1): 156-9, 1987. The paper received a citation count of 49,562 (based on data from Thomson Scientific's *Web of Science*). However, correctly parsing and identifying unique publication names is a difficult task as many databases record those names in cryptic, short-hand forms, e.g., *Analyt. Biochem, Proc. natn. Acad. Sci., J. biol. Chem., J. gen. Physiol., Physiol, Lond., etc.* While many are easily recognized by domain scientists, a computer program would have difficulty parsing them correctly.

- *Institutions:* While researchers publish their research, often the institutions (companies or universities) where they work own their intellectual property. An analysis based on institutional output and productivity can help depict an institution's relative strength and position in the competitive knowledge landscape. Knowledge mapping can help reveal not just the invisible college of researchers, but the "invisible college of institutions." A comparison between basic university research and applied industry invention can also be used to help understand the progression and impact of knowledge creation.
- *Countries or regions:* Similar to institutional analysis, it is often important to analyze publications (especially patents) based on their countries or regions (e.g., Europe vs. Asia) of origin. This kind of analysis is useful for depicting a competitive international landscape and is often relied upon for governmental research policy and funding decisions. For example, the U.S. National Nanotechnology Initiative (NNI) has performed excellent cross-regional analyses for worldwide nanotechnology research, development, and funding.
- *Subject and topics areas:* Academics are often defined by their traditional academic boundaries in colleges or departments. However, a researcher could work in several (often related) subject or topic areas. Academic publication outlets are also defined by their fields of interest and focus. While most academic journals provide a list of topics of interest; some information resources are more comprehensive in their listings. For

example, the USPTO provides a detailed patent classification scheme (USPC), which consists of two levels. The first level contains about 450 categories; while the second contains about 160,000 categories. In addition to these predefined subject categories, important topic-specific keywords, phrases, and concepts can be extracted from the title, abstract, and text body of an article. However, advanced Natural Language Processing (NLP) techniques are needed for such topic identification purposes.

- *Timeline*: All scientific disciplines evolve over time. Most of the online resources for mapping scientific knowledge contain explicit publication dates. Dynamic analysis and visualization of changes in research topics and citation networks can help reveal advances in scientific knowledge.

4. QUESTIONS FOR DISCUSSION

1. What other informal online resources could be used for knowledge mapping?
2. How can the Web and social networking sites help promote the "invisible college" of scholars and researchers?
3. How can multimedia contents (e.g., tables, images, photos, audios, and videos) produced by scholars be analyzed to reveal knowledge generated in a scientific discipline?

Chapter 3

KNOWLEDGE MAPPING: ANALYSIS FRAMEWORK

CHAPTER OVERVIEW

Three types of analysis are often adopted in knowledge mapping research: *text mining, network analysis,* and *information visualization.* Text mining consists of two significant classes of technique: Natural Language Processing (NLP) and content analysis. In NLP, we describe automatic indexing and information extraction techniques that are effective and scalable for concept extraction. In content analysis, clustering algorithms, self-organizing map, multidimensional scaling, principal component analysis, co-word analysis, and PathFinder network are techniques often adopted for knowledge mapping analysis. Network analysis is reviewed based on research in social network analysis (SNA) and complex networks. In SNA, we review research that detects subgroups, discovers patterns of interactions, and identifies roles of individuals. In complex networks, we summarize research in network models, topological properties, and evolving networks. The last section reviews information visualization research of relevance to knowledge mapping. Seven information representation methods are discussed: 1D, 2D, 3D, multidimensional, tree, network, and temporal. Two useful user-interface interaction methods, overview + detail and focus + context, are also presented. We believe these knowledge mapping analysis and visualization methods can be applied to most of the online resources presented earlier.

1. TEXT MINING

Text mining, sometimes alternately referred to as *text data mining*, refers generally to the process of deriving high quality information from text (according to Wikipedia, http://en.wikipedia.org/wiki/Text_mining). Text mining usually involves the process of structuring the input text (usually parsing, along with the addition of some derived linguistic features and the removal of others, and subsequent importation into a database), deriving patterns within the structured data, and finally evaluation and interpretation of the output (Chen & Chau, 2004). Typical text mining tasks include entity and relation extraction, text categorization, text clustering, sentiment analysis, and document summarization (Chen, 2001).

For knowledge mapping research, text mining can be used to identify critical subject and topic areas that are embedded in the title, abstract, and text body of published articles. While most structured fields (such as authors, publication outlets, dates of publication, institutions, etc.) can be parsed from online resources, extracting meanings or semantics from multimedia publications requires advanced computational techniques. Different processing algorithms are needed for different media types, e.g., text (Natural Language Processing), image (color, shape, and texture-based segmentation), audio (indexing by sound and pitch), and video (scene segmentation). Non-text, multimedia content extraction techniques are still under active research and development. Our discussion will be limited to text-based techniques.

1.1 Natural Language Processing

Automatic indexing: Automatic indexing (Salton, 1989) is a method commonly used to represent the content of a document by means of a vector of keywords or terms. The Bag of Words (BOW) representation has often been used as a baseline implementation for information retrieval and text mining research. When implemented using multi-word matching, a Natural Language Processing noun phrasing technique can capture a richer linguistic representation of a document. Most noun phrasing techniques rely on a combination of part-of-speech-tagging (POST) and grammatical phrase-forming rules. This approach has potential to improve precision over other word-based document indexing techniques. Examples of noun phrasing tools include MIT's Chopper, Nptool (Voutilainen, 1997), and Arizona Noun Phraser (Tolle & Chen, 2000).

Information extraction: Information extraction is another computationally effective method to identify important concepts from text documents. It can extract entities of interest (also referred to as entity extraction), such as persons (e.g., "John Doe"), locations (e.g., "Washington, D.C."), and organizations (e.g., "National Science Foundation"), and identify relationships between entities. Other entities that are often extracted from unstructured textual narratives include: dates, times, number expressions, dollar amounts, email addresses, and Web addresses (URLs). Such information can be extracted based on either human-created rules or statistical patterns occurring in text. Most existing information extraction approaches combine machine learning algorithms such as neural networks, decision tree, Hidden Markov Model, and entropy maximization with a rule-based or a statistical approach. The best systems have been shown to achieve more than 90% accuracy in both precision and recall rates when extracting persons, locations, organizations, dates, times, currencies, and percentages from newspaper articles (Chinchor, 1998).

While automatic indexing and information extraction techniques are computationally scalable and feasible for large-scale knowledge mapping research, more advanced and fine-grained computational linguistics techniques are being developed in the NLP community. Sentence-level analysis, including context-free grammar and transformational grammar, can be performed to represent grammatically correct sentences. In addition, semantic analysis based on techniques such as semantic grammar and case grammar can be used to represent semantic (meaning) in sentences and stories. However, most of these full-scale linguistic and semantic analysis techniques lack scalability across different domains and are not yet suitable for large-scale knowledge mapping research (Chen, 2001).

1.2 Content Analysis

Based on automatic indexing or information extraction techniques, documents are often represented as a vector of features (i.e., keywords, noun phases, or entities). Articles that are collected and grouped based on authors, institutions, topic areas, countries, or regions can be analyzed to identify the underlying themes, patterns, or trends. Popular content analysis techniques include: *Clustering Algorithms, Self-Organizing Map (SOM), Multidimensional Scaling (MDS), Principal Component Analysis (PCA), Co-Word Analysis, and PathFinder Network.*

Clustering Algorithms: Everitt (Everitt, 1974) defines a cluster as "a set of entities which are alike, and entities from different clusters are not alike." Clustering algorithms are used to organize (group) similar documents or topics in a hierarchical structure. There are two types of hierarchical cluster analysis: agglomerative and divisive. The agglomerative approach starts with each point as a separate cluster. Each point is merged successively into a larger cluster based on their degree of similarity. Conversely, divisive hierarchical cluster analysis begins with only one large cluster of points. The large cluster is divided successively into smaller clusters based on their degree of similarity. Both approaches produce a dendrogram to represent a hierarchy of points and their associated clusters. Hierarchical agglomerative clustering (HAC) algorithms are the most commonly used method for document clustering (Willett, 1988). One of the most popular HAC algorithms is Ward's clustering (Ward, 1963). Over time, it has been widely used in various domains: astrophysics, pattern recognition, applied statistics, etc. In 1984, Murtagh proposed the Reciprocal Nearest Neighbor (RNN) approach (Murtagh, 1995), which is significantly faster than the original Ward's implementation. The time complexity of the algorithm was reduced from $O(N^3)$ to $O(N^2)$. Roussinov and Chen (Roussinov & Chen, 1999) performed a systematic comparison of the RNN-based Ward's algorithm with the SOM technique for document clustering. Both techniques are computationally efficient for large-scale knowledge mapping applications.

Self-Organizing Map (SOM): The Self-Organizing Map, developed by Kohonen (Kohonen, 1989; Kohonen, 1995), is an unsupervised, two-layered neural network used for clustering and dimension reduction. An advantage of SOM over other clustering algorithms is its ability to visualize high dimensional data using a two-dimensional grid while preserving similarity between data points. It is a technique similar to Multidimensional Scaling (MDS) (Jain & Dubes, 1988). In SOM, each input node corresponds to a dimension (e.g., a keyword). Each output node corresponds to a node in a two-dimensional grid. The network is fully connected in that every output node is connected to every input node with some connection weight. During the training phase, the inputs (e.g., documents represented as vectors of keywords) are presented several times to the SOM to decide their proper placement on the output grid. Connection weights associated with the input and output nodes are adjusted (learned) to ensure that similar inputs are grouped in a close proximity on the two-dimensional grid. After training, all inputs (e.g., documents) can be grouped and placed on a two-dimensional map. Topics (of similar documents) are often represented as regions on the map, where larger regions represent more important topics. The SOM-generated categories were found to be comparable to those generated by human subjects (Orwig et al., 1997). Chen and his team (Chen et al., 1996) developed a multi-layered SOM (ET-Map) to categorize 110,000 Internet Web pages according to their content. Kohonen and his colleagues (Kohonen et al., 2000) adopted SOM to map 6.8 million patent abstracts onto a one-million-node SOM. Several SOM implementations have taken advantage of the sparse input feature vector representation to improve the speed and scalability of their algorithms.

Multidimensional Scaling (MDS): Multidimensional Scaling and Principal Component Analysis (PCA) are two classical and widely used techniques for dimension reduction. They are well known, easy to implement, and computationally efficient. One way to apply MDS is to take a set of p-dimension vectors and approximate them on the two- or three-dimensional Cartesian coordinate space. Beginning with the matrix of distance or dissimilarity between objects, we first compute an initial configuration using Singular Value Decomposition (SVD) (Cox & Cox, 1994; Forsythe et al., 1977; McQuaid et al., 1999). We then measure the information loss between the original matrix and the initial configuration using the STRESS (standardized residual error sum of squares) metric of Kruskal (Kruskal, 1964). It finds a new configuration with smaller information loss than the initial configuration by using an isotonic regression algorithm (Grotzinger & Witzgall, 1984) to obtain fitted distances and a conjugate gradient descent algorithm to optimize. The algorithm repeats until a threshold information loss is reached or until a threshold number of iterations is performed. Similar to SOM for knowledge mapping, MDS produces a two-dimensional display which agrees with human perception of document similarity (McQuaid et al., 1999). Also based on SVD, Latent Semantic Indexing (LSI) uses a term-document matrix, which describes the occurrences of terms in documents (Deerwester et al., 1990). It then transforms the original matrix into a relationship between the terms and (latent) *concepts*, and a relation between the documents and the same concepts. The terms and documents are now indirectly related through the concepts. LSI can be used effectively for concept extraction and association for knowledge mapping.

Principle Component Analysis (PCA): Principal component analysis is central to the study of multivariate data. Although one of the earliest multivariate techniques, it continues to be the subject of much research, ranging from new model-based approaches to algorithmic ideas from neural networks. It is extremely versatile with applications in many disciplines (Jolliffe, 2002). The central idea of PCA is to reduce the dimensionality of a data set in which there are a number of interrelated variables, while retaining as much as possible the variation present in the data set. The reduction is achieved by transforming to a new set of variables, the principal components, which are uncorrelated, and which are ordered so that the first few retain most of the variation present in all of the original variables. Computation of the principal components reduces to the solution of an eigenvalue-eigenvector problem for a positive-semidefinite symmetric matrix (Jollife, 2002). In knowledge mapping, PCA can be used to extract principal components that represent interrelated keywords or topics.

Co-word Analysis: Word-occurrence patterns in text originated in the co-word analysis method developed in the 1980s (Callon et al., 1986). The outcome of co-word analysis is typically depicted as a network of concepts. Given a corpus of *N* documents, each document can be indexed by a set of unique keywords or terms. If two terms, t_i and t_j, appear together in multiple documents, their probability of co-occurrence can be computed using different formulas. A matrix of term co-occurrence defines a network of concepts. In some past research, such a matrix is referred to as a *concept space* (Chen et al., 1996; Chen et al., 1997). The original co-word analysis prunes a concept using a triangle inequality rule on conditional probabilities. If a path that is shorter than the direct path can be found from term t_i to t_j, then the shorter one is chosen. By raising or lowering a threshold, the number of valid links in a network can be decreased or increased.

PathFinder Network (PFNET): According to Börner et al., (Börner et al., 2003), "PathFinder algorithms take estimates of the proximities between pairs of terms as input and define a network representation of the items that preserves only the most important links." The input is pairs of terms, and the pairs are linked as output only if their co-occurrence weights are the highest in their respective vectors. By emphasizing only the most prominent links, PFNET reduces the user's cognitive overload in browsing a network of interrelated concepts (White et al., 2003). In PFNET, paths are required not to violate the triangle inequality $d(a,c) \leq d(a,b) + d(b,c)$, where d is the distance between points a, b, and c. A Spring Embedder algorithm (Kamada & Kawai, 1989) is often used to display the network of terms based on the relative strengths between any pairs of terms. The algorithm aims to provide a layout that avoids crossed links and overlapping nodes. White et al. (White et al., 2003) adopted both SOM and PFNET in creating "localized" mapping of the 24 most relevant terms given a single input term, a medical subject heading, a co-cited author, or a co-cited journal from the *Proceedings of the National Academy of Sciences (PNAS)*, 1971-2002. Mane and Börner (Mane & Börner, 2003) adopted Klienberg's burst detection algorithms, PFNET, and graph layout techniques to generate maps that support the identification of major research topics and trends in *PNAS*, 1982-2001. Both the general co-word analysis and the PFNET algorithm implementation have been shown to be valuable for mapping scientific knowledge.

2. NETWORK ANALYSIS

Recent advances in *social network analysis* and *complex networks* have provided another means for studying the network of productive scholars in the invisible college.

2.1 Social Network Analysis

A collection of methods that are recommended in literature for studying networks are Social Network Analysis (SNA) techniques (McAndrew, 1999; Sparrow, 1991; Xu & Chen, 2005a). Because SNA is designed to discover patterns of interactions between social actors in social networks, it is especially apt for co-authorship network analysis. SNA is capable of detecting subgroups (of scholars), discovering their patterns of interactions, identifying central individuals, and uncovering network organization and structure. It has also been used to study criminal networks (Xu & Chen, 2005a; Xu & Chen, 2005b).

Subgroup Detection: A collaboration or co-authorship network can be partitioned into subgroups consisting of individuals who closely interact with each other. Given a network, traditional data mining techniques such as cluster analysis may be employed to detect underlying groupings that are not otherwise apparent in the data. Burt (Burt, 1976) applied hierarchical clustering methods based on structural equivalence measures (Lorrain & White, 1971) to detect subgroups in a social network. Two nodes are structurally equivalent if they have identical links to and from all other nodes in the network. Since perfectly equivalent nodes rarely exist in reality, this measure is relaxed to be an indicator of the extent to which two nodes are equivalent. With structural equivalence measures between nodes, a hierarchical clustering algorithm partitions a network into subgroups so that members within a group are more similar to each other and members belonging to different groups are more different from each other. Cliques whose members are fully or almost fully connected can also be detected based on clustering results.

Discovery of Patterns of Interactions: Patterns of interactions between subgroups can be discovered using an SNA approach called *blockmodel analysis* (Wasserman & Faust, 1994; Xu & Chen, 2005b). This approach was originally designed to interpret and validate theories of social structures. When used in collaboration or co-author network analysis, it can discover patterns of inter-group interactions and associations and help reveal the overall structure of networks under study. Blockmodeling usually follows clustering, and then determines the presence and absence of an association between a pair of subgroups by comparing the density of the links between these two subgroups with a predefined threshold value. When the link density is greater than the threshold value, an inter-group association presents, indicating that these two subgroups interact with each other constantly and thus have a strong association. Using blockmodeling, a complex network is reduced to a simpler structure by summarizing individual interaction details into interactions between groups, so that the overall structure of the network becomes more salient.

Roles of Individuals: Centrality deals with the roles of individuals in a network. Several measures, such as *degree, betweenness,* and *closeness,* are related to centrality (Wasserman & Faust, 1994). The degree of a particular node is the number of direct links it has; its betweenness is the number of geodesics (shortest paths between any two nodes) passing through it; and the closeness is the total number of all the geodesics between that particular node and every other node in the network. Although these three measures are all intended to describe the importance or centrality of a node, they have different interpretations for the roles network members play. An individual with high degree, for instance, may imply his or her leadership; while an individual with high betweenness may be a gatekeeper or connector in the network.

2.2 Complex Networks

Complex networks of individuals and other entities have been traditionally studied under the random graph theory (Albert & Barabasi, 2002). However, later studies suggested that real-world complex networks (such as collaboration or co-authorship networks) may not be random but may be governed by certain organizing principles. This prompted the study of real-world networks. These studies have explored the topology, evolution and growth, robustness and attack tolerance, and other properties of networks.

Network Models: Three broad models of network topologies have emerged (Albert & Barabasi, 2002): *random graphs*, *small-world* networks, and *scale-free* networks. Random graphs are networks in which any two nodes are connected with a fixed probability p, thus edges are distributed randomly among nodes of the network. Small-world networks are not random networks and have relatively small path lengths despite their often large size (Watts & Strogatz, 1998). In scale-free networks the degrees (number of edges) of nodes follow a power law distribution (Barabasi & Albert, 1999). Some of the networks that have been studied include the World Wide Web (Albert et al., 1999; Kumar et al., 2000), citation networks (Jeong et al., 2000), and co-authorship networks (Newman, 2001a). These networks were found to have similar topological, evolutionary and robustness characteristics (Albert & Barabasi, 2002). They were found to be predominantly small-world and scale-free.

Topological Properties: Topological properties of networks help us study the network as a whole instead of studying the individual constituents. Three concepts dominate the statistical study of the topology of networks:

small-world, *clustering*, and *degree distribution* (Albert & Barabasi, 2002).

Small-world: The small-world concept is based on the fact that large networks often have small path lengths between their nodes. This concept is an important one as it can depict the ease of communications within a network. Communications can range from the spread of disease in human populations to the spread of ideas in a collaboration network. A widely cited example of a small-world network study is the "six degrees of separation" study by psychologist Stanley Milgram, who concluded that there was a path of acquaintances with a typical length of about six between most pairs of people in the United States (Kochen, 1989). The small-world property is measured by the average shortest path length that is obtained by averaging the shortest paths between all pairs of nodes in a network (Albert & Barabasi, 2002). For instance, the average shortest path length between two actors in a network of movie actors (225,226 nodes) was found to be 3.65 (Watts & Strogatz, 1998). The average shortest path length between co-authors in the MEDLINE collection (1.5 million nodes) was found to be 4.6 (Newman, 2001a). There has been research on the phenomenon that leads to the short path lengths in real-world networks. It has been suggested that shortcuts between nodes that otherwise may not be connected reduce the average path length in small-world networks (Watts, 1999). This is especially true in social networks where people are likely to be friends with other individuals outside their immediate friend circle.

Clustering: Cliques that represent circles of friends and acquaintances often form in social networks. For instance, authors often collaborate with the same set of people in a co-authorship network. Cliques also form in networks that do not involve people, for example, related sites on the Web often point to each other through hyperlinks. This inherent tendency to cluster is quantified by the clustering coefficient (Watts & Strogatz, 1998). The clustering coefficient is measured by the ratio of the number of edges that exist in a network to the total number of possible edges (Albert & Barabasi, 2002). Real-world networks tend to have relatively high clustering coefficients as compared to random graphs. The movie actor network had a clustering coefficient of 0.79 (Watts & Strogatz, 1998) and the MEDLINE co-authorship network had a coefficient of 0.066 (Newman, 2001); both values are several orders of magnitude higher then their random counterparts.

Degree distribution: Nodes in a network have different numbers of edges connecting them. The number of edges connected to a node is called its degree. The spread of node degrees is given by a distribution function $P(k)$, which gives the probability that a randomly selected node has exactly 'k' edges (Albert & Barabasi, 2002). The distribution functions of most real-world networks follow power law scaling with exponents ranging from 1.0

to 3.0 Albert & Barabasi, 2002). The movie actor network has a power law degree distribution with an exponent of 2.3 (Watts & Strogatz, 1998). The MEDLINE co-authorship network was found to have an exponent of 1.2 (Newman, 2001a). The degrees of nodes are also used to study the growth and evolution of a network.

Using data from three bibliographic databases in biology, physics, and mathematics, Newman constructed networks in which nodes are scientists, and two scientists are connected if they have coauthored a paper (Newman, 2003). He used various network topological properties to answer a broad variety of questions about collaboration patterns, such as the number of papers authors write, how many people they write them with, what the typical distance between scientists is in the network, and how patterns of collaboration vary between subjects and over time.

Evolving Networks: Most real-world networks are not static and grow due to the addition of nodes and/or links. For instance, the World Wide Web grows exponentially by the addition of new Web pages and a co-authorship network grows by the addition of collaborators. The growth leads to changes in the topological characteristics of the networks. Albert and Barabasi (Albert & Barabasi, 2002) identified two factors in the evolution of a scale-free network: (1) growth: networks expand continuously by adding new nodes and, (2) preferential attachment: new nodes attach preferentially to nodes that are already well connected, an effect called "rich-get-richer." The preferential attachment concept assumes that the probability that a new node will connect to an existing node i depends on the degree of the node i. The higher the degree of i, the higher the probability that new nodes will attach to it. The functional form of preferential attachment ($\Pi(k)$) for a network can be measured by observing the nodes present in the network and their degrees at a particular time, t. After adding new nodes (time = $t+1$), plotting the relative increase as a function of the earlier degree gives the $\Pi(k)$ function (Jeong et al., 2003). Preferential attachment has been studied for citation and co-authorship networks, actor networks and the Internet and has been found to follow the power law distribution (Jeong et al., 2003; Newman, 2001b). In other cases $\Pi(k)$ may grow linearly till a point and then fall off. This usually happens at high degrees, implying that high degree nodes are unable to attract new nodes. For instance, Newman (Newman, 2001b) found that individuals with a large number of collaborators in a co-authorship network did not attract many new ones.

Constraints on the number of links that a node can attract may be due to aging or cost (Amaral et al., 2000). Since the growth of the network may be over time, some high degree nodes might become too old to participate in the network (e.g., actors in a movie network). It might also become too costly for a node to attach to a large number of nodes (e.g., a router in a network slows down when it has too many connections). Constraints on the growth may be domain specific and have been studied in many domains. For instance, in plant-animal pollination networks some animals cannot pollinate certain plants; hence a link cannot be established (Jordano et al., 2003). This is an example of a cost constraint. In criminal networks, trust may restrict the growth of networks. Criminals and terrorists do not include many people in their inner trust circle (Klerks, 2001). In addition, disruption might restrict growth in criminal networks. Individuals may be jailed, wounded or killed and thus not contribute to the growth (Xu & Chen, 2005b).

3. INFORMATION VISUALIZATION

The last step in the knowledge "mapping" process is to make knowledge transparent through the use of various information visualization (or mapping) techniques. *Information representation* and *user-interface interaction* are two dimensions often considered in information visualization research (Zhu & Chen, 2005).

3.1 Information Representation

Shneiderman (1996) proposed seven types of information representation methods including the *1D (one-dimensional), 2D, 3D, multidimensional, tree, network,* and *temporal* approaches. We use this framework as the basis to review related research and present selected examples.

1D representation: The 1D approach represents abstract information as one-dimensional visual objects and displays them on the screen in a linear or a circular manner (Eick et al., 1992; Hearst, 1995). 1D representation has been applied to display either the contents of a single document (Hearst, 1995) or an overview of a collection of documents (Eick et al., 1992). Colors usually represent some attributes of each visual object. For instance, colors indicate type of document in the SeeSoft system (Eick et al., 1992) and depict the location of search terms in a document in Tilebar (Hearst, 1995). Figure 3-1 displays an interface of the "tile bar" that shows the occurrence of search terms in documents. The darkness of each tile indicates the frequency of a search term in a document.

2D representation: A 2D approach represents information as two-dimensional visual objects. Visualization systems based on 2D output of a self-organizing map (SOM) (Kohonen, 1995; Chen et al., 1996; Huang et al., 2003; Huang et al., 2004) belong to this category. Such systems display categories created over a large collection of textual documents, with the layout of each category based on its location in the two-dimensional output of the SOM. Spatial proximity on the interface represents the semantic proximity of the categories created. The challenge in this approach is to help users deal with a large number of categories that have been created for the large volume of textual data. The CancerMap system described in Chen et al. (2003) adopted the SOM and the Arizona Noun Phraser (Tolle & Chen, 2000) to generate a subject map automatically. Figure 3-2 presents two consecutive screen shots, displaying the top-level categories and sub-categories under the category of "Liver Neoplasm." The empirical study described in Chen et al. (2003) indicates that this approach generated a meaningful subject hierarchy to supplement or enhance human-generated hierarchies in digital libraries.

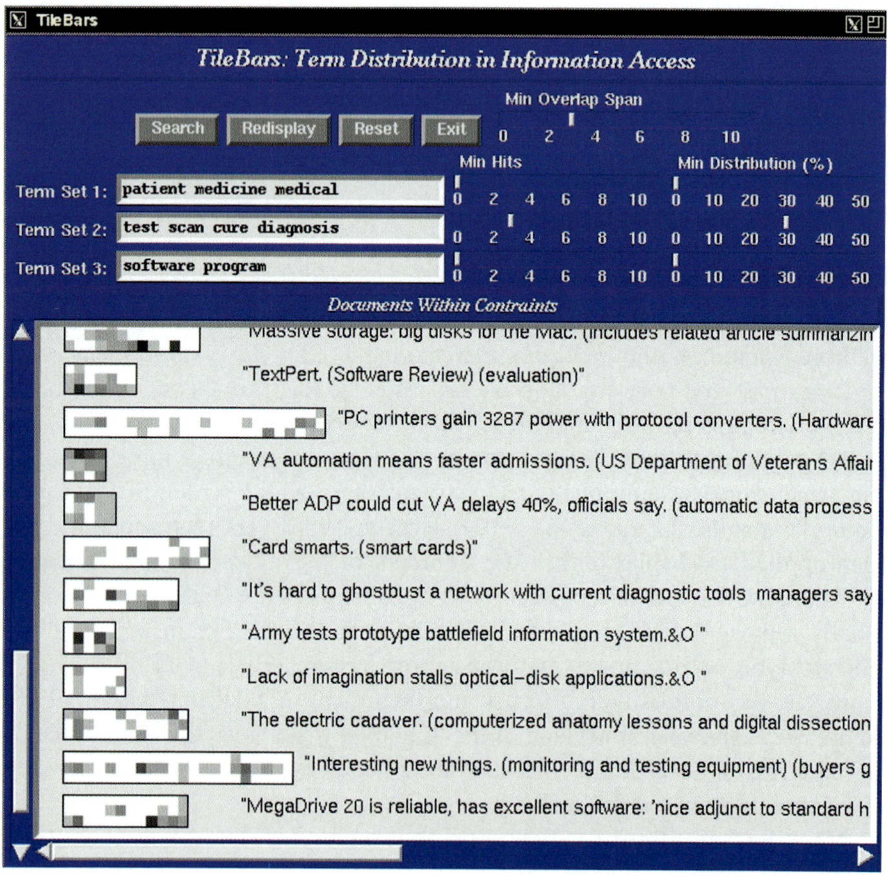

Figure 3-1. TileBar uses 1D representation to show the term-document relevance (http://www.acm.org/sigchi/chi95/Electronic/documnts/papers/mah_fg4.gif @ 1995 ACM, Inc.).

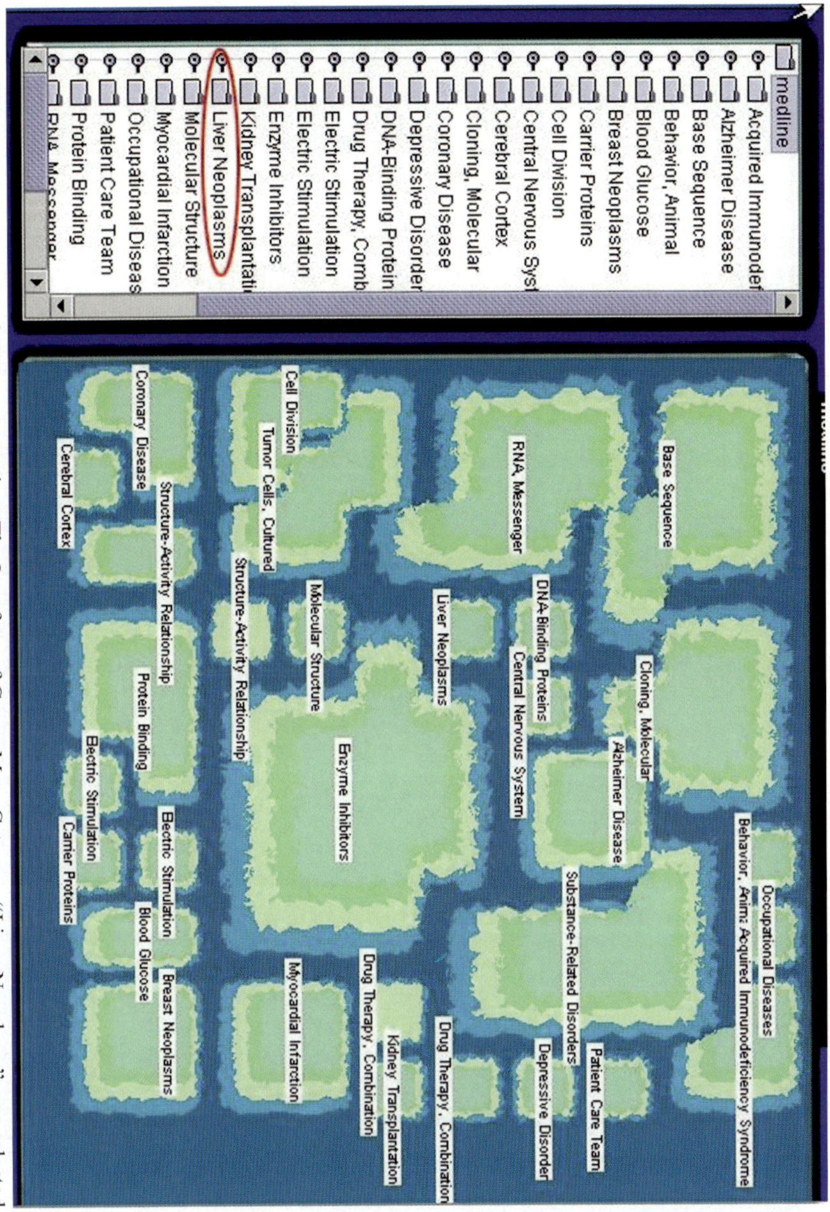

Figurse 3-2a and b. Example of 2D representation: The Interface of CancerMap. Category “Liver Neoplasm” was selected at the top level (3-2a; this page) and the sub-map of “Liver Neoplasm” (3-2b; next page) was displayed.

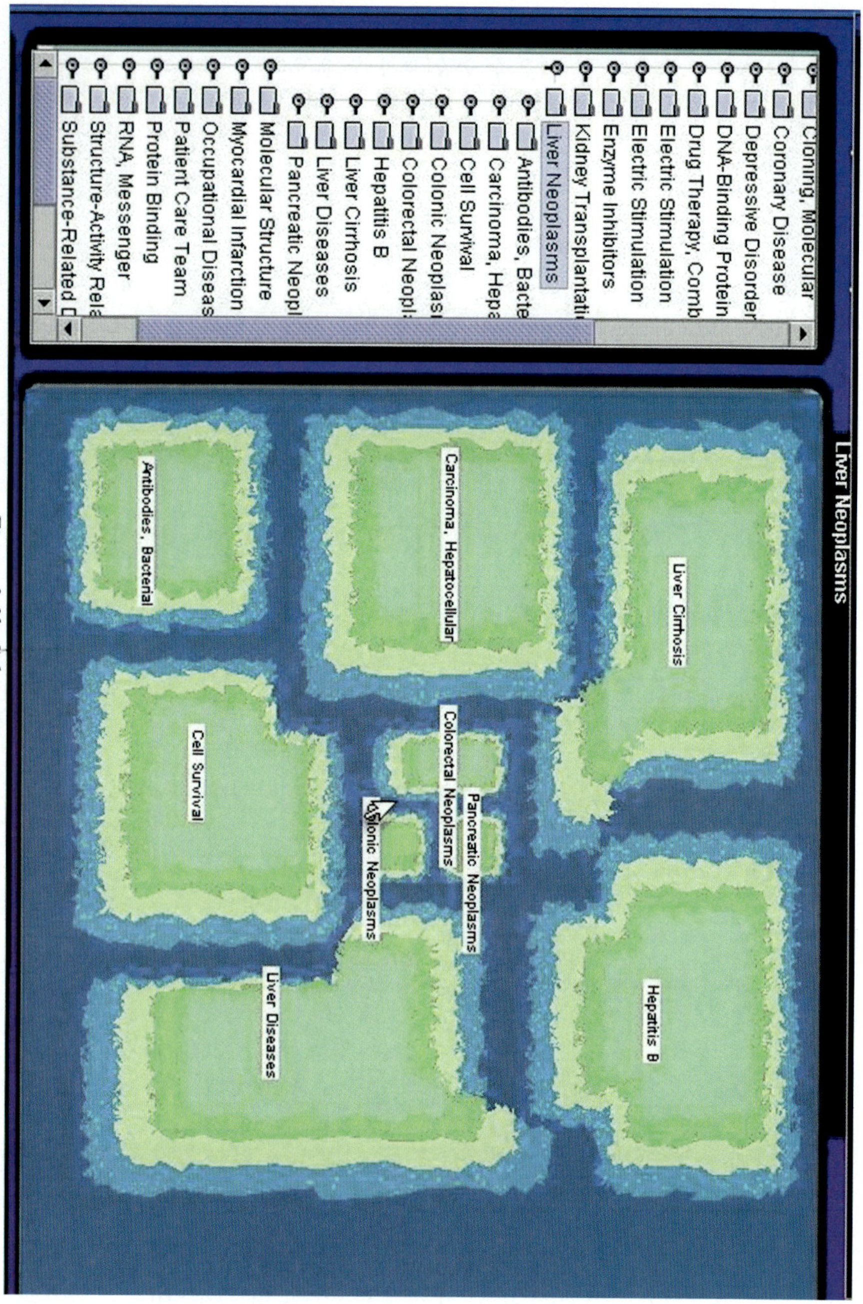

Figure 3-2b. Submap.

3D representation: A 3D approach represents information as three-dimensional visual objects. One example is the WebBook system (Card et al., 1996) that folds Web pages into three-dimensional books. Realistic metaphors such as rooms (Card et al., 1996), bookshelves (Card et al., 1996), or buildings (Andrews, 1995) are employed to depict abstract information. Visualization systems using a 3D version of a tree or network representation also belong to this category. One example is the 3D hyperbolic tree created by (Munzner, 2000) to visualize large-scale hierarchical relationships. Figure 3-3 shows a screenshot of WebBook, where the book metaphor is applied to organize Web pages from the same Web site.

Figure 3-3. Example of 3D representation: The WebBook (http://acm.org/sigchi/chi96/proceedings/papers/Card/skc1txt.html, ©1996 ACM, Inc.).

Multidimensional representation: The multidimensional approach represents information as multidimensional objects and projects them into a three-dimensional or a two-dimensional space. This approach often represents textual documents as a set of key terms that identify the theme of a textual collection. A dimensionality reduction algorithm such as multidimensional scaling (MDS), hierarchical clustering, principal components analysis (PCA), or self-organizing map (SOM) is used to project document clusters or themes into a two-dimensional or three-dimensional space. The SPIRE (Spatial Paradigm for Information Retrieval and Exploration) system presented in Wise et al. (1995) and the VxInsight system in (Boyack et al., 2002) belong to this category. Figures 3-4 and 3-5 display two types of visualization developed for the SPIRE system. The Galaxy (Figure 3-4) clusters 567,437 abstracts of cancer literature based on the semantic similarity, whereas the ThemeView (Figure 3-5) visualizes relationships among topics of a collection of documents. Glyph representation, another type of multidimensional representation, uses graphical objects or symbols to represent data through visual parameters that are spatial (positions x or y), retinal (color and size), or temporal (Chernoff, 1973). It has been applied in various social visualization techniques to describe human behavior during computer-mediated communication (CMC) (Zhu & Chen, 2001; Donath, 2002).

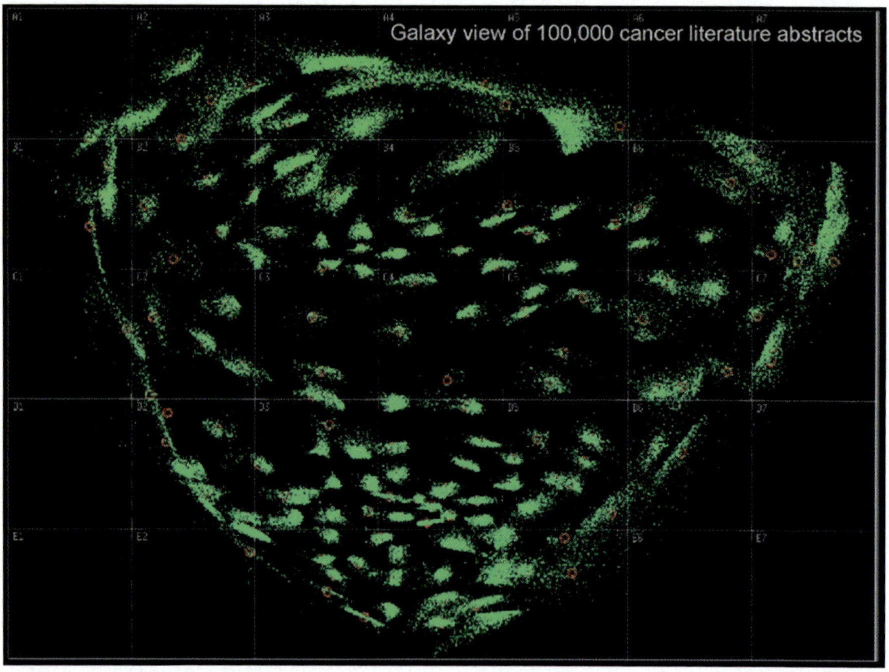

Figure 3-4. Example of multidimensional representation: Galaxy visualization of text documents (http://www.pnl.gov/infoviz/gal_cancer800.gif, reprint with permission).

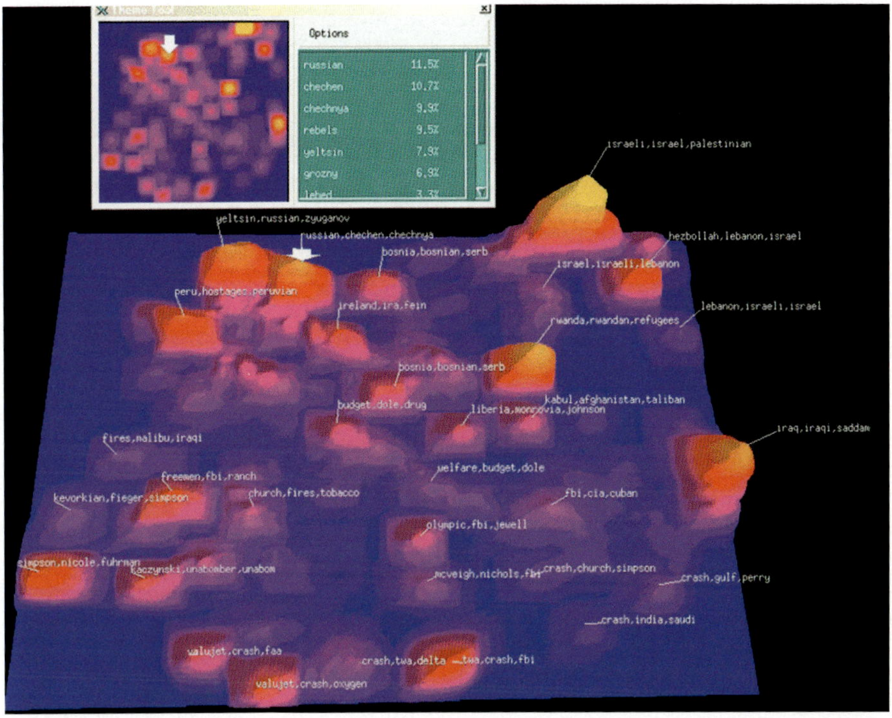

Figure 3-5. Example of multidimensional representation: ThemeView. The height of a peak indicates the strength of a given topic in the collection of documents (http://www.pnl.gov/infoviz/theme_cnn800.gif, reprint with permission).

Tree representation: The tree approach is often used to represent hierarchical relationships. The most common example is an indented text list. Other tree-based systems include the Tree-Map (Johnson & Shneiderman, 1991), the Cone Tree (Robertson et al., 1991), and the Hyperbolic Tree (Lamping et al., 1995). One crucial challenge to this approach is that the number of nodes grows exponentially as the number of tree levels increases. As a consequence, different layout algorithms have been applied. For instance, the Tree-Map (Johnson & Shneiderman, 1991) allocates space according to attributes of nodes, while the Cone Tree (Robertson et al., 1991) takes advantage of the 3D visual structure to pack more nodes on the screen. Figure 3-6 displays the visual interface of the Cat-a-Cone system (Hearst & Karadi, 1997) that applies the 3D Cone Tree to visualize hierarchies in Yahoo. The Hyperbolic Tree (Lamping et al., 1995), on the other hand, projects sub-trees on a hyperbolic plane and puts the plane into the range of display. A 3D version of the hyperbolic tree has also been developed by Munzner (2000) to visualize large-scale hierarchies (Figure 3-7).

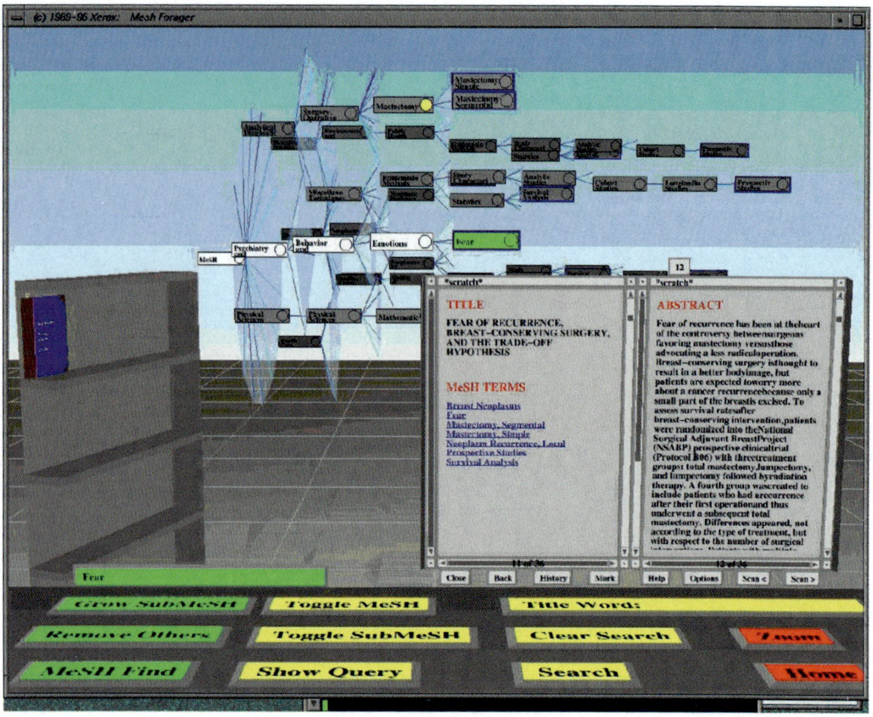

Figure 3-6. Example of tree representation: Cat-a-Cone tree that displays hierarchies in Yahoo (http://www.sims.berkeley.edu/~hearst/cac-overview.html, @ 1997 ACM, Inc.).

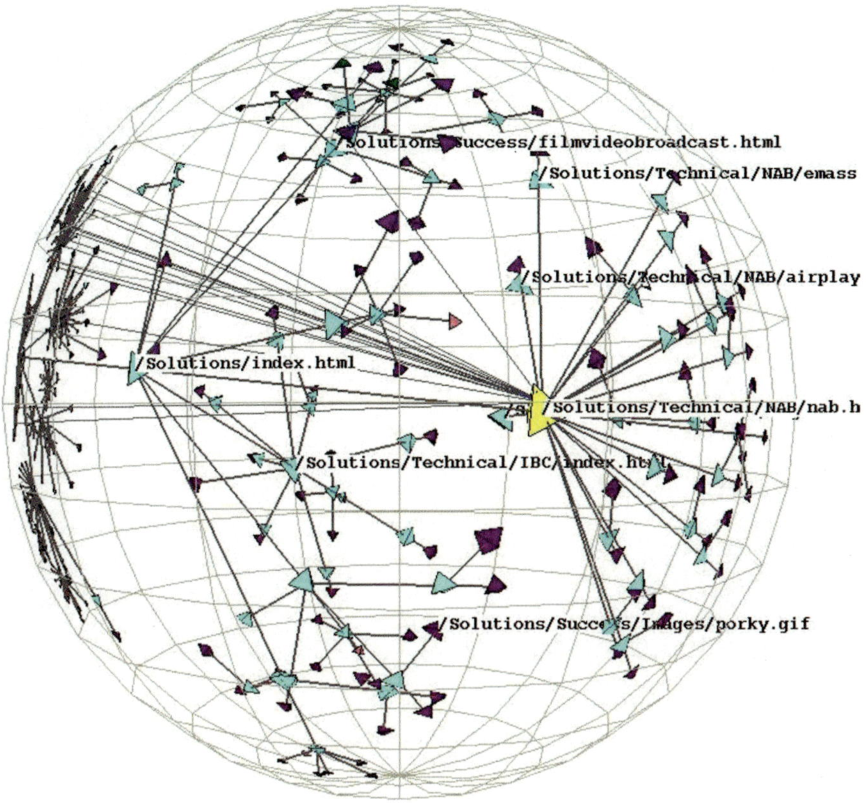

Figure 3-7. Example of tree representation: 3D hyperbolic space (http://graphics.stanford.edu/papers/munzner_thesis/hyp-figs.html, reprint with permission of Tamara Munzner).

Network representation: The network representation method is often applied when a simple tree structure is insufficient for representing complex relationships. Complexity may stem from citations among many academic papers (Mackinlay et al. 1995; Chen & Paul, 2001) or from interconnected Web pages on the Internet (Andrews, 1995). Among various network visualizations that have been created, the Spring Embedder algorithm, originally proposed by Eades (1984), and its variants (Kamada & Kawai, 1989), have become the most popular graph drawing algorithms. Figure 3-8 presents the visualization of co-authorship relationships among 555 scientists using a Spring Embedder algorithm.

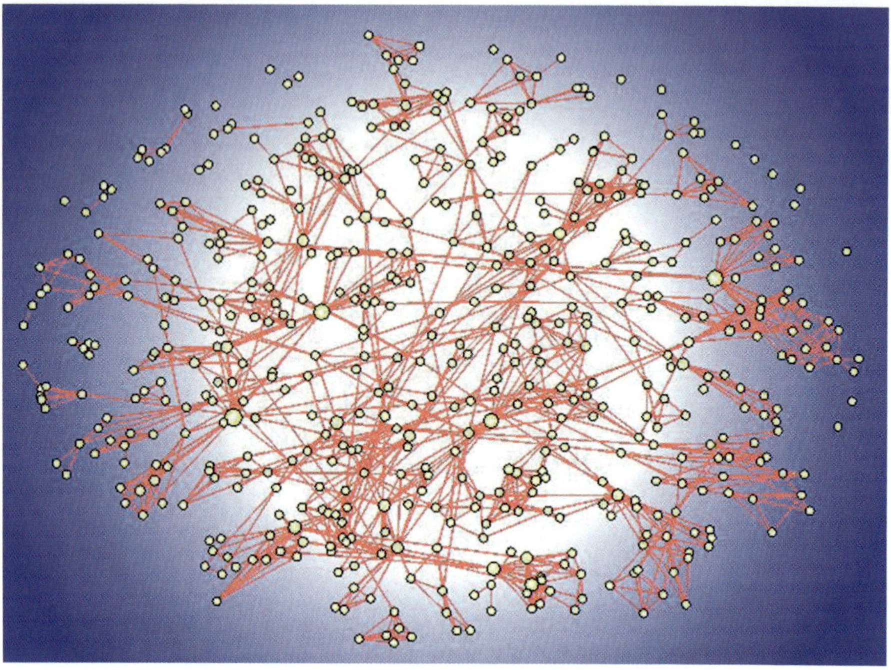

Figure 3-8. Example of network representation: Visualization of a large co-authorship network (http://www.mpi-fg-koeln.mpg.de:80/~lk/netvis/Huge.html, reprint with permission of Lothar Krempel).

Temporal representation: The temporal approach visualizes information based on temporal order. Location and animation are two visual variables commonly used to reveal the temporal aspect of information. Visual objects are usually listed along one axis according to the time when they occurred, while the other axis may be used to display the attributes of each temporal object (Eick et al., 1992; Robertson et al., 1993). For instance, the Perspective Wall (Robertson et al., 1993) lists objects along the x-axis based on time sequence and presents attributes along the y-axis. Animation is another effective way to display temporal information.

The seven types of representation methods turn abstract textual documents into objects that can be displayed. A visualization system usually applies several methods at the same time. For instance, the multi-level ET-Map system created by Chen et al. (Chen et al., 1998) combines both 2D and the tree structure, where a large set of Web pages are partitioned into hierarchical categories based on their content. While the entire hierarchy is organized in a tree structure, each node in the tree is a two-dimensional SOM, on which the sub-categories are graphically displayed.

Many powerful visualization methods also require advanced analysis techniques. For example, the TileBar system (Hearst, 1995) employs the text-tiling analysis algorithm to segment a document, while ThemeView and Galaxy (Wise et al., 1995) use multidimensional scaling (MDS) to cluster and lay out documents on the screen.

3.2 User-Interface Interaction

The "small screen problem" (Robertson et al., 1993) is common to representation methods of any type. To be effective, an information representation method needs to be integrated with user-interface interaction. Recent advances in hardware and software allow quick user-interface interaction, and various combinations of representation methods and user-interface interactions have been employed.

Interaction between an interface and its users not only allows direct manipulation of visual objects displayed, but also allows users to select what is to be displayed and what is not (Card et al., 1999). The two commonly used interaction approaches are: *overview + detail* and *focus + context* (Card et al., 1999).

Overview + Detail: Overview + detail provides multiple views, with the first being an overview that shows overall patterns to users. Details about only the part the user is interested in can then be displayed. These two views can be displayed at the same time or separately. When a detailed view is needed, two types of zooming are usually involved (Card et al., 1999): spatial zooming and semantic zooming. Spatial zooming refers to the process of enlarging selected visual objects spatially to obtain a closer look, whereas semantic zooming provides additional content about a selected visual object by changing its appearance.

Focus + Context: The focus + context technique provides detail (focus) and overview (context) dynamically on the same view. One example is the 3D perception approach adopted by systems like Information Landscape (Andrews, 1995) and Cone Tree (Robertson et al., 1991), where visual objects at the front appear larger than those at the back. Another commonly used focus + context technique is the fisheye view (Furnas, 1986), a distortion technique that acts like a wide-angle lens to amplify part of the focus. The objective is to simultaneously provide neighboring information in reduced detail and supply greater detail on the region of interest. In any focus + context approach, users can change the region of focus dynamically. A well-known visualization system that applies the fisheye technique is the Hyperbolic Tree (Lamping et al., 1995), in which users can scrutinize the focus area and scan the surrounding nodes for a big picture. Other focus +

context techniques include filtering, highlighting, and selective aggregation (Card et al., 1999).

The overview + detail and focus + context user-interface interaction approaches could support knowledge navigation in various knowledge mapping applications.

The presented analysis framework is applied to nanotechnology knowledge mapping in this book.

4. QUESTIONS FOR DISCUSSION

1. What other new analysis and visualization techniques are promising for large-scale knowledge mapping research?
2. How can we adopt selected multimedia indexing techniques (for images and audios) in content analysis and knowledge mapping?
3. How can we adopt selected knowledge mapping techniques to study the informal invisible college of scholars using informal online resources (e.g., via forums and Web blogs)?
4. How can we develop an interactive, user-controlled, highly-visualized system for knowledge mapping using different online resources?

Chapter 4

MAPPING NANOTECHNOLOGY INNOVATIONS VIA THE USPTO DATABASE: A LONGITUDINAL STUDY, 1976-2002

CHAPTER OVERVIEW

Nanotechnology and related areas have seen rapid growth in recent years. The speed and scope of development in the field have made it essential for researchers to be informed on the progress across different laboratories, companies, industries, and countries. This chapter presents results on the basic bibliographic analysis, content map analysis and citation network analysis of U.S. nanotechnology patent documents between 1976 and 2002. The data have been obtained on individual countries, institutions, and technology fields. The top 10 countries with the largest number of nanotechnology patents are the United States, Japan, Germany, France, the United Kingdom, Canada, Switzerland, Italy, Taiwan, and the Netherlands. The fastest growth in the last 5 years has been in chemical and pharmaceutical fields, followed by semiconductor devices. The results demonstrate the feasibility of knowledge mapping techniques for capturing nanotechnology development status.

1. INTRODUCTION

Recent rapid development of nanotechnology promises to bring fundamental changes to a wide range of research fields and industries. The field has been recognized to be critical to a country's future science and technology competence and has recently attracted global research and development interests. The United States announced the National Nanotechnology Initiative (NNI) in 2000 based on a long-term vision (Roco et al., 2000). Both long-term basic research and short-term development related to nanotechnology are being actively explored across many scientific fields and industrial applications. The speed and scope of nanotechnology development make it critical for researchers to be aware of progress in the field across different laboratories, companies, industries, and countries.

As a public information resource, a patent contains rich content regarding technology innovations and is accessible by the general public. A large number of patents are issued everyday and collected systematically worldwide. For example, the United States Patent and Trademark Office (USPTO) has more than 5 million patents, with 3,500 to 4,000 newly granted patents being added into the database each week. Collections of full-text patents over a long period of time are available. The patent documents are also well structured, providing standardized fields such as patent citations, issue date, assignee (the institution to which the patent is assigned), inventors, technology field classification, and country and city of the assignee and inventors, etc. All these special features of patent documents make them a valuable source of knowledge regarding technology development.

In this research we aim to leverage various information analysis and visualization technologies to support domain-specific knowledge mapping from patent documents. The proposed framework is targeted at high-level knowledge landscape analyses, e.g., country-level technology strength comparison, new research field identification, etc. There is substantial past research that uses patent analysis for technology trend assessment (Garfield, 1955; Karki, 1997; Oppenheim, 2000).

In this chapter, we describe the overall design of our research and report results of basic bibliographic analysis, content map analysis, and citation network analysis for three analytical units: countries, institutions, and technology fields.

2. RESEARCH DESIGN

Our overall research objective is to develop a patent analysis framework for knowledge mapping the technology development status of fast-evolving scientific domains, with its first application to nanotechnology.

Analytical Units: Many analytical units have been used in the patent analysis literature. Our proposed analytical units include: geographical regions (e.g., countries, regions, states, cities, etc.); industries/research fields (e.g., genetics, semiconductor, etc.); sectors (e.g., private companies, government organizations, academic institutions, etc.); institutions (e.g., companies, universities, research labs, etc.); individuals; and cross-units (e.g., industries within geographic regions, technology fields within institutions, institutions within industries, etc.).

Types of Analysis: Previous patent analyses can be grouped into three categories:

- Performance evaluation: This analysis type seeks to evaluate an analytical unit's performance in technology development based on various measures of quantity and quality. The quantitative measures indicate the patenting activity level of an analytical unit, e.g., the number of patents, patent growth rate, etc. Measures of quality are mainly based on the citation information. Many citation-based indicators can be used to estimate the impact of patents, cycle time of development, and science linkage.
- Transfer of knowledge: Typical knowledge transfer analysis of patents has focused on the knowledge flow from science literature to patents (Schmoch, 1993). We extend the framework to analyze the knowledge flow between any analytical units. For example, we can analyze the knowledge transfer between countries, sectors, companies, etc. Such analysis will result in a multi-level knowledge transfer network. Both patent citations and journal citations can be used for the construction of the knowledge transfer network.
- Trend analysis: Technology trend analysis is mainly derived from the citation network of patents. The main objective is to use the citation structures together with other indicators, such as patent cycle time, number of patents, number of applicants, etc., to construct a history of the technology development of certain analytical units.

Visualization: Many ideas for visualizing patent data have been proposed in the literature and practiced in the industry. In this research, we leverage our experience in social network and content map visualization for patent analysis. Visualization helps make the patent analysis results more intuitive and convincing.

In this chapter, we present results of basic analysis, content map analysis, and citation network analysis to demonstrate performance evaluation, knowledge transfer analysis, and trend analysis based on USPTO nanotechnology patent documents from the time period of 1976-2002.

3. DATA DESCRIPTION

The data set of nanotechnology-related patents was collected from the patent database of the United States Patent and Trademark Office (USPTO). We use a keyword-based approach to select a subset of the U.S. patents from 1976 to 2002 (U.S. patents prior to 1976 do not have full-text access) that are nanotechnology related.

The nanotechnology search terms were provided by our collaborating domain scientist, as shown in Table 4-1. There are 61,177 patents in the USPTO database that contain such keywords. Most patents were collected by using the "nano*" keyword. We removed patents that contained incidental noise such as "nanosecond" or "nanoliter."

There are 14,417 assignees, 59,683 inventors, and 58 countries in our data set. These patents cover 409 of the 462 first-level United States Patent Classification categories. Examples of such categories are "organic compounds -- part of the class 532-570 series," "drug, bio-affecting and body treating compositions," "chemistry: molecular biology and microbiology," etc. We treat classification category as technology field in our subsequent analysis.

Table 4-1. Nanoscale science and engineering keyword list. (A patent document may contain multiple key phrases listed in the table; thus the total number of unique patent documents is smaller than the total number of collected patent documents presented in the table.)

Terms	Number of Documents
atomic force microscope	1,614
atomic force microscopic	32
atomic force microscopy	1,123
atomic-force-microscope	2
atomic-force-microscopy	2
atomistic simulation	5
biomotor	4
molecular device	102
molecular electronics	195
molecular modeling	1,271
molecular motor	55
molecular sensor	16
molecular simulation	33
nano*	56,865
nanoliter*	25
nanosecond*	258
quantum computing	22
quantum dot*	341
quantum effect*	460
scanning tunneling microscope	946
scanning tunneling microscopic	16

Terms	Number of Documents
scanning tunneling microscopy	666
scanning-tunneling-microscope	23
scanning-tunneling-microscopy	1
self assembled	966
self assembling	590
self assembly	1,136
self-assembled	913
self-assembling	548
self-assembly	1,015
selfassembl*	18
Total	69,263
Unique Total	61,177

4. BASIC BIBLIOGRAPHIC ANALYSIS

Basic analysis evaluates technology development based on basic performance indicators such as the number of issued patents and various citation-based indicators.

4.1 Indicators

We adopted six important patent-relevant performance indicators from Narin (Narin, 2000):

- ***Number of Patents*** indicates the company technology development activity. Definition: The number of patents issued by the U.S. patent system to an analytical unit (a company, a country, or a technology field, etc.).
- ***Cites per Patent*** indicates the impact of an analytical unit's patents. Definition: The average number of the citations received by an analytical unit's patents from subsequent patents.
- ***Current Impact Index (CII)*** indicates the patent portfolio quality. Definition: The number of times the analytical unit's patents issued in the most recent five years had been cited in the current year, relative to the entire patent database. A value of 1 represents average citation frequency. For the analysis results presented in this chapter, the current year was set to 2002.
- ***Technology Independence (TI)*** indicates independence of an analytical unit's technology development. Definition: The number of self-citations divided by the total number of citations.
- ***Technology Cycle Time (TCT)*** indicates speed of invention. Definition: The median age in years of the U.S. patent references cited in an analytical unit's patents.

- ***Science Linkage* (*SL*)** indicates the relationship between an analytical unit's technologies and academic research results. Definition: The average number of scientific papers referenced in an analytical unit's patents.

4.2 Basic Analysis Results for Countries, Institutions, and Technology Fields

4.2.1 Country Analysis

The total numbers of patents issued to the top assignee countries are listed in Table 4-2. Technologically advanced countries, such as the United States, Japan, Germany, and France, had the majority of the nanotechnology patents. The United States was assigned about 70% of the U.S. nanotechnology-related patents between 1976 and 2002.

Table 4-2. Assignee country analysis ("full-text" keyword search; 1976–2002).

Rank	Country Name	Number of patents	Rank	Country Name	Number of patents
1	United States	41,717	11	Australia	317
2	Japan	6,364	12	Rep. of Korea	292
3	Federal Rep. of Germany	1,948	13	Israel	230
4	France	1,947	14	Sweden	223
5	United Kingdom	882	15	Belgium	192
6	Canada	855	16	Denmark	105
7	Switzerland	409	17	Finland	84
8	Italy	347	18	Norway	62
9	China (Taiwan)	346	19	Singapore	52
10	Netherlands	335	20	Austria	47

The yearly numbers of patents of the top 10 countries between 1976 and 2002 are shown in Figure 4-1 and Table 4-3. We observe that the United States, Japan, Germany, France, United Kingdom, Canada, Switzerland, Italy, and Netherlands began research and development activities on nanotechnology in the 1970s. Taiwan started later, in the early 1990s. Because the USPTO database only provides full-text access to patents issued after 1976, our data set may have missed some earlier nanotechnology-related patents.

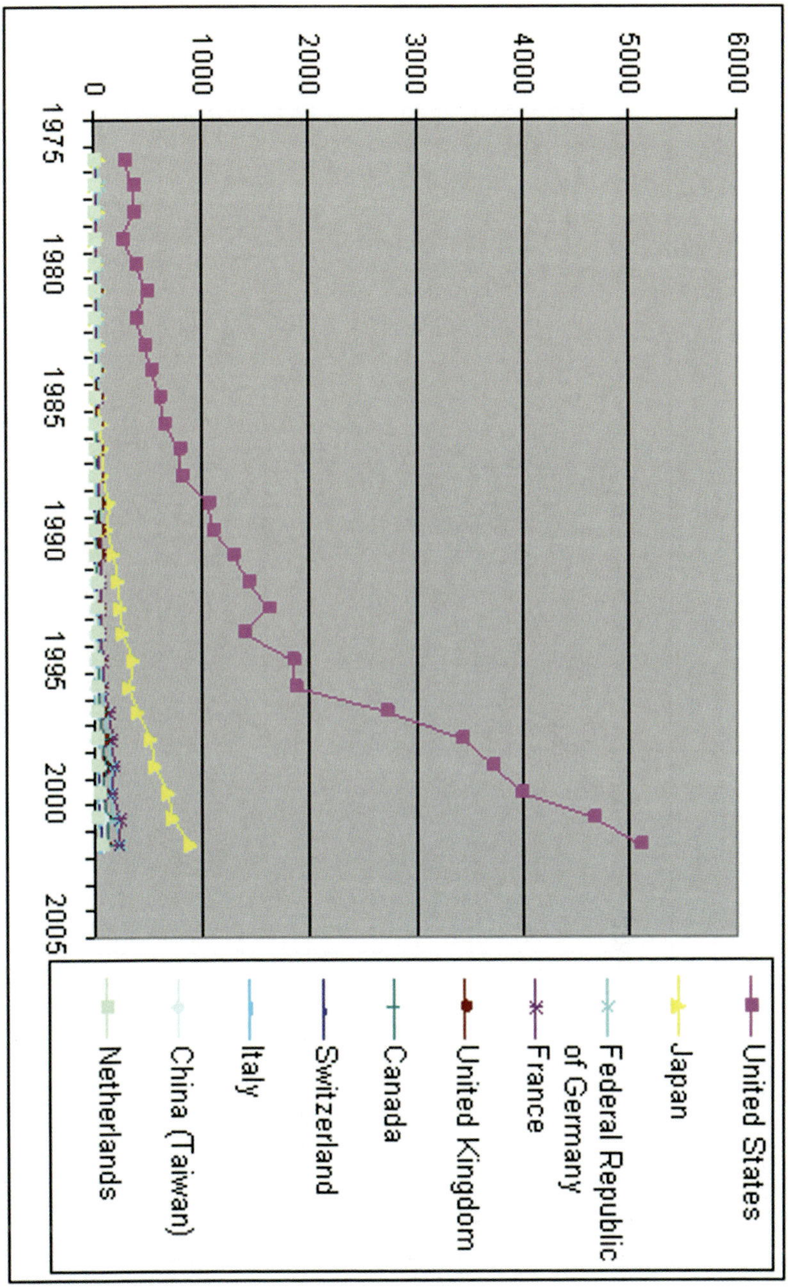

Figure 4-1a. Assignee country analysis by year ("full-text" keyword search; 1976-2002), with the U.S. and Japan.

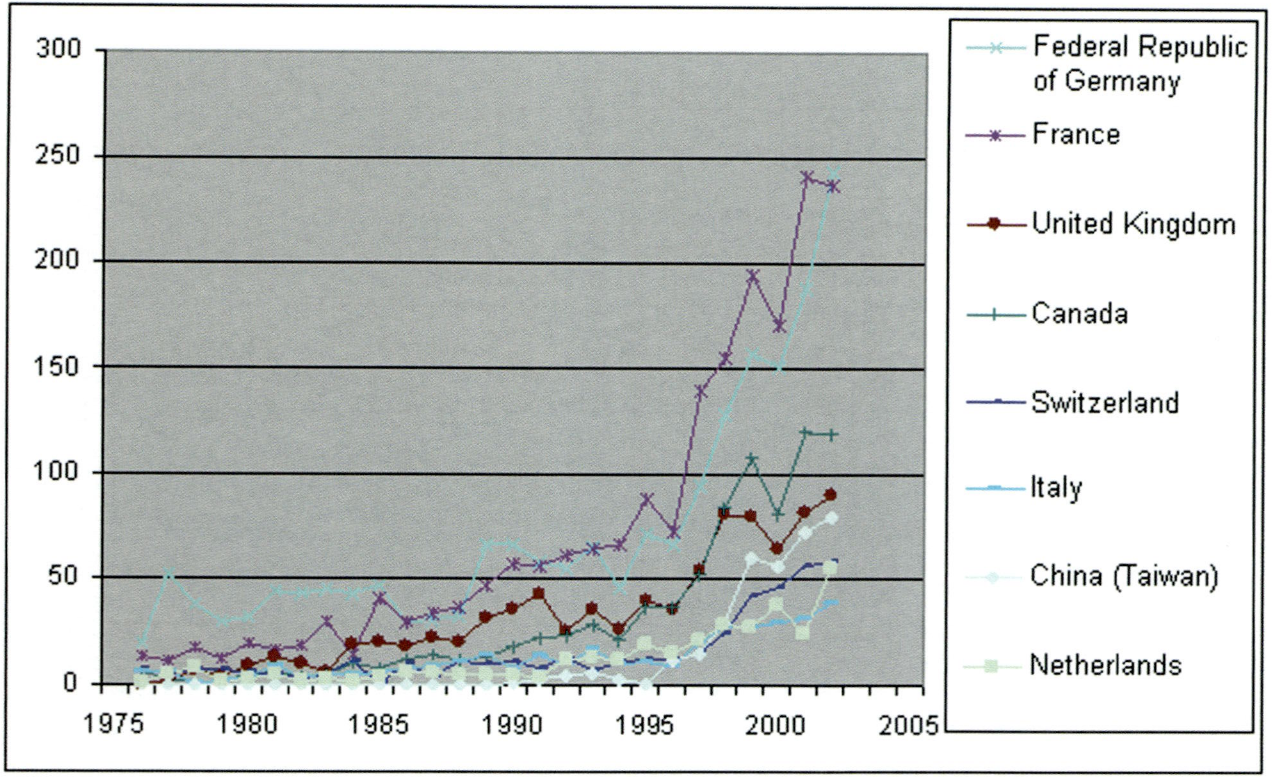

Figure 4-1b. Assignee country analysis by year ("full-text" keyword search; 1976-2002), without the U.S. and Japan.

Figure 4-1. . Assignee country analysis by year ("full-text" keyword search; 1976-2002), with and without the U.S. and Japan.

Table 4-3. Number of patents of assignee countries by year (1976-2002).

Year	USA	Japan	Federal Rep. of Germany	France	UK	Canada	Switzer-land	Italy	China (Taiwan)	Nether-lands
1976	291	38	20	13	0	6	7	6	0	1
1977	363	25	52	11	2	3	5	6	0	5
1978	372	34	38	17	7	1	7	1	0	8
1979	272	22	30	12	4	2	7	2	0	2
1980	386	25	32	20	9	5	6	4	0	3
1981	486	42	44	16	13	6	5	9	0	5
1982	396	28	43	19	10	4	3	4	0	2
1983	466	39	45	30	6	5	5	5	0	3
1984	535	49	43	14	20	10	11	5	0	2
1985	607	44	47	41	21	8	1	6	0	4
1986	663	70	30	30	18	12	10	8	0	6
1987	794	86	31	34	23	14	4	9	0	6
1988	822	87	33	37	21	12	11	11	0	5
1989	1,059	138	67	47	32	13	10	14	0	5
1990	1,113	133	67	58	36	18	11	8	1	5
1991	1,295	185	59	57	43	23	8	14	3	4
1992	1,445	225	55	62	26	24	13	10	5	12
1993	1,632	243	66	65	36	29	7	17	6	13
1994	1,402	273	46	67	27	22	11	10	3	12
1995	1,853	362	72	88	40	37	12	11	1	20
1996	1,866	339	67	73	36	37	12	9	12	15
1997	2,714	421	95	140	54	52	15	21	15	22
1998	3,430	525	128	155	81	84	25	28	30	29
1999	3,710	577	157	194	80	108	42	28	61	28
2000	3,985	696	151	171	65	81	46	30	56	38
2001	4,656	746	188	240	82	120	56	32	73	25

In the analysis on groups of countries, we consider four groups: the United States (US), Japan (JP), European Commission countries (EC) (including Switzerland), and "Other" countries (including Korea, Taiwan, China, Canada, Russia, etc.). The total numbers of nanotechnology-related patents assigned to the four country groups are presented in Table 4-4. The cites per patent measures indicate that U.S. patents have been cited more frequently by the subsequent patents, followed by Japanese patents and European country patents.

Table 4-4. Assignee country group analysis (1976–2002).

Country Group	Number of Patents	Cites Per Patent
US	41,717	1.84
JP	6,364	1.21
EC	6,250	0.91
Others	2,984	0.97

The numbers of patents assigned to the four country groups by year are shown in Figure 4-2a and b. We observe that Japan and the European countries were at approximately the same level of nanotechnology patenting activities.

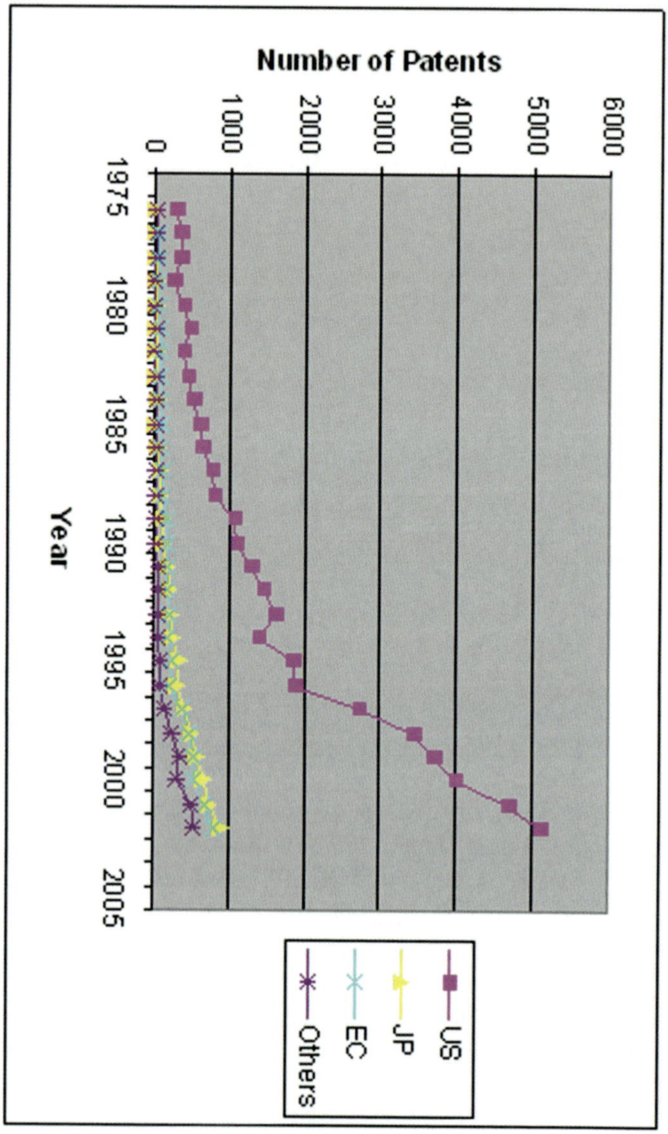

Figure 4-2 a. Assignee country group analysis by years (1976-2002), without U.S.

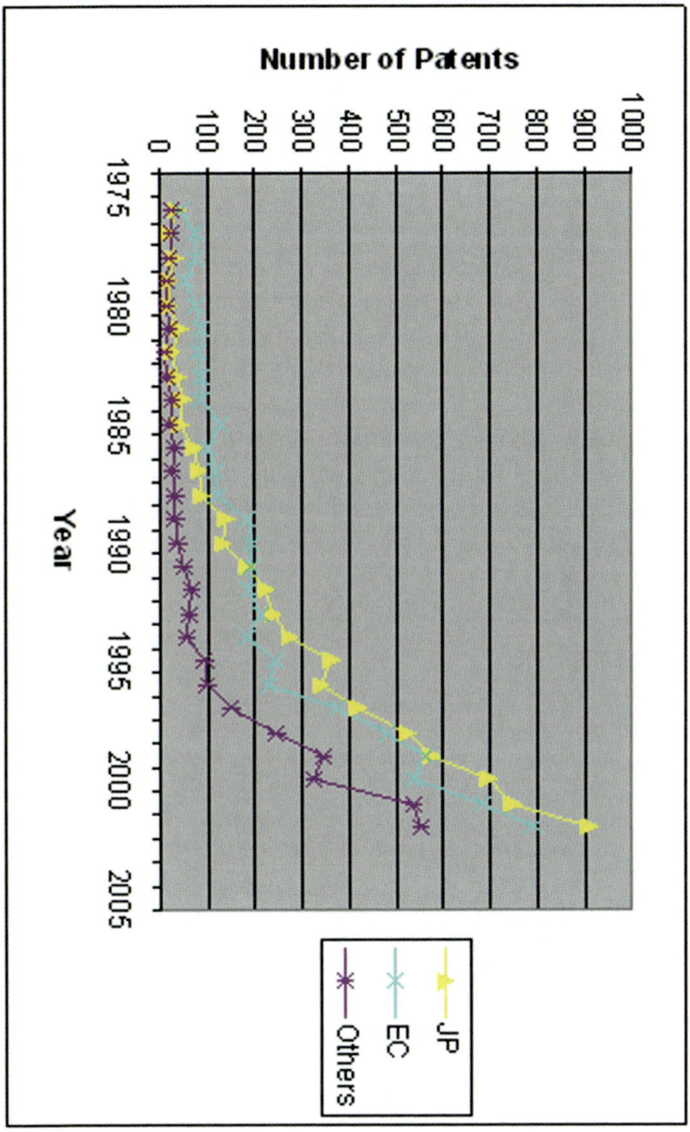

Figure 4-2., Assignee country group analysis by years (1976-2002), without the U.S.

Figure 4-2a and b. Assignee country group analysis by years (1976-2002), with and without the U.S.

4.2.2 Institution Analysis

The top twenty assignees that have received the greatest numbers of nanotechnology patents are shown in Table 4-5. The International Business Machines Corporation (IBM) was issued the greatest number of patents, followed by the Xerox Corporation (Xerox). The average patent age measures (as of 2002) reveal differences in the freshness of the patents assigned to these institutions. We observe that patents issued to Eastman Kodak, Du Pont, General Electric, and the Dow Chemical Company had an average age of over 10 years, while patents issued to the Regents of the University of California, NEC, Micron Technology, and Advanced Micro Devices were of a much "younger" age, under five years. When considering both quantity and freshness of patents assigned, Micron Technology outperformed all other institutions. It had 781 patents issued (the fourth position measured by numbers) with the smallest average patent age (2.53 years), which indicates the company's strong emphasis and potential in this technology area.

Table 4-5. Assignee analysis (1976–2002).

Rank	Assignee Name	Number of Patents	Average Patent Age
1	International Business Machines Corporation	1302	6.74
2	Xerox Corporation	957	7.55
3	Minnesota Mining and Manufacturing Company	807	7.69
4	Eastman Kodak Company	708	10.38
5	Motorola, Inc.	508	7.16
6	The Regents of the University of California	491	4.13
7	NEC Corporation	483	4.42
8	Micron Technology, Inc.	457	2.53
9	Canon Kabushiki Kaisha	408	5.52
10	E. I. Du Pont de Nemours and Company	367	11.45
11	General Electric Company	367	11.54
12	Texas Instruments Incorporated	366	7.77
13	Hitachi, Ltd.	335	6.43
14	The United States of America as represented by the Secretary of the Navy	330	8.63
15	The Dow Chemical Company	327	11.04
16	Kabushiki Kaisha Toshiba	317	5.47
17	Abbott Laboratories	297	6.62
18	Advanced Micro Devices, Inc.	295	2.61
19	Massachusetts Institute of Technology	271	8.28
20	Merck & Co., Inc.	251	8.63

The yearly patenting activities of the top 10 institutions between 1976 and 2002 are shown in Figure 4-3 (the institution names are ordered by the total number of patents issued). Assignees in the United States with early submissions in nanotechnology, include IBM, Xerox, Minnesota Mining and Manufacturing Company (3M), Eastman Kodak, and Motorola. IBM maintained its leading position in most years. Micron Technology and Advanced Micro Devices, Inc. have shown a fast increase in patenting activity in the last several years, rising to the second and third positions. Xerox and 3M, although still in the second and third position respectively in terms of the total number of patents issued, have lagged behind IBM and Micron in recent years. The patenting activities of Xerox, NEC, 3M, and the University of California have remained constant over the last several years.

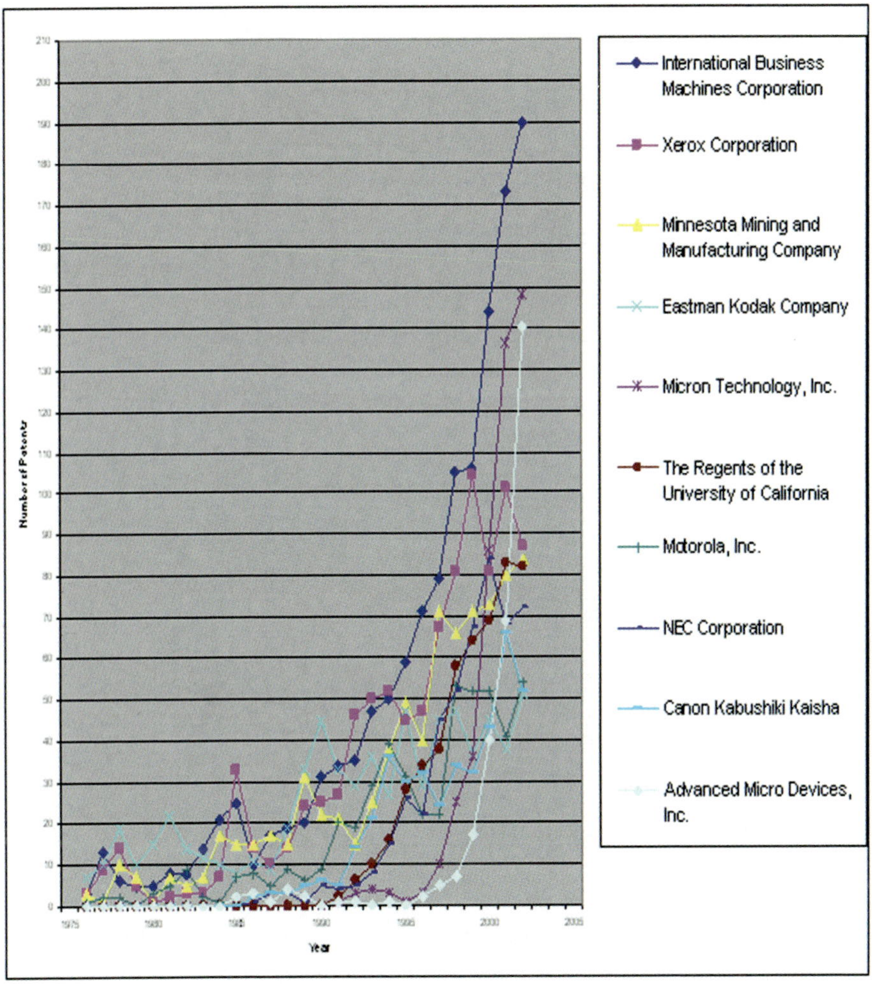

Figure 4-3. Assignee analysis by year (1976-2002).

Cites per patent for assignees are shown in Table 4-6. Patents issued to the Board of Regents, The University of Texas System, Olympus Optical Co., Ltd., and the Dow Chemical Company received the most patent citations: on average each of these institutions' patents was cited more than seven times by subsequent patents. These institutions may have patents of higher quality than other assignees and may possess key technologies of the field.

Table 4-6. Cites per patent by assignee (1976–2002).

Rank	Assignee Name	Cites Per Patent
1	Board of Regents, The University of Texas System	7.95
2	Olympus Optical Co., Ltd.	7.73
3	The Dow Chemical Company	7.43
4	Minnesota Mining and Manufacturing Company	6.17
5	Xerox Corporation	6.05
6	California Institute of Technology	5.92
7	International Business Machines Corporation	5.46
8	PPG Industries, Inc.	5.45
9	Genentech, Inc.	5.04
10	Massachusetts Institute of Technology	4.66

Technology independence is defined by the number of self-citations divided by the total number of citations. The top 10 institutions having the highest technology independence measures are presented in Table 4-7.

Table 4-7. Technology independence analysis (1976–2002).

Rank	Assignee Name	Technology Independence
1	Xerox Corporation	0.2092
2	SmithKline Beecham Corporation	0.1558
3	Eli Lilly and Company	0.1429
4	Canon Kabushiki Kaisha	0.1381
5	The Scripps Research Institute	0.1171
6	Genentech, Inc.	0.1087
7	Dow Corning Corporation	0.0954
8	Olympus Optical Co., Ltd.	0.0944
9	L'oreal	0.0870
10	California Institute of Technology	0.0751

Slow-moving technologies may have longer technology cycle times that are estimated based on the median age in years of the U.S. patent references. It is shown in Table 4-8 that Advanced Micro Devices and Fuji Photo Film Co., Ltd. had the shortest cycle times, which indicate that these institutions' patents cited more recent patents.

Table 4-8. Technology Cycle Time by Assignee (1976–2002).

Rank	Assignee Name	Technology Cycle Time
1	Advanced Micro Devices, Inc.	4
1	Fuji Photo Film Co., Ltd.	4
3	Micron Technology, Inc.	5
4	NEC Corporation	6
4	SmithKline Beecham Corporation	6
4	Intel Corporation	6
4	L'oreal	6
8	Lucent Technologies Inc.	7
8	Sony Corporation	7
8	Applied Materials, Inc.	7
8	Bayer Aktiengesellschaft	7
8	Corning Incorporated	7
8	The Regents of the University of California	7
8	Motorola, Inc.	7
8	Kabushiki Kaisha Toshiba	7
8	3M Innovative Properties Company	7
8	Fujitsu Limited	7
8	The United States of America as represented by the Secretary of the Army	7
8	Sandia Corporation	7
20	California Institute of Technology	8
20	The Procter & Gamble Company	8
20	Xerox Corporation	8
20	Hewlett-Packard Company	8
20	Massachusetts Institute of Technology	8

Institutions at the forefront of a technology field tend to have a stronger Science Linkage (defined by the average number of scientific papers referenced). Academic institutions had higher Science Linkage measures (e.g., California Institute of Technology, the University of Texas System, the University of California, and Massachusetts Institute of Technology) (Table 4-9). On the other hand, high Science Linkage measures of companies like Genentech, Micron Technology, Merck, and Eli Lilly indicate strong connections between these companies' technology development and academic research.

Table 4-9. Science Linkage by assignee (1976–2002).

Rank	Assignee Name	Science Linkage
1	Genentech, Inc.	62.26
2	California Institute of Technology	59.29
3	Board of Regents, The University of Texas System	39.98
4	The Regents of the University of California	29.42
5	Massachusetts Institute of Technology	27.78
6	Micron Technology, Inc.	22.24
7	The Scripps Research Institute	19.40

Rank	Assignee Name	Science Linkage
8	Merck & Co., Inc.	14.19
9	Eli Lilly and Company	13.89
10	Monsanto Company	13.60

4.2.3 Technology Field Analysis

Several technology development indicators of top nanotechnology fields are presented in this section. The top technology fields to which the nanotechnology-related patents were assigned are presented in Table 4-10. "Chemistry: molecular biology and microbiology" and "drug, bio-affecting and body treating compositions" were revealed to be the dominating technology fields.

Table 4-10. Number of patents by technology fields (1976–2002).

Rank	Technology Field	Number of Patents	Average Patent Age
1	435:Chemistry: molecular biology and microbiology	7837	7.28
2	514:Drug, bio-affecting and body treating compositions	6364	7.84
3	424:Drug, bio-affecting and body treating compositions	4760	7.05
4	428:Stock material or miscellaneous articles	3847	8.23
5	250:Radiant energy	3783	10.14
6	530:Chemistry: natural resins or derivatives; peptides or proteins; lignins or reaction products thereof	3772	7.80
7	536:Organic compounds -- part of the class 532-570 series	3701	5.90
8	438:Semiconductor device manufacturing: process	3584	6.31
9	257:Active solid-state devices (e.g., transistors, solid-state diodes)	3480	7.93
10	427:Coating processes	3179	9.10
11	436:Chemistry: analytical and immunological testing	2941	9.87
12	430:Radiation imagery chemistry: process, composition, or product thereof	2883	9.66
13	359:Optics: systems (including communication) and elements	2743	8.77
14	356:Optics: measuring and testing	2556	10.20
15	422:Chemical apparatus and process disinfecting, deodorizing, preserving, or sterilizing	1665	9.05
16	204:Chemistry: electrical and wave energy	1660	9.65
17	252:Compositions	1647	10.48
18	524:Synthetic resins or natural rubbers -- part of the class 520 series	1515	9.01
19	546:Organic compounds -- part of the class 532-570 series	1503	8.62
20	210:Liquid purification or separation	1451	9.48

Figure 4-4 shows trends of the patenting activities in the top 10 technology fields between 1976 and 2002. Names of most active technology fields are listed in the figure in the order of the total number of patents issued. Technology fields that experienced fast growth in recent years include: "chemistry: molecular biology and microbiology," "drug, bio-affecting and body treating compositions," "semiconductor device manufacturing: process," and "organic compounds -- part of the class 532-570 series."

We evaluated technology fields with the highest Current Impact Index measures (Table 4-11) and technology fields with the lowest Technology Cycle Time measures (Table 4-12). "Chemistry: molecular biology and microbiology" has the most patents that are frequently cited. "Semiconductor device manufacturing: process" has the shortest Technology Cycle Time.

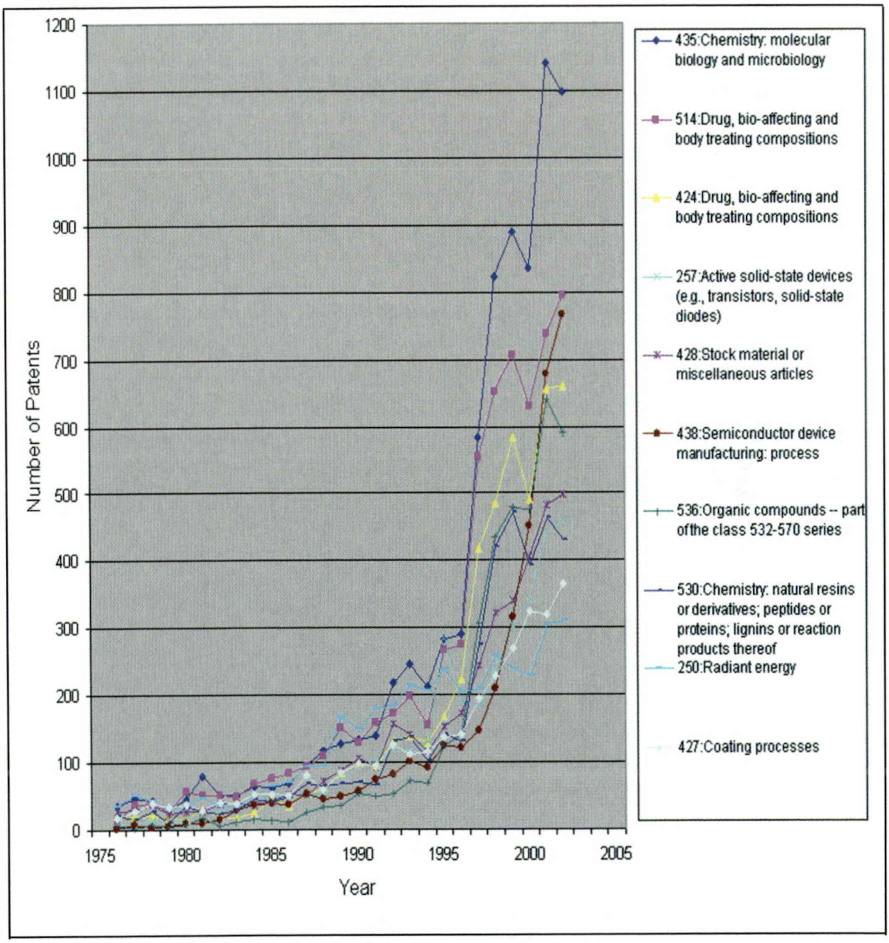

Figure 4-4. Technology field analysis by year (1976-2002).

Table 4-11. Current Impact Index by technology field (2002).

Rank	Technology Field	Current Impact Index
1	435: Chemistry: molecular biology and microbiology	4647
2	428: Stock material or miscellaneous articles	3469
3	438: Semiconductor device manufacturing: process	3021
4	514: Drug, bio-affecting and body treating compositions	2711
5	424: Drug, bio-affecting and body treating compositions	2444
6	427: Coating processes	2306
7	257: Active solid-state devices (e.g., transistors, solid-state diodes)	1812
8	524: Synthetic resins or natural rubbers -- part of the class 520 series	1660
9	536: Organic compounds -- part of the class 532-570 series	1656
10	359: Optics: systems (including communication) and elements	1280
11	436: Chemistry: analytical and immunological testing	1118
12	430: Radiation imagery chemistry: process, composition, or product thereof	1017
13	530: Chemistry: natural resins or derivatives; peptides or proteins; lignins or reaction products thereof	949
14	422: Chemical apparatus and process disinfecting, deodorizing, preserving, or sterilizing	767
15	250: Radiant energy	746
16	385: Optical waveguides	709
17	356: Optics: measuring and testing	612
18	210: Liquid purification or separation	601
19	204: Chemistry: electrical and wave energy	546
20	106: Compositions: coating or plastic	531

Table 4-12. Technology Cycle Time by technology field (1976–2002).

Rank	Technology Field	Technology Cycle Time
1	438: Semiconductor device manufacturing: process	6
2	514: Drug, bio-affecting and body treating compositions	7
2	524: Synthetic resins or natural rubbers -- part of the class 520 series	7
2	546: Organic compounds -- part of the class 532-570 series	7
2	548: Organic compounds -- part of the class 532-570 series	7
2	544: Organic compounds -- part of the class 532-570 series	7
2	549: Organic compounds -- part of the class 532-570 series	7
2	560: Organic compounds -- part of the class 532-570 series	7
9	564: Organic compounds -- part of the class 532-570 series	8
9	324: Electricity: measuring and testing	8
9	510: Cleaning compositions for solid surfaces, auxiliary compositions therefore, or processes of preparing the compositions	8
9	528: Synthetic resins or natural rubbers -- part of the class 520 series	8
9	365: Static information storage and retrieval	8
9	385: Optical waveguides	8
9	423: Chemistry of inorganic compounds	8
9	106: Compositions: coating or plastic	8

Rank	Technology Field	Technology Cycle Time
9	536: Organic compounds -- part of the class 532-570 series	8
9	428: Stock material or miscellaneous articles	8
9	257: Active solid-state devices (e.g., transistors, solid-state diodes)	8

We compared nanotechnology-related patenting activities at the industry level: electronics, materials, chemical/catalysts/pharmaceuticals, and others. We used the United States patent classifications to determine the respective industry in the first-level U.S. patent classifications. The total number of patents issued between 1976 and 2002 and the average number of citations received by the patents in these industries are presented in Table 4-13. The patent development trends of these industries are also presented in Figure 4-5. We can observe that nanotechnology-related research was dominated by the industries of chemical/catalysts/pharmaceuticals and electronics. Significant growth of patenting activity was also observed in the chemical/catalysts/pharmaceuticals industry since 1997.

Table 4-13. Industry analysis (1976–2002).

Industry	Number of Patents	Cites Per Patent
Electronics	8,799	3.53
Materials	4,750	4.37
Chemical/catalyst/pharmaceutical	18,297	4.22
Others	33,016	3.73

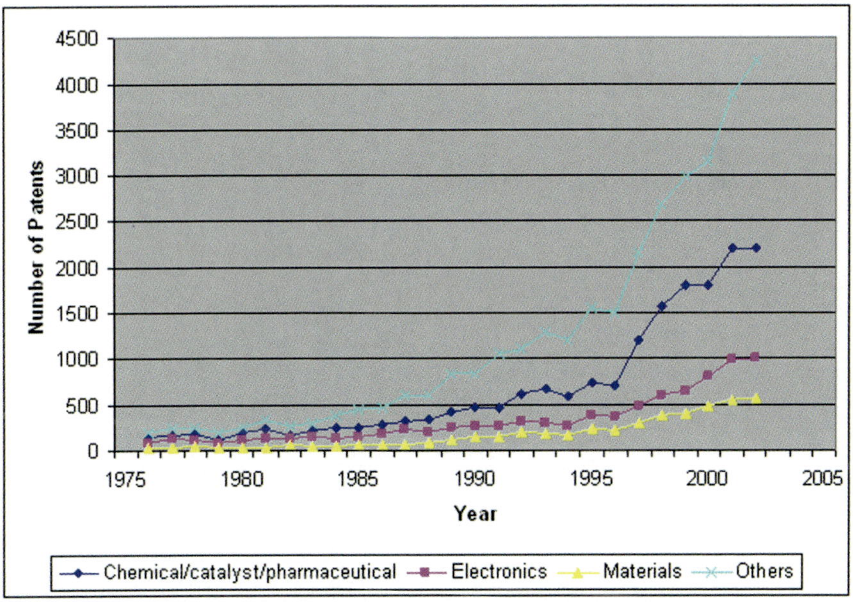

Figure 4-5. Industry analysis by years (1976-2002).

5. PATENT CONTENT MAPS

We applied a content map technology to identify and visualize major research topics in nanotechnology, and the evolution of the major topics was analyzed using a time-sequenced series of content maps.

5.1 Overall Content Map

The hierarchical multi-level Self-Organization Map (SOM) algorithm (Chen, H. et al., 1996; Ong et al., 2003) was used to perform the content analysis of nanotechnology-related patents to discover dominating topic areas. Figures 4-6, 4-7, and 4-8 show three levels of the hierarchical nanotechnology patent content map that was generated based on the titles and abstracts of all the nanotechnology-related patents in our data set.

The topic map display contains two components, a folder tree display on the left-hand side and a hierarchical content map on the right-hand side. The patent documents are organized under technology topics that are represented as nodes in the folder tree and colored regions in the content map. These topics were labeled by representative noun phrases that were identified by the SOM algorithm.

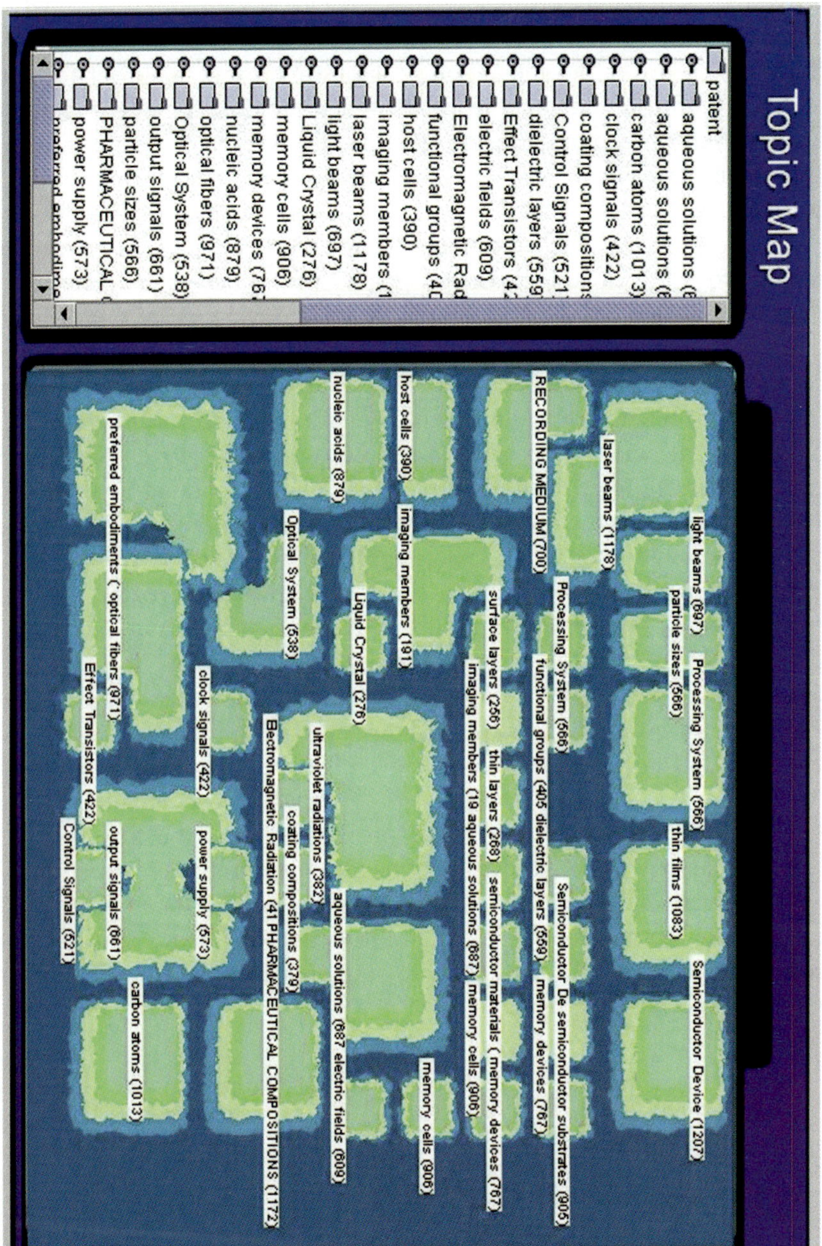

Figure 4-6. First-level technology content map (1976-2002).

Numbers of patent documents that were assigned to the first-level topics are presented in parentheses after the topic labels. Users can either click the folder tree nodes or the content map regions to browse the lower-level topics under a high-level topic. The layers of the colored regions represent the levels of the hierarchies inside the specific regions. The right-hand-side content map display shows all topic regions in the same level under a particular higher-level technology topic region.

In each level of such technology maps, conceptually closer technology topics were positioned closer geographically. Conceptual closeness was derived from the co-occurrence patterns of the technology topics in patent titles and abstracts. The sizes of the topic regions also generally corresponded to the number of patent documents assigned to the topics (Lin et al., 2000). First-level nanotechnology topics are shown in Figure 4-6. We observe that closely related technology topics were positioned in neighborhoods (e.g., "ultraviolet radiations," "coating compositions," "electromagnetic radiation," and "optical systems" in the center of the map).

Technology topics in the lower-level maps were derived from the set of patent documents that belong to a particular higher-level region. As a result, general topics are often found in high-level maps, and more specific technology topics usually appear in low-level maps. The second-level technology topics under "ultraviolet radiations" are shown in Figure 4-7. These topics are more specific technology concepts related to "ultraviolet radiations." Conceptually related topic region neighborhoods can also be observed (e.g., "heat treatments," "transition temperature," "room temperature," and "ultraviolet light" in the center of the map). The third-level technology topic map under "imaging systems" is presented in Figure 4-8. The technology topics identified are more specific than the second-level technology topics. Such a hierarchical technology topic map provides a comprehensive view of the key technology concepts and their relationships in the nanotechnology field.

Figure 4-7. Second-level technology concept map: under the region of "ultraviolet radiations" in the first-level map shown in Figure 4-6.

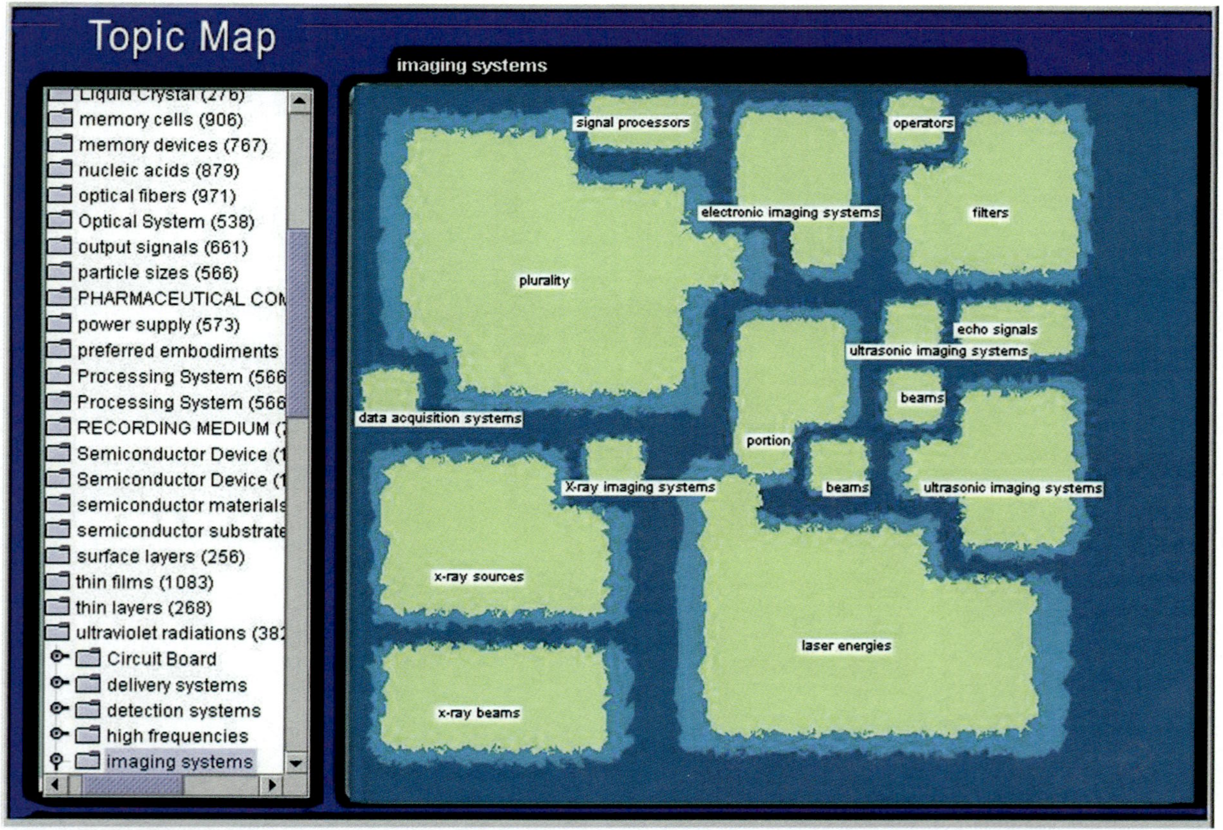

Figure 4-8. Third-level technology concept map, under the region "imaging systems" in the second-level map shown in Figure 4-7.

5.2 Time-series Content Maps

The evolution of major topics in nanotechnology is represented using content maps generated for several time periods:

- 1976-1980 (2,481 patents), map represented in Figure 4-9
- 1981-1985 (3,554 patents), Figure 4-10
- 1986-1990 (6,316 patents), Figure 4-11
- 1991-1995 (11,021 patents), Figure 4-12
- 1996-2000 (23,057 patents), Figure 4-13
- 2001-2002 (14,748patents), Figure 4-14.

The first five time periods are based on a grouping of five years of patents. The last time period is for 2001-2002 due to the importance of more recent patent inventions. By comparing the dominating regions in the top-level content maps of different time periods, we observe some general trends in nanotechnology development.

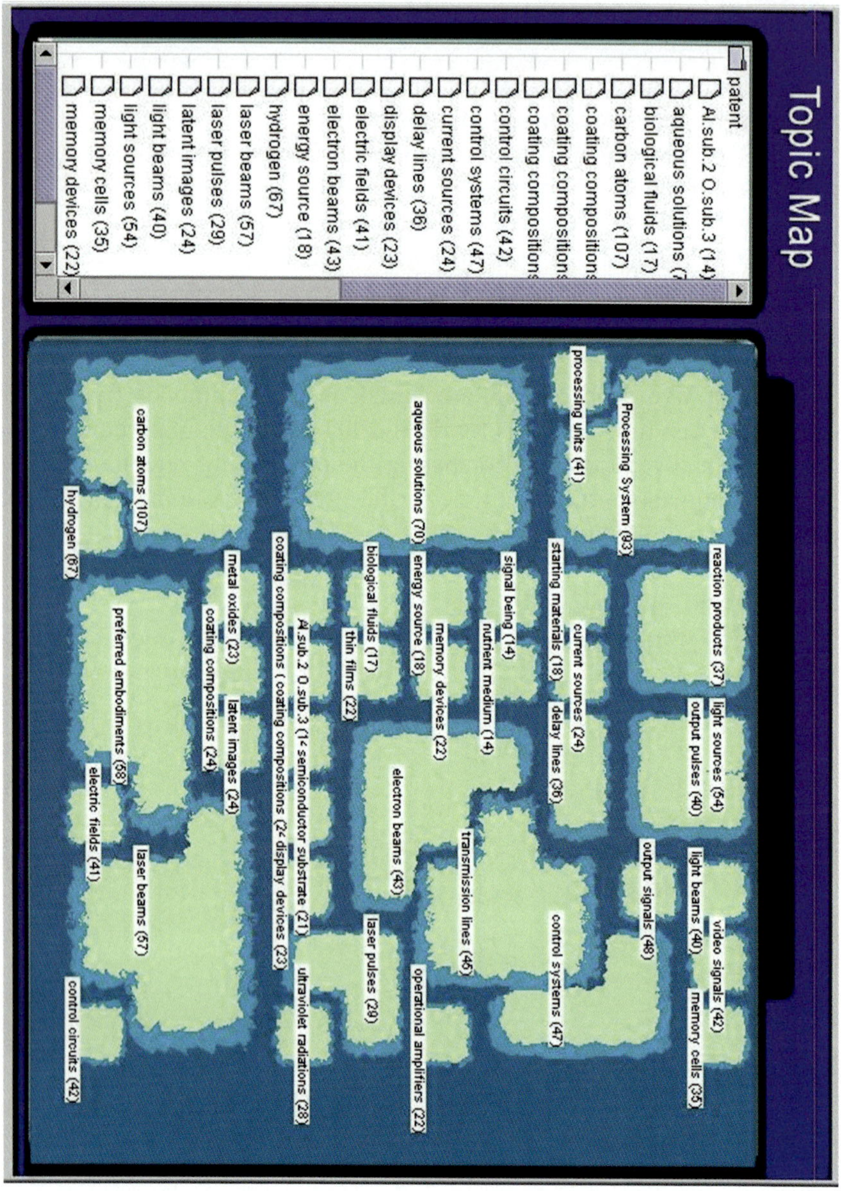

Figure 4-9. Top-layer content map for 1976-1980.

It can be observed that the dominating topic regions between 1976 and 1980 are: "processing systems," "aqueous solutions," "transmission lines," "electron beams," "carbon atoms," "preferred embodiments," and "laser beams." The sizes of these topic regions suggest that they were the key technology topics during the early years of nanotechnology innovation.

Figure 4-10. Top-layer content map for 1981-1985.

During 1981 to 1985, dominating topics of the previous five years (1976-1980), such as “laser beams,” “carbon atoms,” “aqueous solutions,” and “processing systems,” continued to be important. New topics like “control signals,” “control circuits,” “control systems,” and “pharmaceutical compositions” also began to occupy dominating positions.

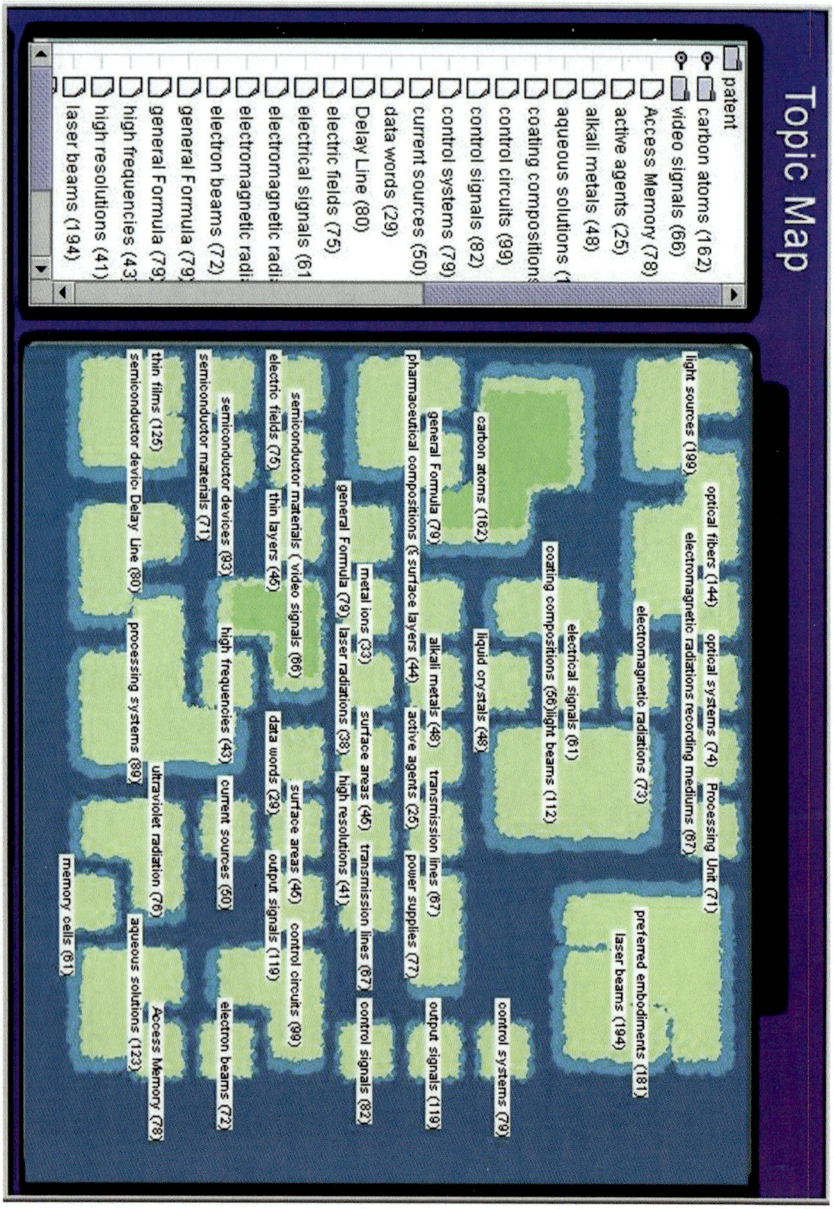

Figure 4-11. Top-layer content map for 1986-1990.

During 1986 to 1990, active technology topics of the previous five years continued to be the central areas of interest. Three new important topics are observed: "light beams," "video signals," and "semiconductor devices."

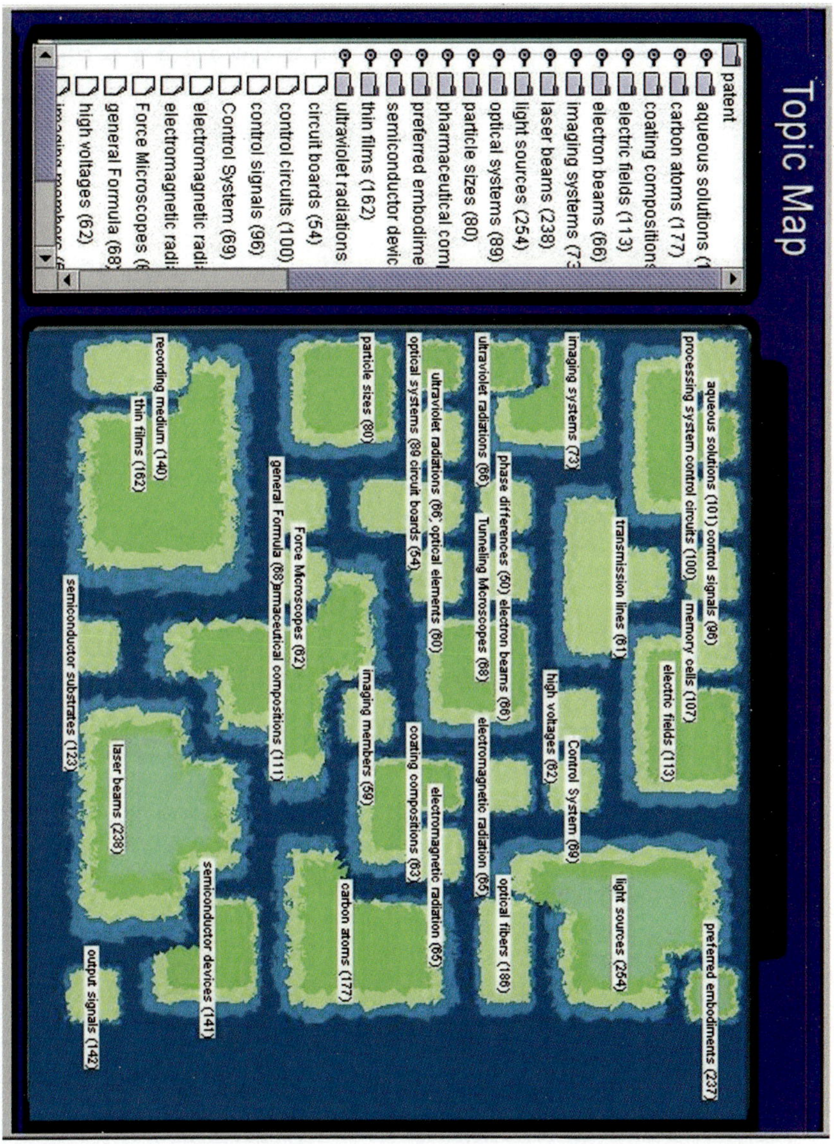

Figure 4-12. Top-layer content map for 1991-1995.

During 1991 to 1995, the most important technology topics were "light sources," "carbon atoms," "pharmaceutical compositions," "thin films," and "laser beams." "Light sources" and "thin films" experienced a remarkable growth and became as important as the other three topics that had long-term dominance in the field. Other important new topics include: "imaging systems," "tunneling microscopes," "coating compositions," and "particle sizes."

Figure 4-13. Top-layer content map for 1996-2000.

During 1996 to 2000, "semiconductor devices" regained the dominating position. New topics like "memory cells," "computer systems," "electromagnetic radiation," "acid sequences," and "nucleic acids" began to appear as major nanotechnology topics.

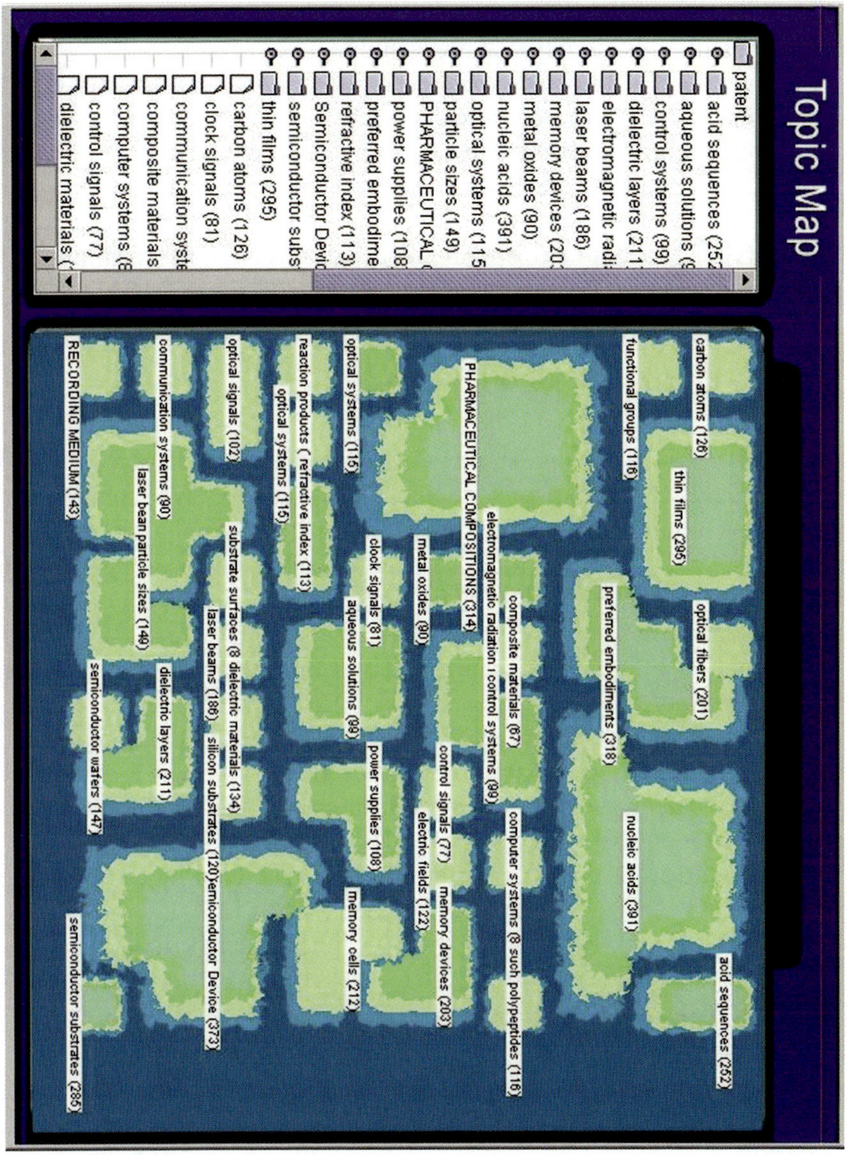

Figure 4-14. Top-layer content map for 2001- 2002.

Major technology topics of the patents issued in the last two years (2001 and 2002) are shown above in Figure 4-14. We did not observe important new topics that occupied the dominating positions. The most important topics continued to be "nucleic acids," "pharmaceutical compositions," "laser beams," and "semiconductor devices." Several new topics can be observed but with smaller region sizes, including "optical systems," "refractive index," "optical signals," "power supplies," and "dielectric layers." These are the most representative technology topics in patents and indicate potential future fields of application.

The overall and time-series nanotechnology content maps presented in this section were generated based on the entire nanotechnology-related patent collection. The SOM content map technology can also be used to analyze and visualize nanotechnology landscapes of individual analytical units by applying it to a specified subset of nanotechnology patent documents, e.g., content maps for individual countries, institutions, or technology fields.

6. CITATION NETWORKS

USPTO patents contain detailed and comprehensive citation information that can be used for knowledge mapping analysis. We computed and summarized citation information for different analytical units: countries, institutions, and technology fields. Based on such information, we applied existing network drawing algorithms to generate a visual representation of their relationships.

The citation networks presented in this section were computed based on all USPTO nanotechnology patents from 1976 to 2002 (Figures 4-15 to 4-17). An arrow represents the direction of the citation. For example, a link such as "Country A → Country B" means that country A's patents cited country B's patents, and the number beside the link represents the total number of citations between them.

The networks presented in this section are generated by an open source graph drawing software, Graphviz, provided by AT&T Labs (Gansner & North, 2000) (available at: http://www.research.att.com/sw/tools/graphviz/).

6.1 Country Citation Network

The nanotechnology-related patent citation networks among countries are presented in this section. The codes and names of the top countries are shown in Table 4-14 (country codes were from USPTO: http://www.uspto.gov/patft/help/helpctry.htm).

The complete citation network among countries is presented in Figure 4-15. A citation link between two countries is present in the map if there are more than three patent citations associated with the link. We observe the following from Figure 4-15:

- The United States (US) dominated the network and the U.S. patents intensively interacted with patents of most other countries.
- Japan (JP) was the second largest patent citation center.
- Other patent citation centers included Germany (DE), France (FR), the United Kingdom (GB), Canada (CA) and Switzerland (CH). There were large amounts of citation activities among the patents of the United States and these countries.
- Regional citation clusters were observed in the citation network, e.g., one large group of linked countries was: Germany, France, United Kingdom, and Canada.

Table 4-14. Country codes and names.

Country Code	Country Name	Country Code	Country Name
AN	Netherlands Antilles	IE	Ireland
AT	Austria	IL	Israel
AU	Australia	IT	Italy
BE	Belgium	JP	Japan
CA	Canada	KR	Republic of Korea
CH	Switzerland	MX	Mexico
CS	Czechoslovakia	NL	Netherlands
DE	Federal Republic	NO	Norway
	of Germany	SE	Sweden
DK	Denmark	SG	Singapore
ES	Spain	TW	China (Taiwan)
FI	Finland	US	United States
FR	France	VE	Venezuela
GB	United Kingdom	VG	British Virgin Islands
HK	Hong Kong	ZA	South Africa

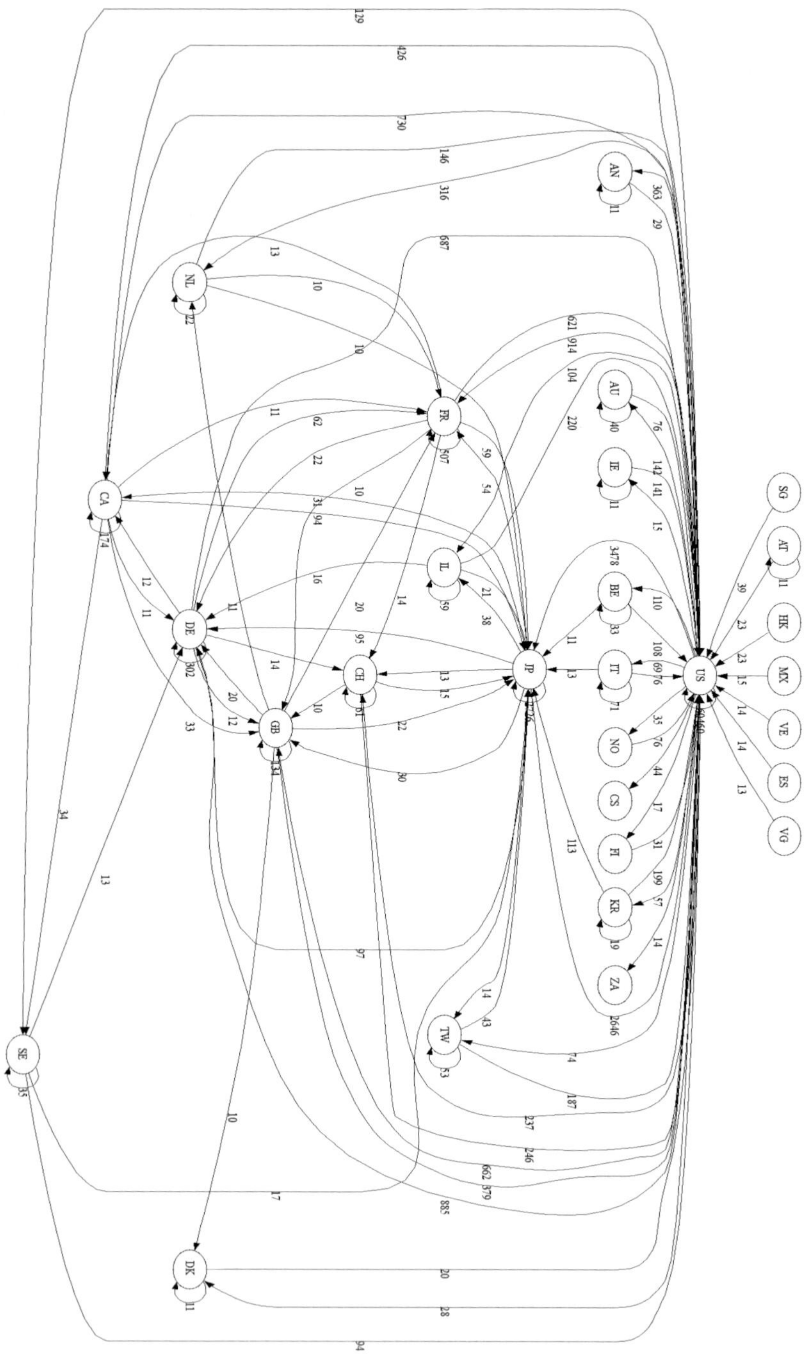

Figure 4-15. Country Citation Network (Minimum Cites: 10).

6.2 Institution Citation Network

The top 50 institutions that own the greatest number of patents in nanotechnology are presented in Table 4-15. We assigned unique institution IDs for analysis and display purposes.

Table 4-15. Top 50 Institutions - IDs and names (1976-2002).

Institution ID	Institution Name
16845	3M Innovative Properties Company
16922	Abbott Laboratories
17089	Advanced Micro Devices, Inc.
17903	AT&T Bell Laboratories
18158	Bayer Aktiengesellschaft
18537	Bell Telephone Laboratories, Incorporated
18641	California Institute of Technology
19393	Canon Kabushiki Kaisha
19486	Digital Equipment Corporation
21088	Dow Corning Corporation
21271	E. I. Du Pont de Nemours and Company
21312	Eastman Kodak Company
21545	Eli Lilly and Company
22326	Fuji Photo Film Co., Ltd.
22357	Fujitsu Limited
22539	Genentech, Inc.
22605	General Electric Company
23244	Hewlett-Packard Company
23330	Hitachi, Ltd.
23417	Honeywell Information Systems Inc.
23489	Hughes Aircraft Company
24073	Intel Corporation
24145	International Business Machines Corporation
24661	Kabushiki Kaisha Toshiba
25085	L'Oreal
25586	LSI Logic Corporation
25610	Lucent Technologies Inc.
25905	Massachusetts Institute of Technology
25959	Matsushita Electric Industrial Co., Ltd.
26268	Merck & Co., Inc.
26501	Micron Technology, Inc.
26602	Minnesota Mining and Manufacturing Company
26650	Mitsubishi Denki Kabushiki Kaisha
26909	Motorola, Inc.
27204	National Semiconductor Corporation
27245	NEC Corporation
28859	PPG Industries, Inc.
29312	RCA Corporation
29701	Rohm and Haas Company

Institution ID	Institution Name
30666	SmithKline Beecham Corporation
30873	Sony Corporation
31877	Texas Instruments Incorporated
32084	The Dow Chemical Company
32311	The Regents of the University of California
32505	The United States of America as Represented by the Secretary of the Air
32508	The United States of America as represented by the Secretary of the Army
32512	The United States of America as represented by the Secretary of the Navy
32524	The United States of America as represented by the United States
33125	U.S. Philips Corporation
34210	Xerox Corporation

The institution citation network is shown in Figure 4-16 (with the minimum number of citations equaling 30). We observe the following:

- International Business Machines Corporation (24145) and Micron Technology, Inc. (26501) were the institutional patent citation centers. Patents of these two companies were cited extensively by patents of other institutions.
- Patents of Kabushiki Kaisha Toshiba (24661), Massachusetts Institute of Technology (25905), Matsushita Electric Industrial Co., Ltd. (25959), Hitachi, Ltd. (23330), Digital Equipment Corporation (19486), and Hewlett-Packard Company (23244) mainly interacted with patents of IBM.
- Patents of RCA Corporation (29312), National Semiconductor Corporation (27204), and Minnesota Mining and Manufacturing Company (26602) mainly interacted with patents of Micron Technology.
- Patents of Texas Instruments Inc. (31877), Advanced Micro Devices, Inc. (17089), Motorola, Inc. (26909), and Intel Corporation (24073) interacted with patents of both IBM and Micron Technology.

There were several other local patent citation networks. Groups of institutions that formed such networks are: (1) Minnesota Mining and Manufacturing Company (3M) (26602 and 16845), the Dow Chemical Company (32084), and U.S. Philips Corporation (33125); (2) Digital Equipment Corporation, Xerox Corporation (34210), and Eastman Kodak Company (21312); and (3) Bayer Aktiengesellschaft (18158) and Lucent Technologies Inc. (25610).

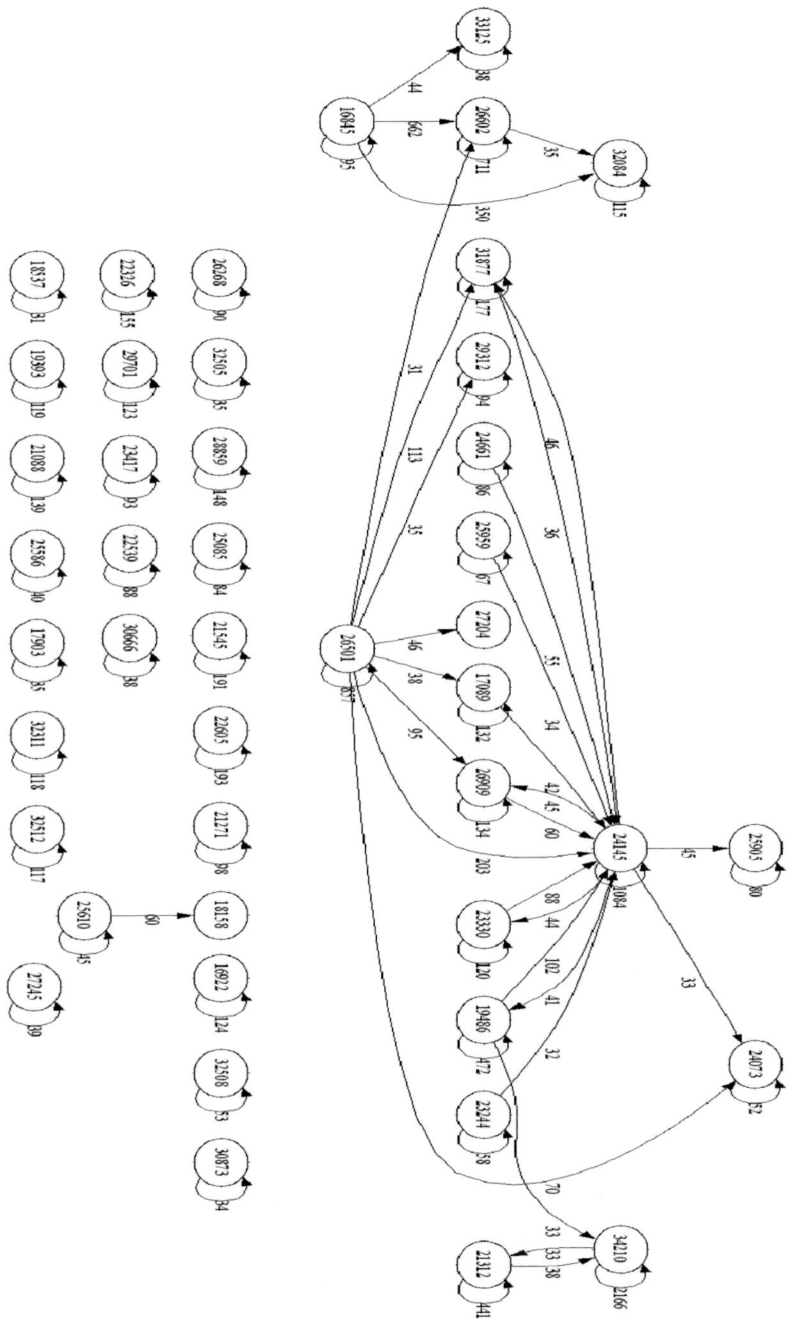

Figure 4-16. Institution Citation Network (Minimum Cites: 30)

6.3 TECHNOLOGY FIELD CITATION NETWORK

The top 50 technology fields that had the largest number of patents are presented in Table 4-16.

Table 4-16. Technology field names.

USClass	Field Name
106	Compositions: coating or plastic
148	Metal treatment
204	Chemistry: electrical and wave energy
205	Electrolysis: processes, compositions used therein, and methods of preparing the compositions
210	Liquid purification or separation
216	Etching a substrate: processes
250	Radiant energy
252	Compositions
257	Active solid-state devices (e.g., transistors, solid-state diodes)
264	Plastic and nonmetallic article shaping or treating: processes
313	Electric lamp and discharge devices
324	Electricity: measuring and testing
326	Electronic digital logic circuitry
327	Miscellaneous active electrical nonlinear devices, circuits, and systems
348	Television
356	Optics: measuring and testing
359	Optics: systems (including communication) and elements
360	Dynamic magnetic information storage or retrieval
361	Electricity: electrical systems and devices
365	Static information storage and retrieval
369	Dynamic information storage or retrieval
370	Multiplex communications
372	Coherent light generators
375	Pulse or digital communications
385	Optical waveguides
422	Chemical apparatus and process disinfecting, deodorizing, preserving, or sterilizing
423	Chemistry of inorganic compounds
424	Drug, bio-affecting and body treating compositions
427	Coating processes
428	Stock material or miscellaneous articles
430	Radiation imagery chemistry: process, composition, or product thereof
435	Chemistry: molecular biology and microbiology
436	Chemistry: analytical and immunological testing
438	Semiconductor device manufacturing: process
501	Compositions: ceramic
502	Catalyst, solid sorbent, or support therefor: product or process of making
514	Drug, bio-affecting and body treating compositions
522	Synthetic resins or natural rubbers -- part of the class 520 series
524	Synthetic resins or natural rubbers -- part of the class 520 series

USClass	Field Name
525	Synthetic resins or natural rubbers -- part of the class 520 series
526	Synthetic resins or natural rubbers -- part of the class 520 series
528	Synthetic resins or natural rubbers -- part of the class 520 series
530	Chemistry: natural resins or derivatives; peptides or proteins; lignins or reaction products thereof
536	Organic compounds -- part of the class 532-570 series
544	Organic compounds -- part of the class 532-570 series
546	Organic compounds -- part of the class 532-570 series
548	Organic compounds -- part of the class 532-570 series
549	Organic compounds -- part of the class 532-570 series
600	Surgery

These technology fields were derived from the first-level United States Patent Classification categories (available at: http://www.uspto.gov/go/classification/ selectnumwithtitle.htm). Some categories have identical names; however, the detailed specifications of such categories are different. The technology field citation network is shown in Figure 4-17 (with minimum number of citations equal to 800). General observations about this network include:

- The fields of "chemistry: natural resins or derivatives; peptides or proteins; lignins or reaction products thereof" (530) and "chemistry: molecular biology and microbiology" (435) were the dominant patent citation centers. The patents of the two fields were cited heavily by other fields.
- Patents of "drug, bio-affecting and body treating compositions" (514), "drug, bio-affecting and body treating compositions" (424), "chemistry: natural resins or derivatives; peptides or proteins; lignins or reaction products thereof" (530) and "chemistry: molecular biology and microbiology" (435) formed an interconnected citation network.
- The patents of "chemistry: analytical and immunological testing" (436) and "organic compounds – part of the class 532-570 series" (536), and "chemistry: natural resins or derivatives; peptides or proteins; lignins or reaction products thereof" (530) and "chemistry: molecular biology and microbiology" (435) formed an interconnected citation network.
- There are several local technology field citation networks. Groups of technology fields that formed such sub-networks are: (1) "active solid-state devices (e.g., transistors, solid-state diodes)" (257) and "semiconductor device manufacturing: process" (438); (2) "coating process" (427) and "stock material or miscellaneous articles" (428); and (3) "radiant energy" (250) and "optics: measuring and testing" (356).

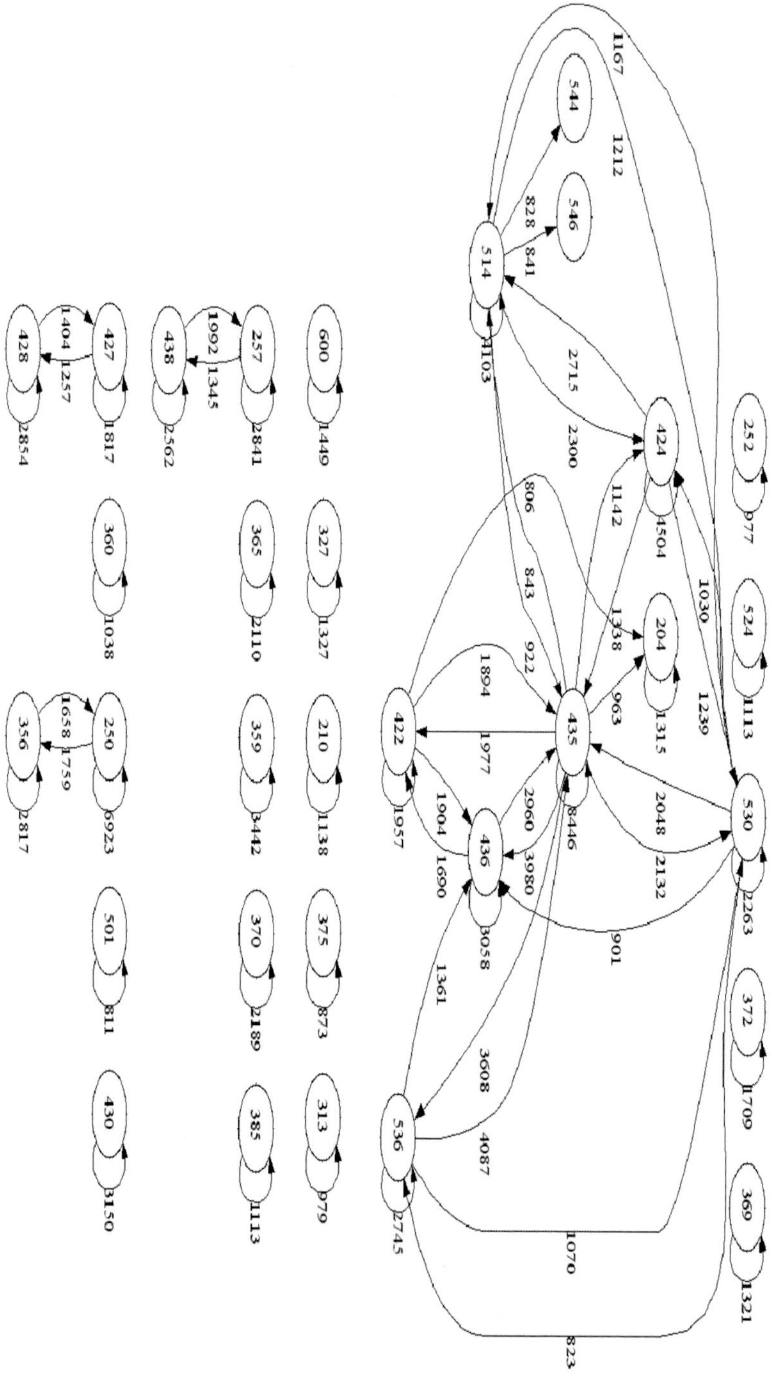

Figure 4-17. Technology Field Citation Network (Minimum Cites: 800)

7. CONCLUSIONS

Several analysis and visualization techniques have been applied on USPTO nanotechnology patents from 1976-2002. Basic analysis, content map analysis, and citation network analysis were conducted on individual countries, institutions, and technology fields.

Technologically advanced countries, such as the United States, Japan, Germany, and France, controlled the majority of nanotechnology patents. The United States was assigned about 70% of the nanotechnology-related patents between 1976 and 2002. While the United States, Japan, Germany, France, United Kingdom, Canada, Italy, and Netherlands began research and development activities reflected in nanotechnology patents since the 1970s, the Republic of Korea and Taiwan started in the early 1990s.

The International Business Machines Corporation (IBM) was issued the largest number of patents, followed by the Xerox Corporation (Xerox). When considering both quantity and freshness of patents assigned, Micron Technology outperformed all other institutions.

"Chemistry: molecular biology and microbiology" and "drug, bio-affecting and body treating compositions" were the dominating technology fields.

Content map analysis shows that dominant topics between 1976 and 1980 were: "processing systems," "aqueous solutions," "transmission lines," "electron beams," "carbon atoms," "preferred embodiments," and "laser beams." Major technology topics of the patents issued in 2001-2002 were: "nucleic acids," "pharmaceutical compositions," "laser beams," and "semiconductor devices." Several new topics can be observed but with smaller region sizes, including "optical systems," "refractive index," "optical signals," "power supplies," and "dielectric layers."

The United States (US) dominated the country citation network and the U.S. patents intensively interacted with patents of most other countries. Japan (JP) was the second largest patent citation center. Other patent citation centers included Germany (DE), France (FR), the United Kingdom (GB), Canada (CA) and Switzerland (CH). International Business Machines Corporation and Micron Technology, Inc. were the institutional patent citation centers. Patents of these two companies were cited extensively by patents of other institutions. The fields of "chemistry: natural resins or derivatives; peptides or proteins; lignins or reaction products thereof" and "chemistry: molecular biology and microbiology" were the dominant patent citation centers. The patents of these two fields were cited heavily by other fields.

This chapter presents a generic framework for patent analysis of nanotechnology. Significant insights were gained about the "invisible college" of inventors and their associated technology fields, institutions, and countries. We believe the proposed framework will be useful for knowledge mapping of other scientific disciplines.

8. QUESTIONS FOR DISCUSSION

1. What are the most important countries in nanotechnology development in the past decade and what are their areas of strength?
2. What institutions are most prominent in nanotechnology innovations in the past decade and what are their areas of strength?
3. Who are the most influential inventors in nanotechnology development in the past decade?
4. What are some of the hottest nanotechnology topics or research areas in the past five years?

Chapter 5

FEDERAL FUNDING AND NANOTECHNOLOGY INNOVATIONS: NSF FUNDING AND USPTO PATENT ANALYSIS, 1991-2002

CHAPTER OVERVIEW

Nanotechnology holds the promise of revolutionizing a wide range of products, processes and applications. It has been recognized by over sixty countries as critical for their development since the beginning of the 21st century. A significant public investment of over $1 billion annually is devoted to nanotechnology research in the United States. This chapter provides an analysis of the National Science Foundation (NSF) funding of nanoscale science and engineering (NSE) and its relationship to innovation as reflected in the United States Patent and Trade Office (USPTO) patent data. Using a combination of bibliometric analysis and visualization tools, we have identified several general trends, the key players, and the evolution of technology topics in the NSF funding and commercial patenting activities. This study documents the rapid growth of innovation in the field of nanotechnology and its correlation to funding. Statistical analysis shows that the NSF-funded researchers and their patents have higher impact factors than other private and publicly funded reference groups. This suggests the importance of fundamental research on nanotechnology development. The number of cites per NSF-funded inventor is about 10 as compared to 2 for all inventors of nanotechnology-related patents recorded at USPTO, and the corresponding Authority Score is 20 as compared to 1.8.

1. INTRODUCTION

Nanotechnology is expected to have broad and fundamental impacts on many sectors of the economy. Advances in nanotechnology have already affected a wide range of research and education areas, and it is changing industry at an accelerated pace. It promises to create interdisciplinary research and development platforms for industry, and to address revolutionary applications such as detecting and treating diseases, efficiently monitoring and protecting the environment, producing and storing energy, and building complex structures for electronic circuits or airplanes. Public funding has played an important role in fostering related research, development and educational activities. In 2000 the United States announced the National Nanotechnology Initiative (NNI) (Roco et al., 2000). In just a few years, nanotechnology has been recognized as a national priority by all industrialized countries and many developing countries, stimulated in part by the NNI. In December 2003 President Bush signed into law the *21st Century Nanotechnology Research and Development Act*, which authorizes $3.7 billion funding for nanotechnology R&D in several agencies for fiscal years 2005-2008. This legislation puts into law the NNI programs and activities, and provides guidance for enhancing innovation and responsible development of the field. The U.S. Nanoscale Science, Engineering and Technology Subcommittee (NSET) of the National Science and Technology Council published its second long-term NNI strategy in 2004 (NSET, 2004).

Increased public funding has been provided for nanotechnology R&D since 1991 and reached a significantly higher level after 2001. It is necessary to understand how funding has impacted the activities in the field. Scientific evaluation of implications of public funding on R&D output and the overall development in a scientific and engineering field is a complex task (Adams and Criliches, 1998). Only sparse literature exists on this topic. Most previous studies have focused on the impact of public funding on scientific publications. Adams and Criliches (1998) have identified a strong correlation between research output and research funding, while Arora and Gambardella (1998) reported a moderate effect of NSF funding on research output. In these studies, the research output is typically measured by the number of published scientific papers and citations these papers receive. A later study by Payne and Siow (2003) showed that $1 million in federal research funding for a research university on average resulted in 11 more articles, 0.2 more patents, and $411,000 more in total faculty salaries. Based on the citations to scientific literature in patent documents, Narin (1998) addressed the impact of research output on the commercial technology development of a field. His study showed that roughly 73% of all papers cited on the front page of U.S. industry patents had their origins in publicly funded projects, indicating the critical impact of funded scientific research on commercial technology development. Narin also concluded that the impact of public funding varies across different technology fields.

This chapter presents an analysis of NSF funding activities in nanotechnology and their relationship to the commercial technology development as reflected in the patent data from the USPTO. Instead of using scientific publication data, as most previous studies did, we analyze funding awards and patent documents, as well as citations and other connections among them. This provides a more direct account of the impact of public funding on technological innovation. We expanded our analysis to the overall development in the nanotechnology field. We apply information visualization techniques to map the relevant technology topic areas and visualize the citation networks. Advanced network analysis techniques are applied to provide more accurate assessment of impacts based on the patent citation network.

The first part of this chapter describes the nanotechnology-related award and patent data sets. Then, the topic areas of the awards and patents are compared using award/patent content map visualization and citation networks for three successive time periods between 1991 and 2002. Finally, we present the impact of NSF-funded researchers on nanotechnology development using the analysis of critical patents and inventors, as well as statistical comparison of the impact of NSF-funded researchers with other groups of inventors.

2. BASIC ANALYSIS OF AWARD AND PATENT DATA

We have used keywords based on the National Nanotechnology Initiative definition of nanotechnology to identify the awards and patents that have partial or full contents in nanotechnology. These keywords are the same as those used in Chapter 4 (Table 4-1). The search has been performed on two public databases: NSF awards (title and abstract) and USPTO patent documents (full-text). The NSF investment in nanotechnology represents about 1/3 of the federal government nanotechnology funding in the United States and about 1/12 of worldwide government nanotechnology funding. Three time periods are considered in this study:

- 1991-1995 (5 years): 1991 corresponds to the beginning of the first program solicitation focused on nanoparticle synthesis and processes at NSF. Other funding programs began during this time interval such as the

National Nanotechnology User Network (NSF, 1994) and instrumentation for nanotechnology (NSF, 2005).
- 1996-2000 (5 years): The U.S. nanotechnology Interagency Working Group was established in 1996, followed by the worldwide study on Nanostructure Science and Engineering in 1997 (Siegel et al., 1999), and NSF program announcements on Functional Nanostructures in 1997-1998 (NSF, 1997).
- 2001-2002 (2 years): These are the first two years of NNI, marked by an increased annual funding of nanotechnology R&D.

The numbers of nanotechnology-related awards and patents per time period are:
- 1991-1995: NSF made 1,146 nanotechnology-related awards and USPTO issued 11,021 nanotechnology patents
- 1996-2000: 2,384 NSF awards and 23,057 USPTO patents
- 2001-2002: 1,733 NSF awards and 14,748 USPTO patents

2.1 Award Data

The NSF funds research and education in science and engineering through awards (grants, contracts, and cooperative agreements). The foundation accounts for about 20% of federal support to academic institutions for basic research (http://www.nsf.gov/home/grants.htm). About 6% of the NSF budget in its 2005-2007 budgets is dedicated to nanotechnology.

The data set for all science and engineering fields includes 122,778 awards in the time period 1991-2002 under 65 NSF Divisions and 638 Programs involving 81,040 investigators. The awards can be accessed at http://www.nsf.gov/awardsearch/index.jsp. A complete list of NSF Divisions, Directorates, and Programs is available at http://www.nsf.gov/home/nsforg/orglist.htm. The keyword search was performed for the title and abstract of each award, which is available on the public Web site. We have identified 5,263 nanotechnology-related NSF awards involving 38 Divisions and 245 Programs between 1991 and 2002. Tables 5-1 and 5-2 present the top NSF Divisions and Programs with the largest numbers of nanotechnology-related awards during 1991–2002 and during the three time intervals. The Division of Materials Research (DMR) was the dominant Division with more than a quarter of the total number of nanotechnology-related awards, followed by the Division of Chemistry (CHE), Division of Chemical and Transportation Systems (CTS), Division of Design, Manufacture and Industrial Innovation (DMI), and Division of Electrical and Communication Systems (ECS). The top 5 NSF Programs funding nanotechnology research during 1991-2002 were: Electronics,

Photonics, and Device Technologies (Program # 1517), Condensed Matter Physics (# 1710), Small Business Phase I (# 5371), Polymers (# 1773), and Major Research Instrumentation (# 1189).

Table 5-1. Top 20 NSF divisions funding nanotechnology research: Number of awards (1991-2002).

Division	Number of grants			
	1991-2002	1991-1995	1996-2000	2001-2002
DMR DIVISION OF MATERIALS RESEARCH	1443	335	686	422
CHE DIVISION OF CHEMISTRY	715	196	326	193
CTS DIV OF CHEMICAL AND TRANSPORT SYSTEMS	533	102	254	177
DMI DIV OF DESIGN, MANUFAC & INDUSTRIAL INNOV	501	82	209	210
ECS DIV OF ELECTRICAL AND COMMUNICATIONS SYS	408	91	153	164
INT OFFICE OF INTERNATL SCIENCE & ENGINEERING	216	56	103	57
DUE DIVISION OF UNDERGRADUATE EDUCATION	191	80	84	27
CCR DIV OF COMPUTER-COMMUNICATIONS RESEARCH	190	3	101	86
CMS DIV OF CIVIL AND MECHANICAL SYSTEMS	168	21	72	75
MCB DIV OF MOLECULAR AND CELLULAR BIOSCIENCE	155	54	69	32
DBI DIV OF BIOLOGICAL INFRASTRUCTURE	98	27	56	15
BES DIV OF BIOENGINEERING & ENVIRON SYSTEMS	97	8	37	52
DMS DIVISION OF MATHEMATICAL SCIENCES	76	5	37	34
EAR DIVISION OF EARTH SCIENCES	65	12	24	29
EEC DIV OF ENGINEERING EDUCATION AND CENTERS	62	9	25	28
PHY DIVISION OF PHYSICS	61	8	24	29
EIA DIVISION OF EXPERIMENTAL & INTEG ACTIVIT	43	2	15	26
OCE DIVISION OF OCEAN SCIENCES	38	17	12	9
IBN DIV OF INTEGRATIVE BIOLOGY AND NEUROSCIE	29	3	18	8
EPS OFFICE OF EXPER PROG TO STIM COMP RSCH	26	2	11	13

Table 5-2. Top 21 NSF programs funding the nanotechnology research: Number of awards (1991-2002).

Program		Number of grants			
		1991-2002	1991-1995	1996-2000	2001-2002
1517	ELECT, PHOTONICS, & DEVICE TEC	319	65	121	133
1710	CONDENSED MATTER PHYSICS	305	83	144	78
5371	SMALL BUSINESS PHASE I	253	51	113	89
1773	POLYMERS	224	65	98	61
1189	MAJOR RESEARCH INSTRUMENTATION	160	0	64	96
1414	INTERFAC TRANS, & THERMODYN PRO	155	36	74	45
1765	MATERIALS THEORY	153	30	73	50
4710	DES AUTO FOR MICRO & NANO SYS	150	0	92	58
1415	PARTICULATE & MULTIPHASE PROCSS	136	23	62	51
7400	UNDERGRAD INSTRM & LAB IMPROVE	124	72	52	0
1972	ELECTROCHEMISTRY & SURFACE CHE	121	37	46	38
1725	NAT'L FACILITIES & INSTRUMNTAT	108	68	40	0
1715	METALS, CERAMICS, & ELEC MATRS	106	67	39	0
1775	ELECTRONIC MATERIALS	100	1	62	37
1750	INSTRUMENT FOR MATERIALS RSRCH	99	1	57	41
1633	SURFACE ENG & MATERIALS DESIGN	94	13	43	38
1762	SOLID-STATE CHEMISTRY	94	4	48	42
1938	CHEMICAL INSTRUMENTATION	82	21	48	13
5373	SMALL BUSINESS PHASE II	69	7	36	26
1108	INSTRUMENTAT & INSTRUMENT DEVP	66	23	37	6
1771	METALS	66	0	40	26

2.2 Patent Data

The USPTO patent documents were extracted using "full-text" search of patent title, abstract, claims, and specifications with our nanotechnology keywords. The resulting data set includes both the patents having nanotechnology products (generally identified by keywords in the "title-claims") and those using the nanoscale science and engineering tools (generally identified in the specifications). The full-text search resulted in 48,826 nanotechnology-related patents issued during 1991 to 2002. There are 11,024 assignees, 90,910 inventors, 79 countries, and 402 first-level United States Patent Classification categories (out of 423 in total associated with these nanotechnology-related patents). Huang et al. (2004) provided a detailed analysis of contributing countries, institutions, and technology fields in nanotechnology for the same USPTO data set. The total numbers of patents issued to top assignee countries between 1991 and 2002 are listed in Table 5-3. The United States produced the majority of the nanotechnology-related patents, followed by Japan, Germany, Canada, and France.

Table 5-3. Top assignee countries for USPTO nanotechnology patents (1991-2002).

Rank	Assignee Country	Number of Patents			
		1991-2002	1991-1995	1996-2000	2001-2002
1	United States	29968	6773	14183	9012
2	Japan	4890	1157	2232	1501
3	Federal Rep. of Germany	3976	998	1927	1051
4	Canada	1321	300	606	415
5	France	1284	269	607	408
6	United Kingdom	551	135	272	144
7	China (Taiwan)	308	15	151	142
8	Israel	294	47	150	97
9	Switzerland	276	47	121	108
10	Australia	254	26	59	169
11	Republic of Korea	248	21	105	122
12	Italy	221	54	101	66
13	Netherlands	209	43	96	70
14	Sweden	154	23	59	72
15	Belgium	125	25	61	39
16	Denmark	92	18	41	33
17	Finland	75	8	36	31
18	Singapore	49	0	14	35
19	Norway	48	6	18	24
20	Austria	41	4	20	17

The numbers of nanotechnology-related patents by the top 20 assignee institutions are shown in Table 5-4. The top five assignees for the interval 1991-2002 are International Business Machines Corporation (IBM), Xerox Corporation, the University of California, NEC Corporation, and Eastman Kodak Company.

Table 5-4. Top assignee institutions in USPTO nanotechnology patents (1991-2002).

Rank	Assignee Name	Number of Patents			
		1991-2002	1991-1995	1996-2000	2001-2002
1	International Business Machines Corp.	1103	226	509	368
2	Xerox Corporation	796	220	385	191
3	The Regents of the University of Calif.	561	70	291	200
4	NEC Corporation	473	59	273	141
5	Eastman Kodak Company	465	173	206	86

Rank	Assignee Name	Number of Patents			
		1991-2002	1991-1995	1996-2000	2001-2002
6	Micron Technology, Inc.	461	13	167	281
7	Motorola, Inc.	447	139	206	102
8	Minnesota Mining and Manufacturing Co.	395	158	236	1
9	Canon Kabushiki Kaisha	389	106	168	115
10	Kabushiki Kaisha Toshiba	313	53	174	86
11	Hitachi, Ltd.	302	84	134	84
12	Advanced Micro Devices, Inc.	298	3	74	221
13	Abbott Laboratories	292	110	141	41
14	Texas Instruments Incorporated	291	74	156	61
15	3M Innovative Properties Company	290	0	103	187
16	General Electric Company	289	98	100	91
17	Matsushita Electric Industrial Co., Ltd.	262	99	115	48
18	The U.S.A. as represented by the Secretary of the Navy	257	88	108	61
19	Genentech, Inc.	245	36	147	62
20	E. I. Du Pont de Nemours and Co.	239	127	87	25

The technology fields correspond to the first-level United States Patent Classification categories (available at: http://www.uspto.gov/go/classification/selectnumwithtitle.htm). Some categories have identical names; however, the detailed specifications of such categories are different. We used the category name as well as their assigned U.S. Patent Classification ID number to label each technology field. The top technology fields of the nanotechnology-related patents are presented in Table 5-5. The first five fields are: "Chemistry: molecular biology and microbiology," "Drug, bio-affecting and body treating compositions" (514 and 424), "Organic compounds – part of the class 532-570 series," and "Semiconductor device manufacturing: process."

Table 5-5. Top U.S. patent classification first-level nanotechnology fields (1991-2002).

Rank		Technology Field	Number of Patents			
			1991-2002	1991-1995	1996-2000	2001-2002
1	435	Chemistry: molecular biology and microbiology	6779	1097	3447	2235
2	514	Drug, bio-affecting and body treating compositions	5322	955	2840	1527
3	424	Drug, bio-affecting and body treating compositions	4197	666	2221	1310
4	536	Organic compounds -- part of the class 532-570 series	3447	368	1856	1223
5	438	Semiconductor device manufacturing: process	3185	478	1261	1446
6	530	Chemistry: natural resins or	3172	579	1700	893

Rank		Technology Field	Number of Patents 1991-2002	1991-1995	1996-2000	2001-2002
		derivatives; peptides or proteins; lignins or reaction products thereof				
7	428	Stock material or miscellaneous articles	3136	660	1494	982
8	257	Active solid-state devices (e.g., transistors, solid-state diodes)	2922	758	1264	900
9	250	Radiant energy	2777	1020	1141	616
10	427	Coating processes	2434	585	1162	687
11	359	Optics: systems (including communication) and elements	2162	546	979	637
12	436	Chemistry: analytical and immunological testing	2151	597	985	569
13	430	Radiation imagery chemistry: process, composition, or product thereof	2113	569	971	573
14	356	Optics: measuring and testing	1825	553	891	381
15	422	Chemical apparatus and process disinfecting, deodorizing, preserving, or sterilizing	1298	325	576	397
16	204	Chemistry: electrical and wave energy	1217	310	559	348
17	546	Organic compounds -- part of the class 532-570 series	1207	254	637	316
18	524	Synthetic resins or natural rubbers -- part of the class 520 series	1169	282	538	349
19	252	Compositions	1147	308	532	307
20	210	Liquid purification or separation	1099	309	498	292

By comparing the major NSF Divisions' awards and the patent technology fields, we observe that the dominant positions of the NSF Divisions covering material science, chemistry, chemical engineering, and device design and manufacturing are consistent with the importance of related technology fields reflected in the patent data. However, the major technology fields related to biology and pharmaceutical research in nanotechnology-related patents do not have corresponding NSF Divisions. Most of the U.S. federal biological and pharmaceutical research is under the jurisdiction of the National Institutes of Health (NIH).

2.3 Trend Analysis

Figure 5-1 shows that the growth of the number of nanotechnology-related awards at NSF is similar to the growth of the number of nanotechnology-related patents after 1991.

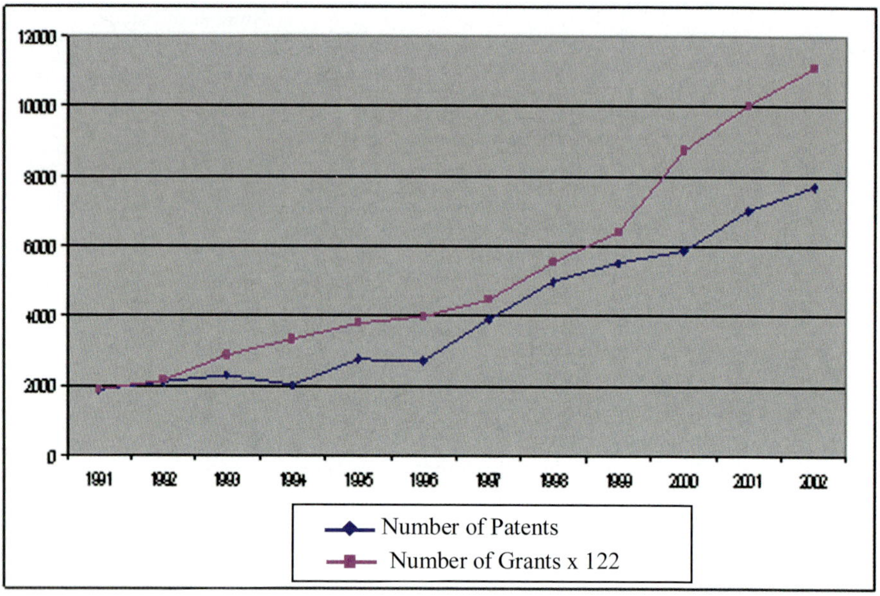

Figure 5-1. Trend analysis: the numbers of nanotechnology patents and NSF awards (1991-2002).

Four of the top five NSF Divisions, including the Division of Materials Research (DMR), Division of Chemistry (CHE), Division of Design, Manufacture and Industrial Innovation (DMI), and Division of Chemical and Transport Systems (CTS), had consistent prominent presence in nanotechnology funding activities after 1994 (Figure 5-2). The DMR had a slight drop in funding activities in 2002, while the other three continued a general increasing trend. The Division of Electrical and Communications Systems (ECS) had substantial increases in nanotechnology funding after 1998.

Figure 5-3 illustrates the patenting activity trends in the top 20 technology fields. The technology fields are listed in the order of decreasing total number of patents issued between 1991 and 2002. The fastest growths in patenting activities in recent years are in "Chemistry: molecular biology and microbiology," "Drug, bio-affecting and body treating compositions," "Semiconductor device manufacturing: process," and "Organic compounds -- part of the class 532-570 series."

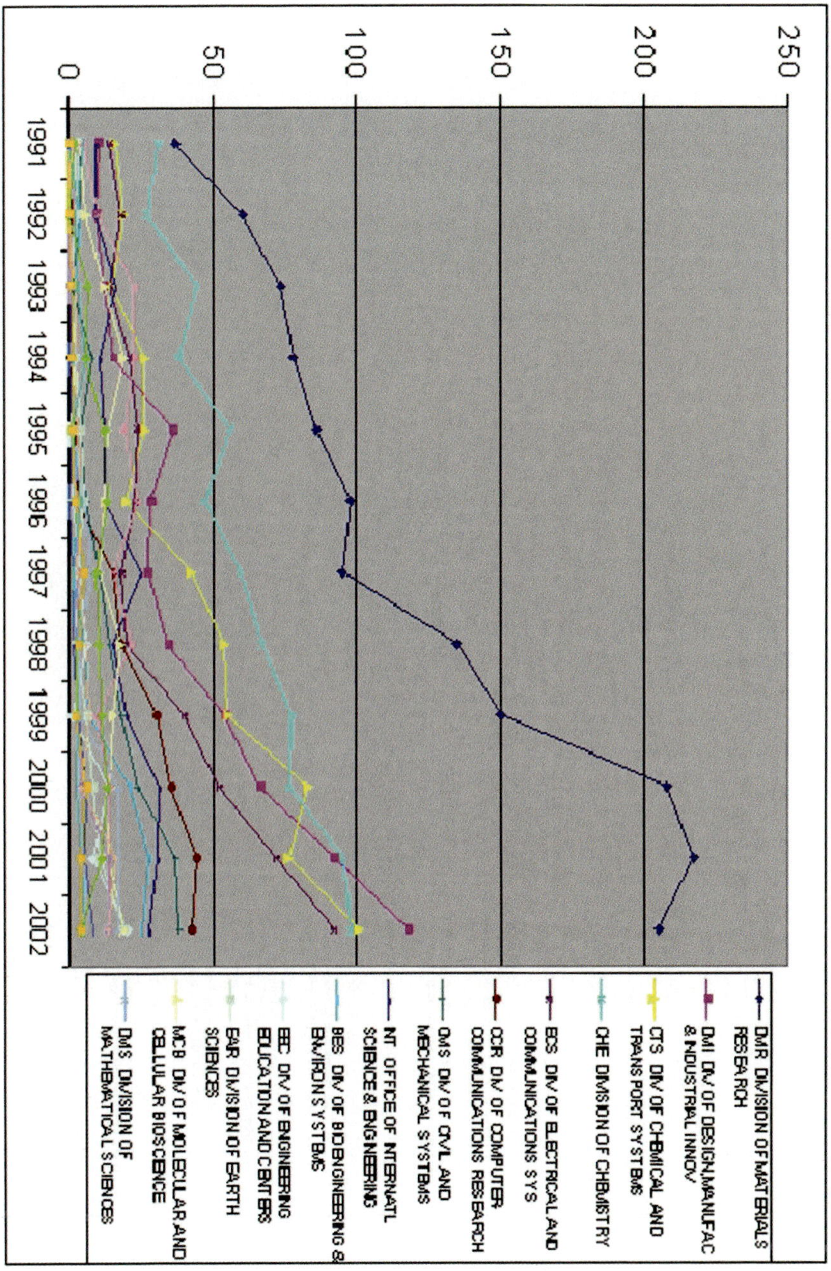

Figure 5-2. Trend analysis: number of awards per NSF division (1991-2002).

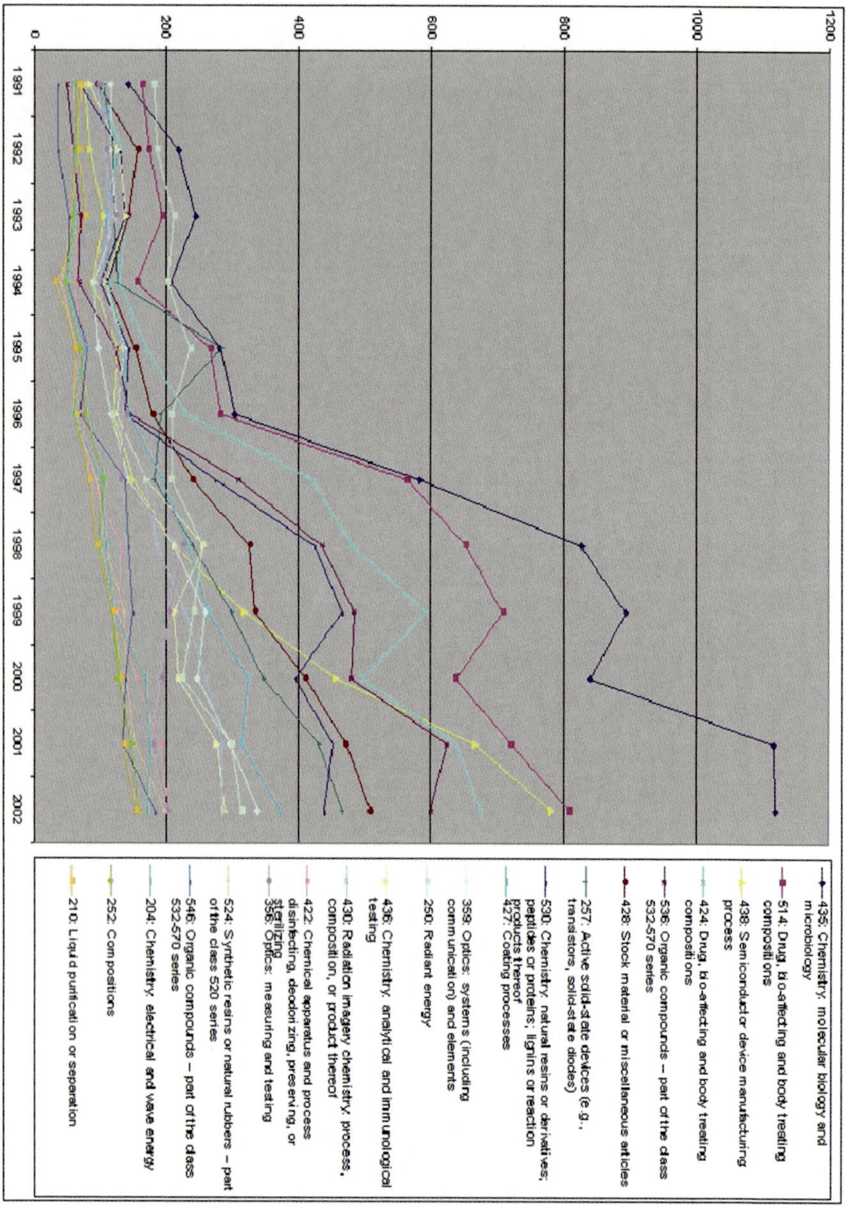

Figure 5-3. Trend analysis: Number of patents per technology field (1991-2002).

2.4 Linking NSF Award and Patent Data

The key linkage between the NSF awards and USPTO patents is the set of nanotechnology patent inventors who are also principal investigators of NSF awards. They are referred to as *"PI-inventors"* in this chapter. We identified *307 PI-inventors* by matching their names and their institutions in the award and patent data sets. These PI-inventors were associated with 760 nanotechnology-related patents and 628 NSF awards.

Table 5-6 shows the top PI-inventors with the largest numbers of nanotechnology-related patents. Thomas J. Pinnavaia of Michigan State University (also Claytec, Inc.) topped the list with 30 patents, followed by George M. Whitesides of Harvard University (also with several companies), who had filed 24 nanotechnology-related patents. The remaining top 6 PI-inventors measured by number of nanotechnology-related patents were: Stuart M. Lindsay of Arizona State University (also Molecular Imaging Corporation), Mark E. Thompson of the University of Southern California, Sanford A. Asher of University of Pittsburgh, and Charles R. Cantor of Boston University (also Genelabs Technologies, Inc).

Table 5-6. Top PI-inventors with the largest numbers of patents (1991-2002).

Rank	PI-inventors	Institution	No. of patent(s)
1	Pinnavaia, Thomas J.	Board of Trustees of Michigan State Univ.	30
2	Whitesides, George M.	Harvard Univ.	24
3	Lindsay, Stuart M.	Arizona Board of Regents acting on behalf of Arizona State Univ.	18
4	Thompson, Mark E.	The Trustees of Princeton Univ.	14
5	Asher, Sanford A.	Univ. of Pittsburgh	14
6	Cantor, Charles R.	Boston Univ.	14
7	Maris, Humphrey J.	Brown Univ. Research Foundation	13
8	Fejer, Martin M.	Board of Trustees, Leland Stanford Junior Univ., Stanford Univ.	12
9	Bard, Allen J.	Board of Regents, The Univ. of Texas System	12
10	Hansma, Paul K.	Digital Instruments, Inc.	11
11	Klabunde, Kenneth J.	Kansas State Univ. Research Foundation	11
12	Minne, Stephen C.	Board of Trustees of the Leland Stanford Jr. Univ.	10
13	Searson, Peter C.	Candescent Technologies Corporation	10
14	Newkome, George R.	The Univ. of South Florida	9
15	Kear, Bernard H.	Rutgers the State Univ. of New Jersey	9
16	Bowers, John E.	CIENA Corporation	8
17	Rothschild, Kenneth J.	Amber Gen. Inc.	8
18	McCandlish, Larry E.	Exxon Research & Engineering Company	8
19	Lindsey, Jonathan S.	North Carolina State Univ.	8
20	Zare, Richard N.	Board of Trustees of the Leland Stanford Junior Univ.	7
21	Lieber, Charles M.	President and Fellows of Harvard College	7

Table 5-7 lists the top PI-inventors in terms of number of nanotechnology-related awards. Henry I. Smith from Louisiana State University and MIT topped the list with 12 nanotechnology-related awards. The next five top PI-inventors are: Charles M. Lieber from Harvard, Cyrus

R. Safinya from the University of California, David J. Sellmyer from the University of Nebraska, Allen J. Bard from the University of Texas, and Kenneth J. Klabunde from Kansas State University. By comparing Tables 5-6 and 5-7, we observe that the top PI-inventors with the largest numbers of nanotechnology-related patents do not appear to overlap with the top PI-inventors with the largest numbers of nanotechnology-related awards. Only Allen J. Bard appeared near the top of both lists, 9th in Table 5-6 and 5th in Table 5-7. Matching the complete lists in Tables 5-6 and 5-7 we identify only five PI-inventors as overlaps, including Stuart M. Lindsay, Paul K. Hansma, Allen J. Bard, Kenneth J. Klabunde, and Charles M. Lieber. These findings would indicate that most PI-inventors either focus on fundamental research projects (such as those funded by NSF) or commercial technology development.

Table 5-7. Top PI-inventors with the largest numbers of NSF awards (1991-2002).

Rank	PI-inventors	Institution	Number of grants
1	Smith, Henry I.	Board of Supervisors of Louisiana State Univ. and Agricultural and Mechanical College	12
2	Lieber, Charles M.	President and Fellows of Harvard College	9
3	Safinya, Cyrus R.	The Regents of the Univ. of California	9
4	Sellmyer, David J.	Board of Regents of the Univ. of Nebraska	8
5	Bard, Allen J.	Board of Regents, The Univ. of Texas System	8
6	Klabunde, Kenneth J.	Kansas State Univ. Research Foundation	8
7	Dzenis, Yuris A.	Board of Regents, Univ. of Nebraska-Lincoln	8
8	Requicha, Aristides A. G.	Univ. of Southern California	7
9	Awschalom, David D.	The Regents of the Unversity of California	7
10	Hansma, Paul K.	Digital Instruments, Inc.	7
11	Chou, Stephen Y.	Regents of the Univ. of Minnesota	7
12	Lindsay, Stuart M.	Arizona Board of Regents acting on behalf of Arizona State Univ.	7
13	Colvin, Vicki L.	The Regents of the Univ. of California	7
14	Ruoff, Rodney S.	Washington Univ.	6
15	Westervelt, Robert M.	President and Fellows at Harvard College	6
16	Hamilton, Andrew D.	Univ. of Pittsburgh	6
17	Tour, James M.	Univ. of South Carolina	6
18	Moore, Jeffrey S.	The Board of Trustees of the Univ. of Illinois	6
19	Stucky, Galen D.	GRT, Inc.	6

3. CONTENT MAP ANALYSIS

Previous evaluation studies of R&D funding have been mostly based on publication and citation counts. In this chapter, we rely on advanced text analysis and visualization techniques to comparatively evaluate the technology topics of the nanotechnology-related awards and patents.

A topic map contains two components: a folder tree displayed on the left-hand side and a hierarchical content map on the right-hand side. The award/patent documents are organized under technology topics, which are represented as nodes in the folder tree and colored regions in the content map. These topics were labeled by representative noun phrases using a Natural Language Processing tool, the Arizona Noun Phraser, which identifies the key noun phrases based primarily on linguistic patterns (Tolle and Chen, 2000). Numbers of patent documents that were assigned to the first-level topics are presented in parentheses after the topic labels. The layers of the colored regions represent the levels of the hierarchies inside the specific regions. The mapping is based on the multi-level self-organization map algorithm (Chen et al., 1996; Ong et al., 2004) developed by the Arizona Artificial Intelligence Lab.

The award and patent content maps for the three time periods of 1991-1995, 1996-2000, and 2001–2002 are presented. For content maps of 1996-2000 and 2001-2002 we visualize the changes of topic areas from the previous time period using different colors to indicate the growth rate of a topic area. The growth rate of a topic area was computed as the ratio between the number of documents in the current time period and that of the previous time period. A base growth rate was computed as the ratio between the total number of documents in the current time period and that of the previous time period. A topic region with a growth rate similar to the base growth rate was assigned a green color that is consistent with the region color in the 1991-1995 content maps. The higher (or lower) the growth rate of a topic region is, the warmer (or colder) the color assigned.

We also analyzed the distribution of NSF Divisions and Directorates covering the topic areas in the award content map. For each topic region "*i*" we counted the number of awards v_{ij} in each Division/Directorate "*j*." We present the dominant NSF Divisions/Directorates for each topic region in a table. Divisions/Directorates of a topic region that had more than $\mu_i + \sigma_i$ awards were considered dominant Divisions/Directorates, where μ_i and σ_i are the mean and standard deviation of v_{ij}.

3.1 Content Map Analysis for 1991-1995

Figures 5-4 and 5-5 show nanotechnology-related award and patent content maps from 1991 to 1995, respectively.

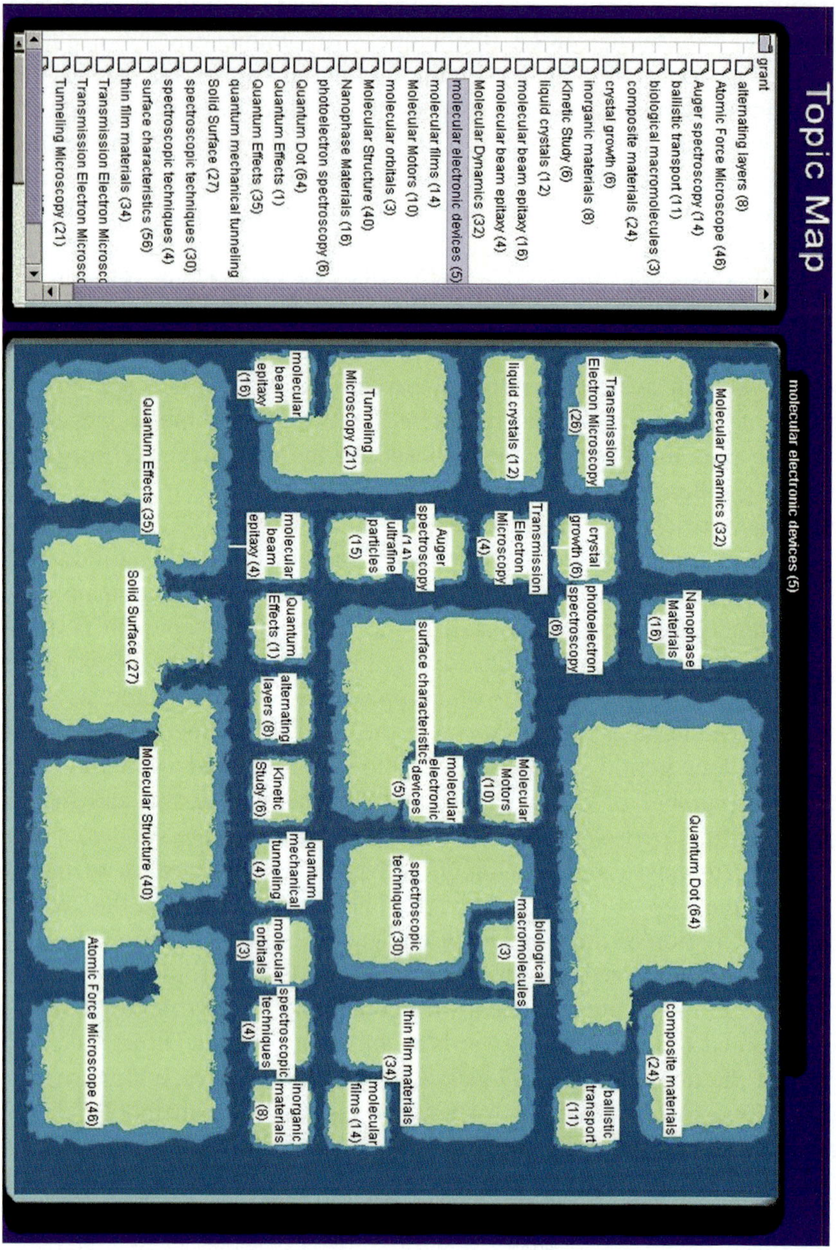

Figure 5-4. Nanotechnology-related NSF award content map (1991-1995).

Figure 5-4 shows that from 1991 to 1995, the nanotechnology-related NSF awards were concentrated in several technology topics including "Quantum Dot," "surface characteristics," "Atomic Force Microscope," and "Molecular Structure."

Figure 5-5 shows that nanotechnology-related patents cover broader technology topics, including "light sources," "carbon atoms," "pharmaceutical compositions," "thin films," and "laser beams." The topic labels are noun phrases extracted using the Arizona Noun Phraser as described previously. However, phrases with capitalization as well as morphological and inflectional variations were treated as the same phrase for the patent/award content representation used by the self-organizing map algorithm.

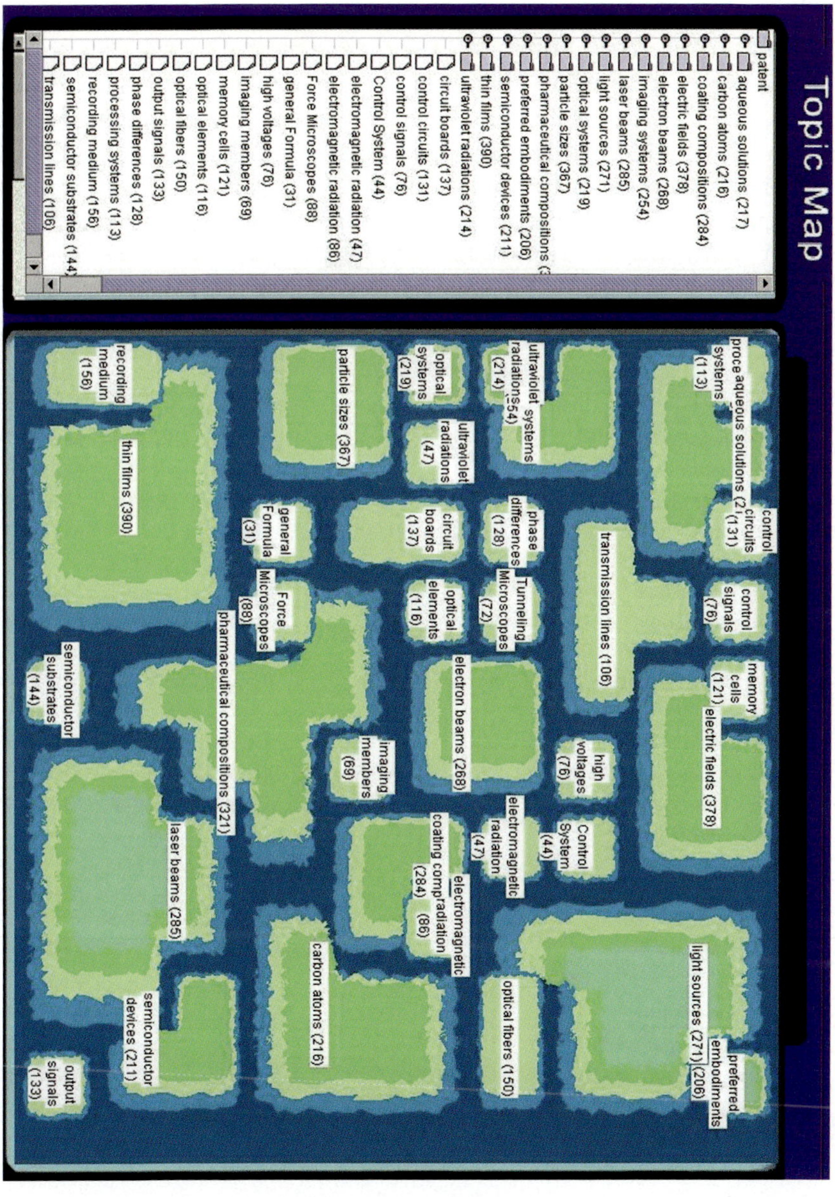

Figure 5-5. Nanotechnology-related USPTO patent content map (1991-1995).

Table 5-8 summarizes the distribution of awards per NSF Divisions and Directorates between 1991 and 1995. The numbers in the parentheses after the dominant Divisions/Directorates indicate the number of awards under that Division/Directorate within the topic region. "Quantum Dot" was the largest region in the nanotechnology-related award content map for this time period. Sixty-four awards were grouped into this area with an average funding amount of $209,000. The Division of Materials Research (DMR) had a large contribution with 30 out of 64 awards. The Directorate for Mathematical and Physical Sciences (MPS) was the most active Directorate of the region with 37 out of 64 awards. Most other topic areas, such as "Quantum Effects," "Atomic Force Microscope," and "spectroscopic techniques," were similarly dominated by the DMR and MPS. There were 10 topic areas that had multiple dominant Divisions. For example, "Molecular Structure" had 12 awards under the Division of Undergraduate Education (DUE) and 8 awards under the Division of Chemistry (CHE). Overall, DMR under MPS dominated the majority of topic areas of the nanotechnology-related awards from 1991 to 1995.

Table 5-8. NSF Division/Directorate distribution of awards per topic area (1991-1995).

Region Size	Region Label	No. of Grants	Average Amount	Major Division	Major Directorate
13	Quantum Dot	64	209,000	DMR(30)	MPS(37)
9	Molecular Structure	40	233,000	DUE(12);CHE(8)	MPS(15)
8	Quantum Effects	36	817,000	DMR(16)	MPS(26)
7	surface characteristics	56	456,000	CHE(19);DMR(16)	MPS(35)
7	Atomic Force Microscope	46	160,000	DMR(17)	MPS(20)
7	Solid Surface	27	201,000	CHE(7)	ENG(13)
6	spectroscopic techniques	34	514,000	DMR(11)	MPS(15)
5	thin film materials	34	239,000	CHE(7);CTS(6)	MPS(12); ENG(9)
5	Molecular Dynamics	32	262,000	DMR(9);CHE(8)	MPS(18)
5	Tunneling Microscopy	21	219,000	DMR(7);CHE(6)	MPS(13)
4	Transmission Electron Microscopy	30	234,000	DMR(13);CHE(11)	MPS(24)
4	composite materials	24	400,000	DMR(11)	MPS(11)
2	molecular beam epitaxy	20	364,000	DMR(10)	MPS(11)
2	Nanophase Materials	16	177,000	DMR(6)	MPS(6); ENG(5)
2	liquid crystals	12	2,034,000	DMR(8)	MPS(10)
1	ultrafine particles	15	190,000	DMR(5)	MPS(7)
1	Auger spectroscopy	14	408,000	DMR(7)	MPS(9)
1	molecular films	14	347,000	CHE(4)	MPS(7)
1	ballistic transport	11	223,000	DMR(8)	MPS(8)

Region Size	Region Label	No. of Grants	Average Amount	Major Division	Major Directorate
1	Molecular Motors	10	125,000	MCB(6)	BIO(6)
1	alternating layers	8	202,000	DMR(4)	MPS(6)
1	inorganic materials	8	172,000	CHE(3);DMR(3)	MPS(6)
1	crystal growth	6	301,000	DMR(5)	MPS(5)
1	Kinetic Study	6	297,000	DMR(2);CHE(2); DUE (2)	MPS(4)
1	photoelectron spectroscopy	6	271,000	DMR(4)	MPS(4)
1	molecular electronic devices	5	350,000	CHE(3)	MPS(3)
1	quantum mechani-cal tunneling	4	423,000	ECS(2)	ENG(2)
1	biological macromolecules	3	76,000	MCB(1);DBI(1); CTS (1)	BIO(2)
1	molecular orbitals	3	148,000	DUE(2)	EHR(2)

3.2 Content Map Analysis for 1996-2000

Figure 5-6 presents the award content map for the interval 1996-2000. It displays many new award topics (shown as red regions), for instance, "Organic Molecules," "Electron Microscope," "optical fibers," "polymeric materials," and "crystal structures." The major topic areas of "Tunneling Microscopy," "Nanostructured Materials" (shown as red regions because of large growth rates) and "Transmission Electron Microscopy" and "Quantum Dots" (shown as orange and yellow regions) continued to have a higher growth rate than the base rate. The topic area "Composite Materials" (green region) had the base growth rate. Other topic areas, including "Molecular Structure," "Molecular Dynamics" and "Thin Film Materials" (shown as blue and purple regions), had a lower growth rate than the base growth rate. In general, the microscope-related topics continue to dominate nanotechnology-related NSF awards. The detailed topic region growth rates and base growth rate are also presented in Table 5-9.

In the table, some regions' grant numbers are listed as "N/A." These topics appeared in only one of the content maps and had too few grants in the other time period to be reported by the content map. Since these regions have such low numbers of grants, they can be considered either new topics or disappearing topics between the two time periods. Table 5-9 arranges topics in three sections. The first section contains the new topics, which are sorted in descending order by the number of grants in the second time period. The second section contains topics that existed in both time periods and are sorted in descending order by their growth/decline rate. The last section lists topics that disappeared from the previous time period and are sorted in ascending order by the number of grants in the first time period.

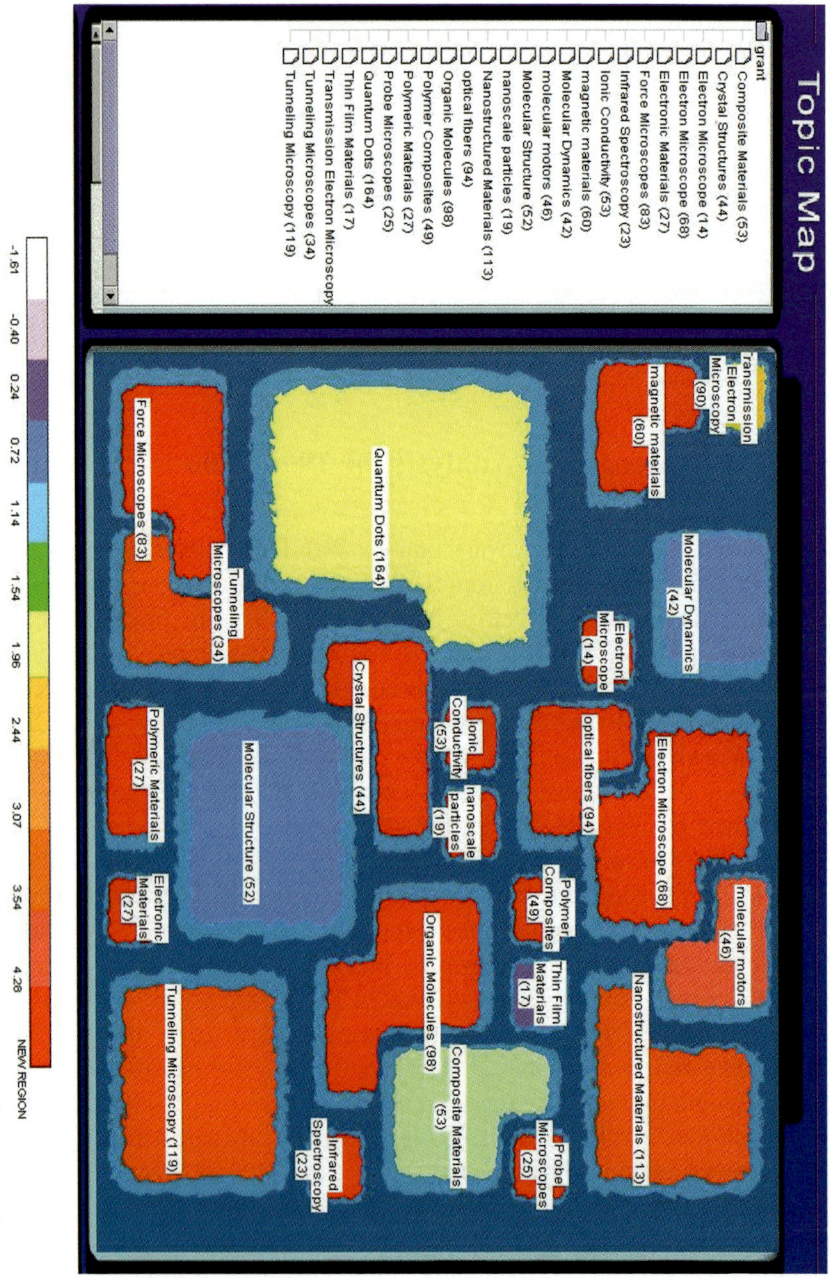

Figure 5-6. Nanotechnology-related NSF award content map (1996-2000). The color scale shows the rate of increase of the number of awards published in the two time periods in the respective topic.

Table 5-9. Nanotechnology-related NSF award change details (1996 – 2000). * "N/A" indicates that the number of awards in that region is too small to be shown on the content map

Region Label	# of Grants in the region (1996-2000)*	# of Grants in the region (1991-1995)*	Growth Rate
Organic Molecules	98	N/A	N/A
optical fibers	94	N/A	N/A
Force Microscopes	83	N/A	N/A
Electron Microscope	82	N/A	N/A
magnetic materials	60	N/A	N/A
Ionic Conductivity	53	N/A	N/A
Polymer Composites	49	N/A	N/A
Crystal Structures	44	N/A	N/A
Electronic Materials	27	N/A	N/A
Polymeric Materials	27	N/A	N/A
Probe Microscopes	25	N/A	N/A
Infrared Spectroscopy	23	N/A	N/A
nanoscale particles	19	N/A	N/A
Tunneling Microscopes	153	21	6.29
Nanostructured Materials	113	16	6.06
molecular motors	46	10	3.60
Transmission Electron Microscopy	90	26	2.46
Quantum Dots	164	64	1.56
Composite Materials	53	24	1.21
Molecular Dynamics	42	32	0.31
Molecular Structure	52	40	0.30
Thin Film Materials	17	34	-0.50
biological macromolecules	N/A	3	N/A
molecular orbitals	N/A	3	N/A
quantum mechanical tunneling	N/A	4	N/A
molecular electronic devices	N/A	5	N/A
crystal growth	N/A	6	N/A
Kinetic Study	N/A	6	N/A
photoelectron spectroscopy	N/A	6	N/A
alternating layers	N/A	8	N/A
inorganic materials	N/A	8	N/A
ballistic transport	N/A	11	N/A
liquid crystals	N/A	12	N/A
Auger spectroscopy	N/A	14	N/A
molecular films	N/A	14	N/A
ultrafine particles	N/A	15	N/A
Nanophase Materials	N/A	16	N/A
molecular beam epitaxy	N/A	20	N/A
Solid Surface	N/A	27	N/A
spectroscopic techniques	N/A	34	N/A
Quantum Effects	N/A	36	N/A
Atomic Force Microscope	N/A	46	N/A
surface characteristics	N/A	56	N/A
Baseline Growth Rate			1.337

Table 5-10 presents the major Divisions/Directorates contributing to various topic areas during 1996-2000. The DMR Division under the MPS Directorate was still the dominant source of funding while CHE became another important Division. The Engineering (ENG) Directorate also funded a significant number of awards in many small award topic areas. A significant difference from the previous time period is that more than half of the topic areas now had multiple funding divisions, indicating more diversified funding sources for nanotechnology-related projects and increasing the role of engineering (ENG).

Table 5-10. NSF Division/Directorate distribution of awards per topic area (1996-2000).

Region Size	Region Label	No. of Grants	Average Amount	Major Division	Major Directorate
17	Quantum Dots	164	238,000	DMR(85)	MPS(99)
9	Tunneling Microscopy	119	449,000	DMR(64)	MPS(88)
9	Molecular Structure	52	500,000	DMR(14);CHE(7)	MPS(23)
8	Electron Microscope	82	322,000	DMR(35)	MPS(42); ENG(32)
7	Nanostructured Materials	113	953,000	DMR(50)	MPS(58)
6	Organic Molecules	98	323,000	CHE(26);CTS(23); DMR(22)	MPS(48)
5	Force Microscopes	83	229,000	DMR(24);CTS(15)	ENG(34); MPS(27)
5	Composite Materials	53	248,000	DMR(17);DMI(11)	ENG(27)
4	Crystal Structures	44	251,000	DMR(15);CHE(15)	MPS(30)
4	Molecular Dynamics	42	276,000	DMR(8);CHE(8); ACI(6)	MPS(19)
4	Tunneling Microscopes	34	246,000	DMR(23)	MPS(26)
3	optical fibers	94	255,000	DMR(27)	MPS(41); ENG(36)
3	magnetic materials	60	395,000	DMR(31)	MPS(43)
3	molecular motors	46	253,000	MCB(14)	BIO(19)
2	Transmission Electron Microscopy	90	295,000	DMR(38)	MPS(62)
2	Polymeric Materials	27	1,164,000	DMR(20)	MPS(23)
1	Ionic Conductivity	53	261,000	DMR(14);CTS(11); CHE(10)	MPS(24); ENG(22)
1	Polymer Composites	49	273,000	DMI(8);CHE(8); DMR(7)	ENG(22)
1	Electronic Materials	27	326,000	DMR(13)	MPS(22)
1	Probe Microscopes	25	280,000	DMR(7);DUE(4)	MPS(9)
1	Infrared Spectroscopy	23	205,000	CHE(7);DMR(7)	MPS(14)
1	nanoscale particles	19	184,000	DMR(5)	MPS(9)
1	Thin Film Materials	17	1,634,000	CHE(4);DMI(4)	ENG(8); MPS (8)

Figure 5-7 presents the patent content map of the same time period. The dominant patent topics were “semiconductor devices,” “memory cells,” “thin films," "acid sequences," "nucleic acids," and "laser beams." New topic areas include "memory devices," "dielectric layers," "semiconductor wafers," "electronic devices," "metal oxides," "composite materials," "supply voltages," "power supplies," "clock signals," "computer systems," "communication systems," "acid sequence," "refractive indexes," "nucleic acids," and "disk drives."

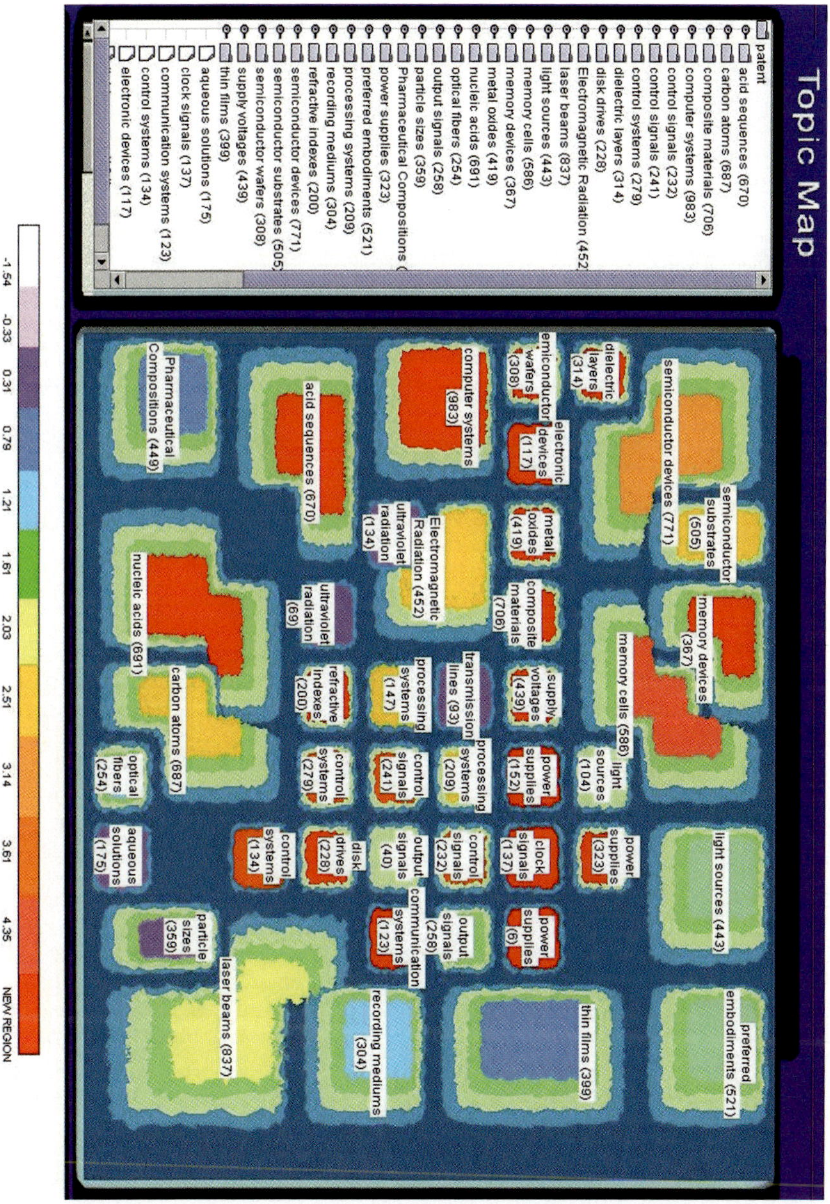

Figure 5-7. Nanotechnology-related USPTO patent content map (1996 – 2000).

Table 5-11 presents the detailed growth rate information. Similar to the topic areas in the award content map, most patent topic areas were new or had a higher growth rate than the base rate, indicating that the patents issued in 1996-2000 were dominated by new topic areas.

Table 5-11. Growth rates for nanotechnology-related USPTO patents (1996-2000).

Region Label	# of Patents in the region (1996 - 2000)*	# of Patents in the region (1991 - 1995)*	Growth Rate
computer systems	983	N/A	N/A
composite materials	706	N/A	N/A
nucleic acids	691	N/A	N/A
acid sequences	670	N/A	N/A
power supplies	481	N/A	N/A
supply voltages	439	N/A	N/A
metal oxides	419	N/A	N/A
memory devices	367	N/A	N/A
dielectric layers	314	N/A	N/A
semiconductor wafers	308	N/A	N/A
disk drives	228	N/A	N/A
refractive indexes	200	N/A	N/A
clock signals	137	N/A	N/A
communication systems	123	N/A	N/A
electronic devices	117	N/A	N/A
control systems	413	44	8.39
control signals	473	76	5.22
memory cells	586	121	3.84
semiconductor devices	771	211	2.65
semiconductor substrates	505	144	2.51
Electromagnetic Radiation	452	133	2.40
carbon atoms	687	216	2.18
processing systems	356	113	2.15
laser beams	837	285	1.94
preferred embodiments	521	206	1.53
output signals	298	133	1.24
light sources	547	271	1.02
recording mediums	304	156	0.95
optical fibers	254	150	0.69
Pharmaceutical Compositions	449	321	0.40
thin films	399	390	0.02
particle sizes	359	367	-0.02
transmission lines	93	106	-0.12
aqueous solutions	175	217	-0.19
ultraviolet radiation	203	261	-0.22
general Formula	N/A	31	N/A
imaging members	N/A	69	N/A
Tunneling Microscopes	N/A	72	N/A
high voltages	N/A	76	N/A
Force Microscopes	N/A	88	N/A
optical elements	N/A	116	N/A
phase differences	N/A	128	N/A
control circuits	N/A	131	N/A
circuit boards	N/A	137	N/A
optical systems	N/A	219	N/A

3.3 Content Map Analysis for 2001-2002

Figure 5-8 presents the award content map for 2001-2002, with the growth rates per topic area presented in Table 5-12.

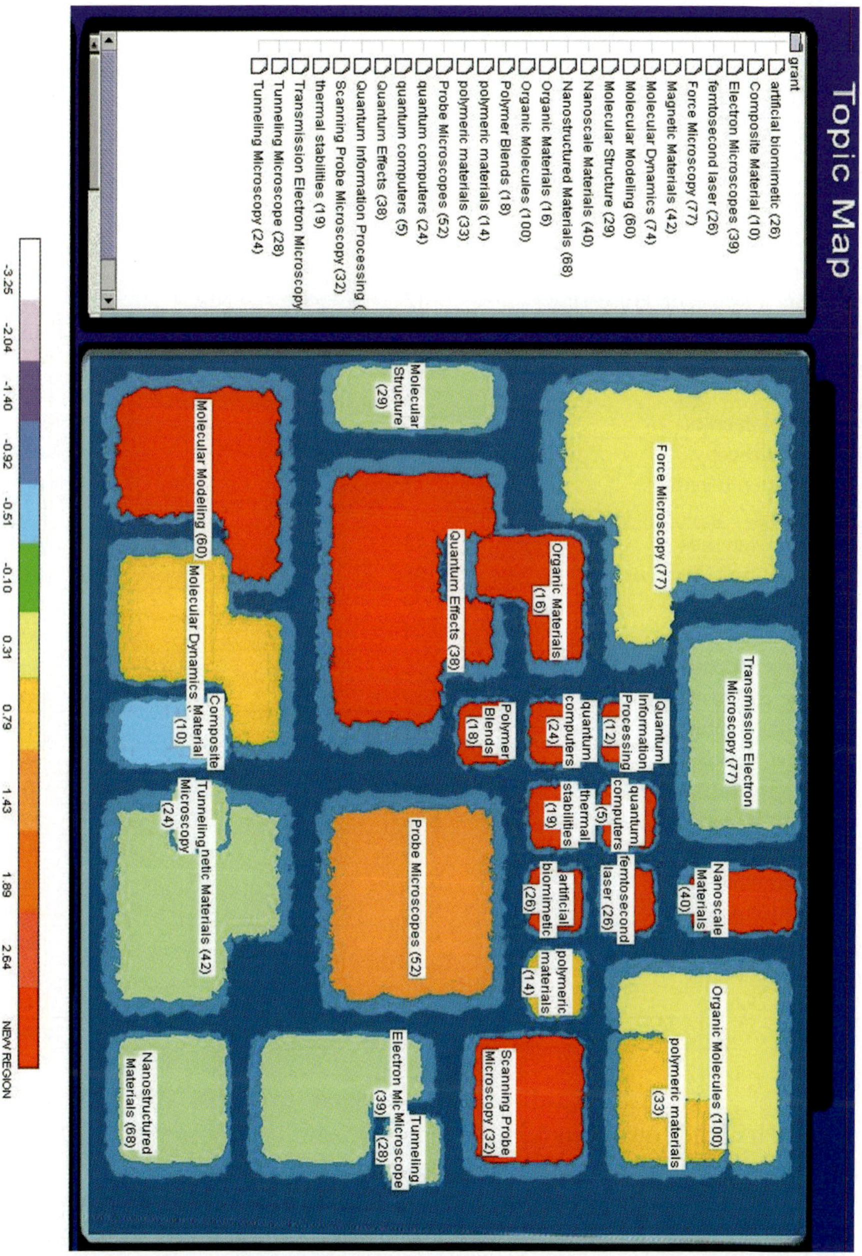

Figure 5-8. Nanotechnology-related NSF award content map (2001-2002).

Table 5-12. Growth rate of nanotechnology-related NSF awards per detailed topic areas in 2001-2002 as compared to 1996-2000.

Region Label	# of Grants in the region (2001-2002)*	# of Grants in the region (1996-2000)*	Growth Rate
Molecular Modeling	60	N/A	N/A
Nanoscale Materials	40	N/A	N/A
Quantum Effects	38	N/A	N/A
Scanning Probe Microscopy	32	N/A	N/A
quantum computers	29	N/A	N/A
artificial biomimetic	26	N/A	N/A
femtosecond laser	26	N/A	N/A
thermal stabilities	19	N/A	N/A
Polymer Blends	18	N/A	N/A
Organic Materials	16	N/A	N/A
Quantum Information Processing	12	N/A	N/A
Probe Microscopes	52	25	1.08
Molecular Dynamics	74	42	0.76
polymeric materials	47	27	0.74
Organic Molecules	100	98	0.02
Force Microscopy	77	83	-0.07
Transmission Electron Microscopy	77	90	-0.14
Magnetic Materials	42	60	-0.30
Nanostructured Materials	68	113	-0.40
Molecular Structure	29	52	-0.44
Electron Microscopes	39	82	-0.52
Tunneling Microscopy	52	153	-0.66
Composite Material	10	53	-0.81
Thin Film Materials	N/A	17	N/A
nanoscale particles	N/A	19	N/A
Infrared Spectroscopy	N/A	23	N/A
Electronic Materials	N/A	27	N/A
Crystal Structures	N/A	44	N/A
molecular motors	N/A	46	N/A
Polymer Composites	N/A	49	N/A
Ionic Conductivity	N/A	53	N/A
optical fibers	N/A	94	N/A
Quantum Dots	N/A	164	N/A
Baseline Growth Rate			*-0.305*

We observe that award topics continued to focus on microscopy-, molecular- and quantum-related topics. New topics include "Quantum Effects," "Molecular Modeling," "Scanning Probe Microscopy," "Organic Materials," "Nanoscale Materials," "Polymer Blends," "quantum computers," "thermal stabilities," "artificial biomimetic," "Quantum Information Processing," and "femtosecond laser." Some of these new topic region labels were variations of or closely related to topic areas in previous time periods, such as "Molecular Modeling," "Organic Materials," and "Polymer Blends." Several of these new topic areas turned out to truly represent new directions in the field, such as "quantum computers," "artificial biomimetic," and "femtosecond laser."

Table 5-13 presents the major Division/Directorate per topic area in 2001-2002. Continuing the trend from 1996-2000, the Engineering Directorate had a more prominent presence in most major topic areas and become an important funding source for nanotechnology research, comparable to the MPS Directorate. At the same time, the DMR Division still dominated most topic areas, but many other dominant funding Divisions contributed significantly as well, including CTS, DMI, CHE, MCB, CMS, and ECS.

Table 5-13. Major NSF Division/Directorate per topic area (2001-2002).

Region Size	Region Label	No. of Grants	Average Amount	Major Division	Major Directorate
12	Force Microscopy	77	421,000	MPS(27); ENG(27)	DMR(23); CTS(11)
10	Quantum Effects	38	360,000	MPS(26)	DMR(14)
9	Probe Microscopes	52	1,021,000	ENG(28)	DMR(16); DMI(10)
7	Molecular Modeling	60	426,000	ENG(16); MPS(15)	CTS(11); CHE(10); MCB(7)
7	Magnetic Materials	42	810,000	MPS(28)	DMR(20)
6	Organic Molecules	100	511,000	MPS(46); ENG(40)	DMR(29); ECS(17); CHE(14)
6	Transmission Electron Microscopy	77	411,000	MPS(48)	DMR(33)
6	Molecular Dynamics	74	549,000	ENG(40)	DMR(16); CTS(14); DMI(10); CMS(9)
5	Nanostructured Materials	68	1,310,000	ENG(31); MPS(28)	DMR(23)
5	Electron Microscopes	39	433,000	MPS(18); ENG(17)	DMR(17)
4	Polymeric materials	47	547,000	MPS(23)	DMR(17)
4	Scanning Probe Microscopy	32	422,000	MPS(23)	DMR(13); CHE(9)
3	Molecular Structure	29	1,036,000	MPS(16)	CHE(9); DMR(7)
3	Organic Materials	16	230,000	ENG(6)	ECS(3); CHE(3); OCE(3); DMR(2)

Region Size	Region Label	No. of Grants	Average Amount	Major Division	Major Directorate
2	Nanoscale Materials	40	771,000	MPS(15); ENG(12)	DMR(11); ECS(9)
2	Quantum computers	29	625,000	MPS(13)	DMR(7); ECS(6)
2	Composite Material	10	333,000	ENG(6)	ECS(3); CTS(2); DMR(2)
1	Tunneling Microscope	28	909,000	MPS(21)	DMR(19)
1	Artificial biomimetic	26	477,000	MPS(19)	CHE(9); DMR(9)
1	Femtosecond laser	26	401,000	ENG(17)	DMI(7); DMR(5)
1	Tunneling Microscopy	24	360,000	MPS(12)	DMR(7)
1	Thermal stabilities	19	258,000	ENG(11)	DMI(6)
1	Polymer Blends	18	214,000	MPS(13)	DMR(10)
1	Quantum Information Processing	12	457,000	MPS(6); ENG(6)	DMR(4); ECS(4)

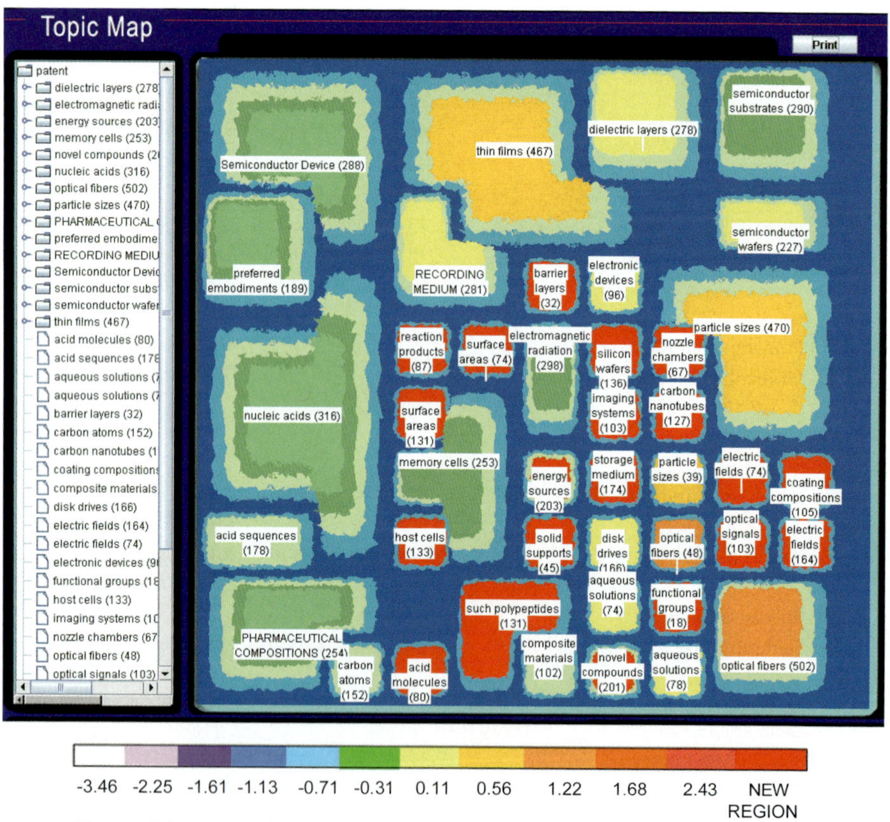

Figure 5-9. Nanotechnology-related USPTO patent content map (2001-2002).

The most important patent topics continued to be "optical fibers," "particle sizes," "thin film," "semiconductor wafers," "semiconductor substrates," "pharmaceutical compositions," "nucleic acids," and "semiconductor devices" during 2001-2002 (Figure 5-9). The detailed topic area changes are presented in Table 5-14. Compared to the previous time period, there were fewer new topic areas during 2001-2002. These new topics included "electric fields," "surface areas," "energy sources," "silicon wafers," "(such) polypeptides," "carbon nanotubes," "coating compositions," and "imaging systems."

Table 5-14. Growth rate of nanotechnology-related USPTO patents per detailed topic area in 2001-2002 as compared to 1996-2000.

Region Label	# of Patents in the region (2001-2002)*	# of Patents in the region (1996-2000)*	Growth Rate
electric fields	238	N/A	N/A
surface areas	205	N/A	N/A
energy sources	203	N/A	N/A
novel compounds	201	N/A	N/A
storage medium	174	N/A	N/A
silicon wafers	136	N/A	N/A
host cells	133	N/A	N/A
such polypeptides	131	N/A	N/A
carbon nanotubes	127	N/A	N/A
coating compositions	105	N/A	N/A
imaging systems	103	N/A	N/A
reaction products	87	N/A	N/A
acid molecules	80	N/A	N/A
nozzle chambers	67	N/A	N/A
solid supports	45	N/A	N/A
barrier layers	32	N/A	N/A
functional groups	18	N/A	N/A
optical fibers	550	254	1.17
particle sizes	509	359	0.42
Thin Film	467	399	0.17
aqueous solutions	178	175	0.02
RECORDING MEDIUM	281	304	-0.08
dielectric layers	278	314	-0.11
electronic devices	96	117	-0.18
semiconductor wafers	227	308	-0.26
disk drives	166	228	-0.27
electromagnetic radiation	298	452	-0.34
semiconductor substrates	290	505	-0.43
pharmaceutical compositions	254	449	-0.43
nucleic acids	316	691	-0.54
memory cells	253	586	-0.57
SEMICONDUCTOR DEVICE	288	771	-0.63
preferred embodiments	189	521	-0.64

Region Label	# of Patents in the region (2001-2002)*	# of Patents in the region (1996-2000)*	Growth Rate
optical signals	103	298	-0.65
acid sequences	178	670	-0.73
carbon atoms	152	687	-0.78
composite materials	102	706	-0.86
clock signals	N/A	137	N/A
communication systems	N/A	123	N/A
computer systems	N/A	983	N/A
control signals	N/A	473	N/A
control systems	N/A	413	N/A
laser beams	N/A	837	N/A
light sources	N/A	547	N/A
memory devices	N/A	367	N/A
metal oxides	N/A	419	N/A
power supplies	N/A	481	N/A
processing systems	N/A	356	N/A
refractive index	N/A	200	N/A
supply voltages	N/A	439	N/A
transmission lines	N/A	93	N/A
ultraviolet radiation	N/A	203	N/A
Baseline Growth Rate			-0.51

For both the award and patent topic maps, the numbers of new topics were smaller in 2001-2002 than in 1996-2000. This is mainly because the content maps are only for a two-year interval as compared to the previous interval of five years (1996-2000). The annual rate of change increased in 2001-2002 as compared to the previous interval.

3.4 Award/Patent Map Topic Associations

We visualize the award-patent links per topic for of all three time periods using a "topic association network" (Figure 5-10). Here, the award and patent topics identified in the content maps are shown as circles and triangles, respectively. The green, orange, and red colors represent the three time periods, 1991-1995, 1996-2000, and 2001-2002, respectively. Links were allowed only between award and patent topics to indicate associations between an award topic and its related patent topics. These links were created by consulting a nanotechnology expert. We generated two lists of award and patent topics that appeared in the content maps. The expert was asked to identify the relevant patent topics for each award topic. In Figure 5-10 we placed nodes that represent the same topics in different time periods close to each other. For example, the patent topic "electromagnetic radiation" in the upper-right corner of Figure 5-10 appeared in the patent content maps of all three time periods. It was associated with the award topic "Infrared Spectroscopy" in 1996-2000.

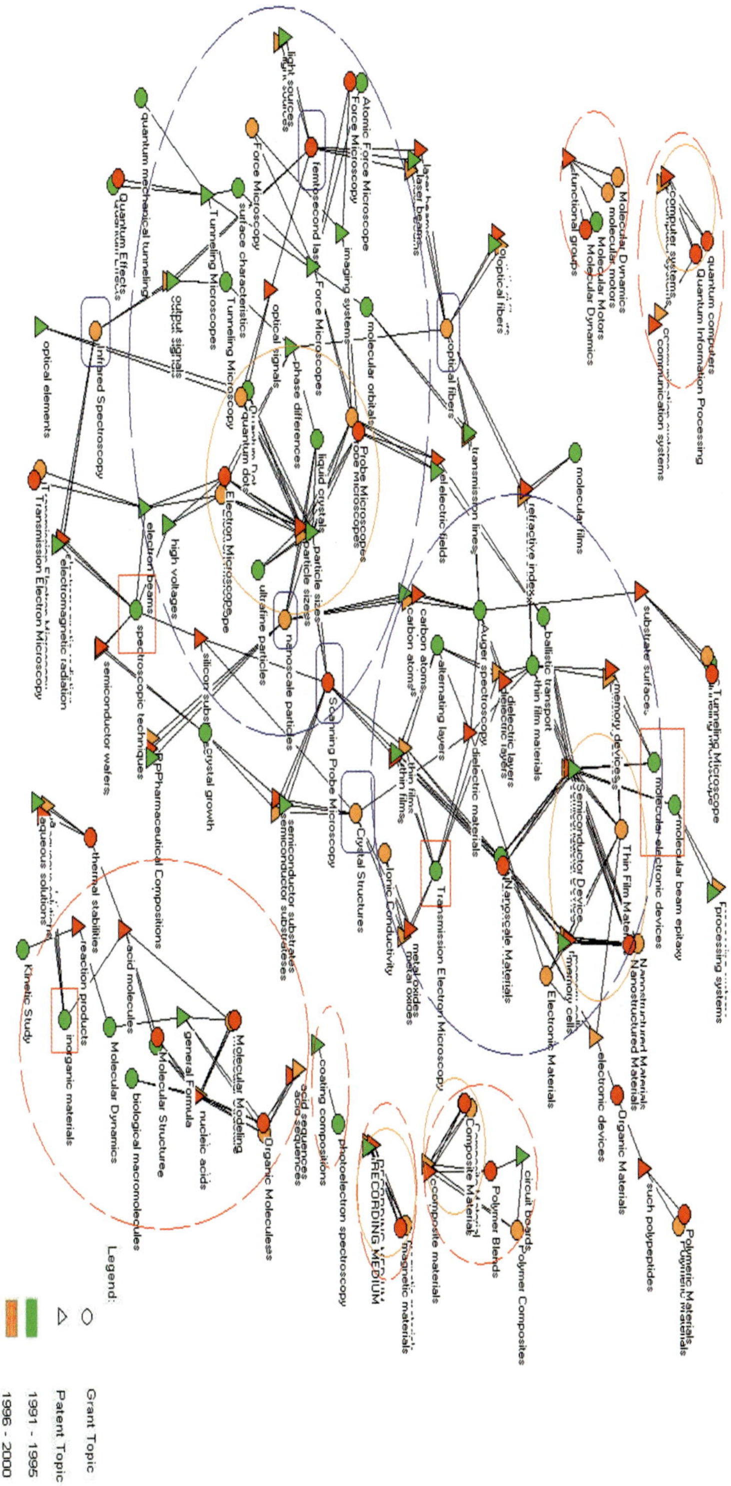

Figure 5-10. Nanotechnology-related award/patent topic associations: longitudinal links for 1991-1995, 1996-2000 and 2001-2002.

Figure 5-10 shows that most nanotechnology award topics are relevant to nanotechnology patent topics. A large connected component of the award/patent map consists of material- and optics-related topics. There are also several isolated topic areas, which are circled with red dotted lines in the figure. The two most important isolated topic association clusters were in the lower-right part of Figure 5-10, corresponding to topics related to organic molecular and composite materials. Several dense topic association clusters (marked using circles of blue dotted lines) also appear in the large connected component. These may represent the central topic areas of NSF-funded nanotechnology research and their impacted patent areas. Most topics in these clusters are optics and material topics. We observe several local award/patent topic association patterns:

- *Pattern I (marked with red squares),* an award topic followed by several associated patent topics in later time periods: This pattern may indicate that a new research funded by the NSF eventually invoked substantial industry efforts. Such award topics include "molecular electronic devices," "Transmission Electron Microscopy," "inorganic materials," "molecular beam epitaxy," and "spectroscopic techniques."
- *Pattern II (marked with orange circles),* consistent award and patent topic associations throughout all time periods: This pattern may indicate that basic research and industry development evolve in parallel in these areas. Such award topics include "quantum dots," "probe microscopes," "nanostructured materials," and "magnetic materials."
- *Pattern III (marked with blue squares)*, a patent topic followed by award topics in the later time periods: Such patterns may indicate that basic research was initiated by early industry developments in related areas. Such award topics include "nanoscale particles," "infrared spectroscopy," "femtosecond laser," "optical fibers," "crystal structures," and "Scanning Probe Microscopy."

4. CRITICAL PATENT/INVENTOR ANALYSIS

An analysis of patents and inventors that have the highest impact in nanotechnology research and development is presented based on patent citation. We refer to these high-impact patents and inventors in different subfields as "critical patents/inventors." In this section, the nanotechnology-related patent data from 1976-1990 are included in our analysis for a more comprehensive assessment. Details regarding the complete nanotechnology-related patent data set can be found in our previous work (Huang et al., 2003; Huang et al., 2004).

4.1 Measures

A simple measure of impact used in bibliometric and patent analysis is the number of times an author, paper or patent is cited by others. The *citation count* was initially introduced for evaluating academic publications. Citation implies an acknowledgement of authority on the part of the citing author to the cited one, and an author's citation level reflects the community's perceived value of their contribution to the field. This idea is supported by a substantial amount of academic literature on citation indexing. Garfield's 1955 vision of an interdisciplinary science citation index introduced the concept of citation as an impact factor indicator and the concept has since been widely applied.

Similarly, the citations among patents also indicate the authority of patents and inventors in a technology field. USPTO patents are required by law to cite all important prior works on the pertinent topic. These citations eventually determine the scope of the monopoly power of that patent. Although debates on the nature and quality of the citations in patent documents are still ongoing, patent citation analysis has been widely applied in practical technology analysis of various domains (e.g., Huang et al., 2003; Narin, 2000). The number of citations received by a patent has been the main measure in evaluating impact and quality of patents, inventors, and the technology competence of institutions and countries. In our study, we define a *number of cites measure for a patent* as the number of later patents from which this patent receives citations and a *number of cites measure for an inventor* as the sum of the number of cites of all his/her patents.

In addition to using this simple measure of number of cites, we also assign inventors and patents an *Authority Score* based on the Hypertext Induced Topic Selection (HITS) algorithm (Kleinberg, 1998), which was intended to identify important Web pages based on hyperlink citation structure. Following the formulation of the original HITS algorithm, two types of scores are defined for each patent, an Authority Score and a Hub Score. A patent with a high Authority Score has a significant impact/influence on the follow-up patents. A high Hub Score, on the other hand, indicates that a patent has cited many critical patents and describes a new technology that integrates many important prior innovations. The Authority and Hub Scores mutually reinforce each other. Using our patent citation data set, we initialize the Authority Scores as the number of times the patents are cited by others and the Hub Scores as the number of times the patents cite others. The two scores are then computed following an iterative updating process:

$$\text{Authority Score}(p) = \sum_{q \text{ has cited } p} \text{Hub Score}(q)$$

$$\text{Hub Score}(q) = \sum_{q \text{ has cited } p} \text{Authority Score}(p) \qquad (1)$$

The Authority Score we use for our study is obtained with three iterations of score updating. The simple number of cites measure can be viewed as a local measure of impact, while the Authority Score provides a global impact measure that accounts for transitive citations. Similar to the Authority Score of patents, we define the Authority Score of an inventor as the sum of the Authority Scores of his/her patents.

4.2 Subfield Analysis

We identify the highest ranked PI-inventors and PI-inventors' patents based on both the number of cites and the Authority Score in nanotechnology subfields. For each subfield, the scores of a PI-inventor are derived by summing the scores of the patents of a PI-inventor in that subfield.

The top 20 nanotechnology fields from 1976 to 2002 have been analyzed. Table 5-15 presents the PI-inventors' patents that received the highest number of cites per subfield. For 10 of these subfields the top PI-inventor patents rank in the top 20. Considering the much smaller number of PI-inventors (307) relative to the entire set of nanotechnology-related patent inventors (123,752), the rankings of these PI-inventors demonstrate their significant impacts on nanotechnology development.

The highest ranked PI-inventor patents are: (1) "Formation of microstamped patterns on surfaces and derivative articles" (patent number 5512131) filed by the University of California, which received 40 citations and was ranked 4th in the subfield of "Semiconductor device manufacturing: process;" (2) "Atomic force microscope with optional replaceable fluid cell" (patent number 4935634) filed by Columbia University, which received 58 citations and was ranked 5th in the subfield "Radiant energy."

The highest rank of PI-inventor patents in the second largest nanotechnology subfield "Drug, bio-affecting and body treating compositions" was only 268th. Other subfields in which PI-inventors' patents had low rankings were: "Optics: systems (including communication) and elements" (the highest rank of PI-inventor patent is 72) and "Coating processes" (the highest rank of PI-inventor patent is also 72). These fields represent technology areas where NSF-funded research had substantially lower impact.

Table 5-15. Critical patent analysis: number of cites.

Rank	US_Class	Name	No. of patents	Highest ranked PI-inventor patent measured by Number of Cites
1	435	Chemistry: molecular biology and microbiology	7946	29th: [5620850] "Molecular recognition at surfaces derivatized with self-assembled monolayers." President and Fellows of Harvard College. NumCites = 41
2	514	Drug, bio-affecting and body treating compositions	6183	268th: [5157032] "Mixed ligand complexes and uses thereof as binding agents and uses thereof as binding agents and probes to DNA." The Trustees of Columbia University in the City of New York. NumCites = 6.
3	424	Drug, bio-affecting and body treating compositions	4683	53rd: [4721669] "Chemical probes for left-handed DNA and chiral metal complexes as Z-specific anti-tumor agents." The Trustees of Columbia Univ. in the City of New York. NumCites = 23
4	250	Radiant energy	4657	**5th: [4935634] "Atomic force microscope with optional replaceable fluid cell." The Regents of the Univ. of CA. NumCites = 58**
5	428	Stock material or miscellaneous articles	3939	**16th: [4728591] "Self-assembled nanometer lithographic masks and templates and method for parallel fabrication of nanometer scale multi-device structures." Trustees of Boston Univ. NumCites = 43**
6	257	Active solid-state devices (e.g., transistors, solid-state diodes)	3933	30th: [5751018] "Semiconductor nanocrystals covalently bound to solid inorganic surfaces using self-assembled monolayers." The Regents of the Univ. of CA. NumCites = 20
7	438	Semiconductor device manufacturing: process	3877	**4th: [5512131] "Formation of microstamped patterns on surfaces and derivative articles." President and Fellows of Harvard College. NumCites = 40**
8	536	Organic compounds -- part of the class 532-570 series	3756	**13th: [4500707] "Nucleosides useful in the preparation of polynucleotides." Univ. Patents, Inc. NumCites = 35**
9	530	Chemistry: natural resins or derivatives; peptides or proteins; lignins or reaction products thereof	3753	**10th: [5620850] "Molecular recognition at surfaces derivatized with self-assembled monolayers." President and Fellows of Harvard College. NumCites = 41**
10	359	Optics: systems (including communi-cation) and elements	3404	1) 72nd: [4187336] "Non-iridescent glass structures." Roy G. Gordon. NumCites = 14. // 2) 72nd: [5036220] "Nonlinear optical radiation generator and method of controlling regions of ferroelectric

Rank	US_Class	Name	No. of patents	Highest ranked PI-inventor patent measured by Number of Cites
				polarization domains in solid state bodies." Leland Stanford Junior Univ. NumCites = 14
11	427	Coating processes	3265	**18th: [5512131] "Formation of microstamped patterns on surfaces and derivative articles." President and Fellows of Harvard College. NumCites = 40**
12	436	Chemistry: analytical and immunological testing	3027	24th: [5472881] "Thiol labeling of DNA for attachment to gold surfaces." University of Utah Research Foundation. NumCites = 34
13	430	Radiation imagery chemistry: process, composition, or product thereof	2983	**15th: [4728591] "Self-assembled nanometer lithographic masks and templates and method for parallel fabrication of nanometer scale multi-device structures." Trustees of Boston Univ. NumCites = 43**
14	356	Optics: measuring and testing	2957	25th: [4675300] "Laser-excitation fluorescence detection electrokinetic separation." The Board of Trustees of the Leland Stanford Junior Univ. NumCites = 24
15	365	Static information storage and retrieval	2310	48th: [5268862] "Three-dimensional optical memory." The Regents of the Univ. of CA. NumCites = 15
16	327	Miscellaneous active electrical nonlinear devices, circuits, and systems	2286	42th: [5475341] "Sub-nanoscale electronic systems and devices." Yale University. NumCites = 9
17	204	Chemistry: electrical and wave energy	1864	**20th: [5202004] "Scanning electrochemical microscopy." Digital Instruments, Inc. NumCites = 25**
18	422	Chemical apparatus and process disin-fecting, deodorizing, preserving, or sterilizing	1829	**12th: [5620850] "Molecular recognition at surfaces derivatized with self-assembled monolayers." President and Fellows of Harvard College. NumCites = 41**
19	372	Coherent light generators	1775	**13th: [5036220] "Nonlinear optical radiation generator and method of controlling regions of ferroelectric polarization domains in solid state bodies." Leland Stanford Junior Univ. NumCites = 14**
20	252	Compositions	1680	100th: [5324457] "Devices and methods for generating electrogenerated chemiluminescence." Board of Regents, The Univ. of TX System. NumCites = 6

Table 5-16 presents the PI-inventors' patents that received the highest Authority Scores in the top 20 nanotechnology subfields. In 12 of these subfields top PI-inventor patents rank in the top 20. The highest ranked PI-inventor patents are different from the ones ranked by number of cites in Table 5-15. These patents include: (1) "Self-assembled nanometer lithographic masks and templates and method for parallel fabrication of nanometer scale multi-device structures" (patent number 4728591) filed by Boston University, which was ranked 1st in both the field "Stock material or miscellaneous articles" and the field "Radiation imagery chemistry: process, composition, or product thereof;" (2) "Scanning electrochemical microscopy" (patent number 5202004) filed by Digital Instruments (Allen J. Bard of the University of Texas), which was ranked 1st in the field "Chemistry: electrical and wave technology." Patent 4935634 of Columbia University was still highly ranked by the Authority Score; it was ranked 4th in the field "Radiant energy" (higher than its ranking by number of cites). However, patent 5512131 of the University of California was ranked much lower by the Authority Score than by number of cites. In the field "Semiconductor device manufacturing: process" the highest ranked PI-inventor patent by the Authority Score was filed by Stanford University and was ranked 9th. Patents of PI-inventors had relatively low rankings in the following fields: "Drug, bio-affecting and body treating compositions" (58th), "Coherent light generators" (69th), and "Compositions" (53rd).

PI-inventors' patents generally had higher rankings as measured by the Authority Score than by the number of cites, indicating that PI-inventors' patents had relatively higher indirect impact on industrial nanotechnology development. The titles of the highly ranked PI-inventor patents also match nicely with the content map analysis results. These critical PI-inventors represented the dominant topic areas of NSF-funded nanotechnology research.

Table 5-16. Critical patent analysis: Authority Score.

Rank	US_Class	Name	No. of patents	Highest ranked PI-inventor patent measured by Authority Score
1	435	Chemistry: molecular biology and microbiology	7946	32nd: [5106729] "Method for visualizing the base sequence of nucleic acid polymers." Arizona Board of Regents acting on behalf of Arizona State University. Authority Score = 88.90502
2	514	Drug, bio-affecting and body treating compositions	6183	40th: [5157032] "Mixed ligand complexes and uses thereof as binding agents and probes to DNA." The Trustees of Columbia University in the City of New York. Authority Score = 6.20386

Rank	US_Class	Name	No. of patents	Highest ranked PI-inventor patent measured by Authority Score
3	424	Drug, bio-affecting and body treating compositions	4683	58th: [5639473] "Methods for the preparation of nucleic acids for in vivo delivery." Vivorx Pharmaceuticals, Inc. Authority Score = 8.721781
4	250	Radiant energy	4657	**4th: [4935634] "Atomic force microscope with optional replaceable fluid cell." The Regents of the University of California. Authority Score =490.1931**
5	428	Stock material or miscellaneous articles	3939	**1st: [4728591] "Self-assembled nanometer lithographic masks and templates and method for parallel fabrication of nanometer scale multidevice structures." Trustees of Boston University. Authority Score =165.5829**
6	257	Active solid-state devices (e.g., transistors, solid-state diodes)	3933	**16th: [5093699] "Gate adjusted resonant tunnel diode device and method of manufacture." Texas A & M University System. Authority Score = 16.50901**
7	438	Semiconductor device manufacturing: process	3877	**9th: [5618760] "Method of etching a pattern on a substrate using a scanning probe microscope." The Board of Trustees of the Leland Stanford, Jr. University. Authority Score = 30.16755**
8	536	Organic compounds – part of the class 532–570 series	3756	**8th: [4500707] "Nucleosides useful in the preparation of polynucleotides." University Patents, Inc. Authority Score = 152.8311**
9	530	Chemistry: natural resins or derivatives; peptides or proteins; lignins or reaction products thereof	3753	**16th: [5620850] "Molecular recognition at surfaces derivatized with self-assembled monolayers." President and Fellows of Harvard College. Authority Score = 28.0039**
10	359	Optics: systems (including communication) and elements	3404	21st: [5479024] "Method and apparatus for performing near-field optical microscopy." The Regents of the University of California. Authority Score =16.1751
11	427	Coating processes	3265	**20th: [5512131] "Formation of microstamped patterns on surfaces and derivative articles." President and Fellows of Harvard College. Authority Score = 18.20165**
12	436	Chemistry: analytical and immunological testing	3027	**20th: [5106729] "Method for visualizing the base sequence of nucleic acid polymers." Arizona Board of Regents acting on behalf of Arizona State University. Authority Score = 88.90502**
13	430	Radiation imagery chemistry: process, composition, or product thereof	2983	**1st: [4728591] "Self-assembled nanometer lithographic masks and templates and method for parallel fabrication of nanometer scale multi-device structures." Trustees of Boston University. Authority Score = 165.5829**

Rank	US_Class	Name	No. of patents	Highest ranked PI-inventor patent measured by Authority Score
14	356	Optics: measuring and testing	2957	**19th: [5479024] "Method and apparatus for performing near-field optical microscopy." The Regents of the University of California. Authority Score = 16.1751**
15	365	Static information storage and retrieval	2310	47th: [5228001] "Optical random access memory." Syracuse University. Authority Score = 6.262227
16	327	Miscellaneous active electrical nonlinear devices, circuits, and systems	2286	24th: [5475341] "Sub-nanoscale electronic systems and devices." Yale University. Authority Score = 2.072026
17	204	Chemistry: electrical and wave energy	1864	**1st: [5202004] "Scanning electrochemical microscopy." Digital Instruments, Inc. Authority Score = 436.3328**
18	422	Chemical apparatus and process disin-fecting, deodorizing, preserving, or sterilizing	1829	23rd: [5620850] "Molecular recognition at surfaces derivatized with self-assembled monolayers." President and Fellows of Harvard College. Authority Score = 28.0039
19	372	Coherent light generators	1775	69th: [5036220] "Nonlinear optical radiation generator and method of controlling regions of ferroelectric polarization domains in solid state bodies." Leland Stanford Junior University. Authority Score = 0.4296076
20	252	Compositions	1680	53rd: [5324457] "Devices and methods for generating electrogenerated chemiluminescence." Board of Regents, Univ. of TX System. Authority Score=1.09404.

The PI-inventors who had the highest number of cites in each of the top 20 nanotechnology fields are listed in Table 5-17. In three of these fields the top PI-inventors were ranked in the top 20. George M. Whitesides (Harvard University) and Paul K. Hansma (University of California) were the top-ranked PI-inventors. George M. Whitesides is arguably the most influential PI-inventor by this measure. He was highly ranked in four fields, including: "Semiconductor device manufacturing process" (ranked 13th with 120 citations) and "Chemistry: natural resins or derivatives; peptides or proteins; lignins or reaction products thereof" (ranked 13th with 216 citations). Paul K. Hansma received 427 citations and was ranked 5th in "Radiant energy."

NSF PI-inventors had low rankings in the following fields: "Drug, bio-affecting and body treating compositions" (486th), "Coherent light generators" (246th), and "Optics: measuring and testing" (223rd).

Table 5-17. Critical inventor analysis: number of cites.

Rank	US_class	Name	No. of inventors	Highest ranked PI-inventor measured by Number of Cites
1	435	Chemist: molecular biology and microbiology	10971	28th: [Whitesides, George M.] President and Fellows of Harvard College. Number of Cites = 468
2	514	Drug, bio-affecting and body treating compositions	8526	486th: [Barton, Jacqueline K.] The Trustees of Columbia University in the City of New York. Number of Cites = 18
3	424	Drug, bio-affecting and body treating compositions	6442	94th: [Suslick, Kenneth S.] University of Illinois, VivoRX Pharmaceuticals, Inc. Number of Cites = 106
4	250	Radiant energy	6238	**5th: [Hansma, Paul K.] The Regents of the University of California. Number of Cites = 427**
5	428	Stock material or miscellaneous articles	6039	33rd: [Gordon, Roy G.] Harvard College. Number of Cites = 194
6	257	Active solid-state devices (e.g., transistors, solid-state diodes)	5955	34th: [Reed, Mark A.] Yale University. Number of Cites = 127
7	438	Semiconductor device manufacturing: process	5808	**13th: [Whitesides, George M.] President and Fellows of Harvard College. Number of Cites = 120**
8	536	Organic compounds – part of the class 532–570 series	5220	26th: [Caruthers, Marvin H.] University Patents, Inc. Number of Cites = 133
9	530	Chemistry: natural resins or derivatives; peptides or proteins; lignins or reaction products thereof	5108	**13th: [Whitesides, George M.] President and Fellows of Harvard College. Number of Cites = 216**
10	359	Optics: systems (including communication) and elements	4747	81st: [Gordon, Roy G.] President and Fellows of Harvard College. Number of Cites = 56
11	427	Coating processes	4369	63rd: [Gordon, Roy G.] President and Fellows of Harvard College. Number of Cites = 114
12	436	Chemistry: analytical and immunological testing	4223	26th: [Bard, Allen J.] The University of Texas System, Board of Regents, Digital Instruments, Inc., IGEN International Inc. Number of Cites = 322 // 26th: [Whitesides, George M.] President and Fellows of Harvard College. Number of Cites = 322
13	430	Radiation imagery chemistry: process, composition, or product thereof	3916	223rd: [Zaidi, Saleem H.] The University of New Mexico. Number of Cites = 46
14	356	Optics: measuring and testing	3518	65th: [Maris, Humphrey J.] Brown University Research Foundation. Number of Cites = 49
15	365	Static information storage and retrieval	3200	121st: [Birge, Robert R.] Syracuse University. Number of Cites = 35

Rank	US_class	Name	No. of inventors	Highest ranked PI-inventor measured by Number of Cites
16	327	Miscellaneous active electrical nonlinear devices, circuits, and systems	2868	86th: [Reed, Mark A.] Yale University. Number of Cites = 18
17	204	Chemistry: electrical and wave energy	2482	35th: [Lindsay, Stuart M.] Arizona State University, Molecular Imaging Corporation, The United States of America as represented by the Secretary of the Navy. Number of Cites = 84 // 35th: [Zare, Richard N.] The Board of Trustees of the Leland Stanford Junior College. Number of Cites = 84
18	422	Chemical apparatus and process disinfecting, deodorizing, preserving, or sterilizing	2438	66th: [Whitesides, George M.] President and Fellows of Harvard College. Number of Cites = 101
19	372	Coherent light generators	2260	246th: [Fejer, Martin M.] Stanford University. Number of Cites = 15
20	252	Compositions	2044	54th: [Bard, Allen J.] University of Texas. Number of Cites = 36

Table 5-18 presents the PI-inventors who had the highest ranking measured by the Authority Score. In six of these fields the top PI-inventors were ranked in the top 20. Paul K. Hansma was again one of the most important PI-inventors. He was highly ranked in three fields, including "Radiant energy" (ranked 4th). The other most influential PI-inventor was Noel A. Clark of Boston University, who was ranked 4th in the field "Stock material or miscellaneous articles." PI-inventors had relatively lower Authority Scores in the fields of "Coherent light generators" (159th) and "Chemical apparatus and process disinfecting, deodorizing, preserving, or sterilizing" (97th).

Table 5-18. Critical inventor analysis: Authority Score.

Rank	US_Class	Name	Number of inventors	Highest ranked PI-inventor measured by Authority Score
1	435	Chemistry: molecular biology and micro-biology	10971	52nd: [Lindsay, Stuart M.] Arizona State University, Molecular Imaging Corporation. AuthScore = 355.6201
2	514	Drug, bio-affecting and body treating composi-tions	8526	78th: [Barton, Jacqueline K.] Columbia University. AuthScore = 18.61158
3	424	Drug, bio-affecting and body treating composi-tions	6442	69th: [Suslick,Kenneth S.] University of Illinois, VivoRX Pharmaceuticals, Inc. AuthScore = 47.45637
4	250	Radiant energy	6238	**4th: [Hansma, Paul K.] The University of California. AuthScore = 4208.508**

Rank	US_Class	Name	Number of inventors	Highest ranked PI-inventor measured by Authority Score
5	428	Stock material or miscellaneous articles	6039	**4th: [Clark, Noel A.] Boston University. AuthScore = 662.3314 // 4th: [Rothschild, Kenneth J.] AmberGen Inc, Boston University. AuthScore = 662.3314**
6	257	Active solid-state devices (e.g., transistors, solid-state diodes)	5955	35th: [Kirk, Wiley P.] Texas A & M University System, Texas Instruments Incorporated. AuthScore = 99.05406
7	438	Semiconductor device manufacturing: process	5808	**14th: [Minne, Stephen C.] Stanford University, NanoDevices, Inc. AuthScore = 90.50266**
8	536	Organic compounds – part of the class 532–570 series	5220	**9th: [Caruthers, Marvin H.] University Patents, Inc. AuthScore = 466.1678**
9	530	Chemistry: natural resins or derivatives; peptides or proteins; lignins or reaction products thereof	5108	29th: [Whitesides, George M.] Harvard College. AuthScore = 120.134
10	359	Optics: systems (including communication) and elements	4747	82th: [Hansma, Paul K.] The University of California. AuthScore = 16.1751
11	427	Coating processes	4369	61st: [Clark, Noel A.] Boston University. AuthScore = 42.54337 // 61st: [Rothschild, Kenneth J.] AmberGen Inc., Boston University. AuthScore = 42.54337
12	436	Chemistry: analytical and immunological testing	4223	82nd: [Lindsay, Stuart M.] Arizona State University, Molecular Imaging Corporation. AuthScore = 88.90502
13	430	Radiation imagery chemistry: process, composition, or product thereof	3916	**9th: [Clark, Noel A.] Boston University. AuthScore = 165.5829 // 9th: [Rothschild, Kenneth J.] AmberGen Inc., Boston University. AuthScore = 165.5829**
14	356	Optics: measuring and testing	3518	87th: [Hansma, Paul K.] The University of California. AuthScore = 16.1751
15	365	Static information storage and retrieval	3200	59th: [Birge, Robert R.] Syracuse University. AuthScore = 18.98621
16	327	Miscellaneous active electrical nonlinear devices, circuits, and systems	2868	34th: [Reed, Mark A.] Yale University. AuthScore = 4.144051
17	204	Chemistry: electrical and wave energy	2482	**7th: [Lindsay, Stuart M.] Arizona State University, Molecular Imaging Corporation. AuthScore = 1410.155**
18	422	Chemical apparatus and process disin-	2438	97th: [Whitesides, George M.] Harvard College. AuthScore = 34.00076

Rank	US_Class	Name	Number of inventors	Highest ranked PI-inventor measured by Authority Score
		fecting, deodorizing, preserving, or sterilizing		
19	372	Coherent light generators	2260	159th: [Fejer, Martin M.] Stanford University. AuthScore = 0.45
20	252	Compositions	2044	43rd: [Bard, Allen J.] University of Texas, Digital Instruments. AuthScore = 5.014

The strongest and weakest nanotechnology fields for the PI-inventors and their patents were consistent across the results presented in Tables 5-15 to 5-18, confirming again the findings in the award and patent content maps discussed previously.

4.3 Sample Patent Citation Networks

To further illustrate the relative impact of PI-inventors, we selected three representative nanotechnology fields and visualized the patent citation networks of these fields. For the nanotechnology fields, we selected a minimum number of cites and obtained a manageable set of core patents to visualize. The inventors of these core patents (represented by circles in the network) were identified. Inventors who had at least two patents in the core set were shown in the network as triangles. In these networks, links between patents represent the patent citation. A link with the form "patent A → patent B" represents that patent A had been cited by patent B. Links between inventors and patents represent the patent authorship, i.e., "inventor A → patent B" represents that A is one of the inventors of patent B. We used the color yellow to mark the PI-inventors and their patents in these networks. We also selectively show the keywords in the titles of the core patents.

The citation network for the largest nanotechnology field "Chemistry: molecular biology and microbiology" is shown in Figure 5-11. Only patents with more than 5 cites are shown. The keywords in the titles of the most important patents include: "Detectable molecules, method of preparation and use," "process for amplifying nucleic acid sequences," and "fluid handling in mesoscale analytical devices." Overall we found that PI-inventors and their patents form a closely linked cluster within the largest connected component of the citation network. This PI-inventor cluster includes Allen J. Bard of the University of Texas, Jacqueline K. Barton of Columbia University, and George M. Whitesides of Harvard University.

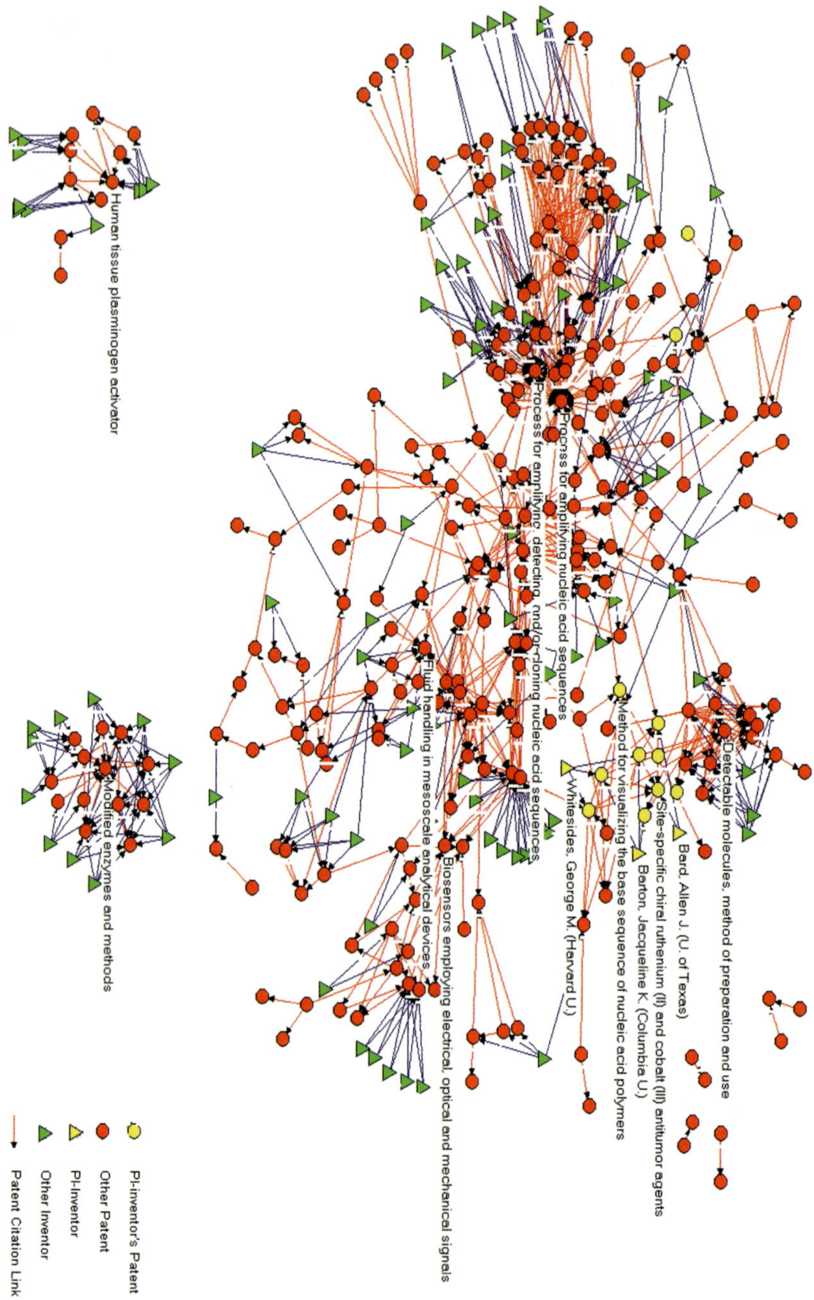

Figure 5-11. Patent citation network: "Chemistry: molecular biology and microbiology."

Figure 5-12 displays the citation network for the field "Radiant energy," in which PI-inventors had the strongest positions. The patents shown were cited more than 14 times. The network shows that the core patents in this field form one dense connected component without lower level subclusters. "Atomic force microscope," "scanning ion conductance microscope," and "scanning tunneling microscope" are the major topics. PI-inventors' patents were among the top-cited patents covering the central topic of the field, microscope-related topics, e.g., "atomic force microscope with optional replaceable fluid cell" and "high resolution atomic force microscope." Paul K. Hansma and Allen J. Bard appeared in this network as well. Stuart M. Lindsay of Arizona State University and Joseph W. Lyding of the University of Illinois also appeared in the network.

Figure 5-13 shows the citation network for a relatively small field among the top 20 nanotechnology fields: "Chemistry: electrical and wave energy." Patents shown were cited more than once. Important topics in this field include: "capillary gel electrophoresis columns," "laser-excitation fluorescence detection electrokinetic separation," and "silicon semiconductor wafer for analyzing micronic biological samples." The network shows that three patents by Jonathan V. Sweedler and Richard N. Zare of Stanford University occupied the central positions in the largest connected component. Two other PI-inventors' patents were in a smaller isolated cluster: Stuart M. Lindsay of Arizona State University and Allen J. Bard of the University of Texas.

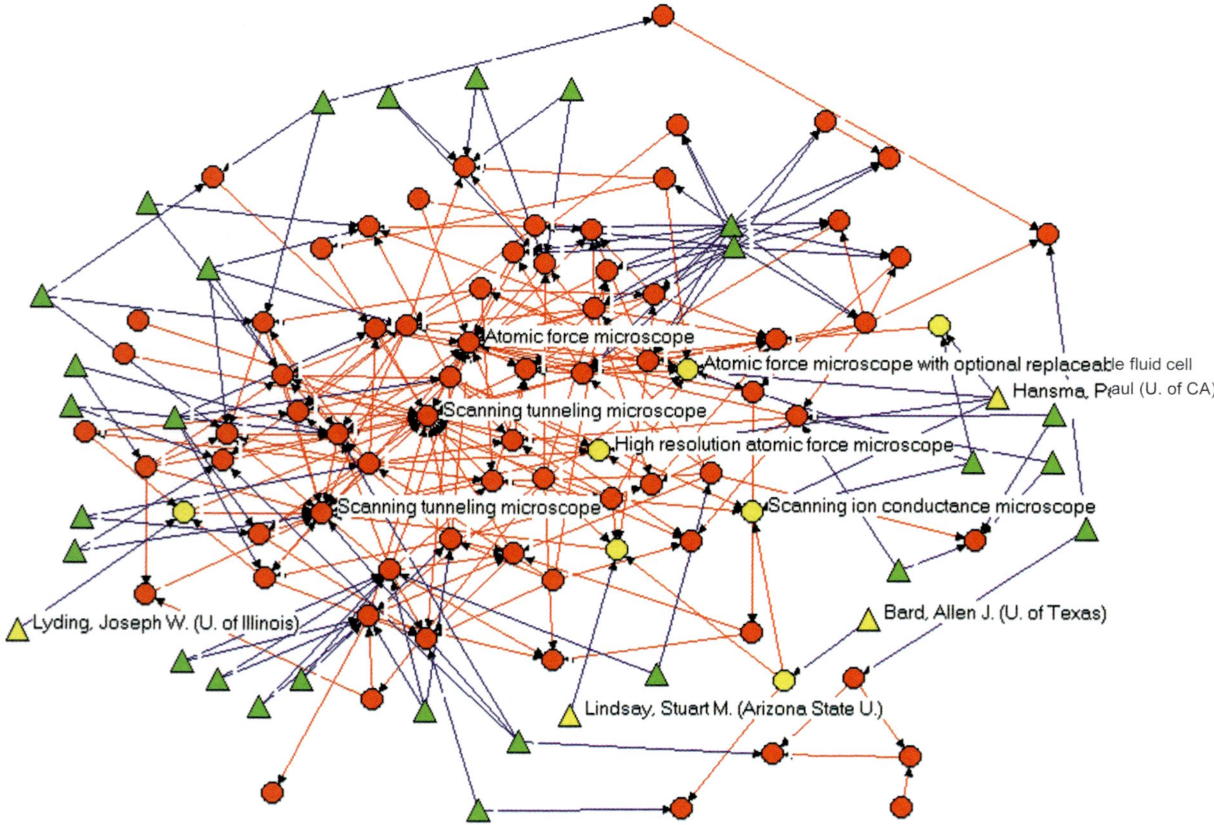

Figure 5-12. Patent citation network: "Radiant energy."

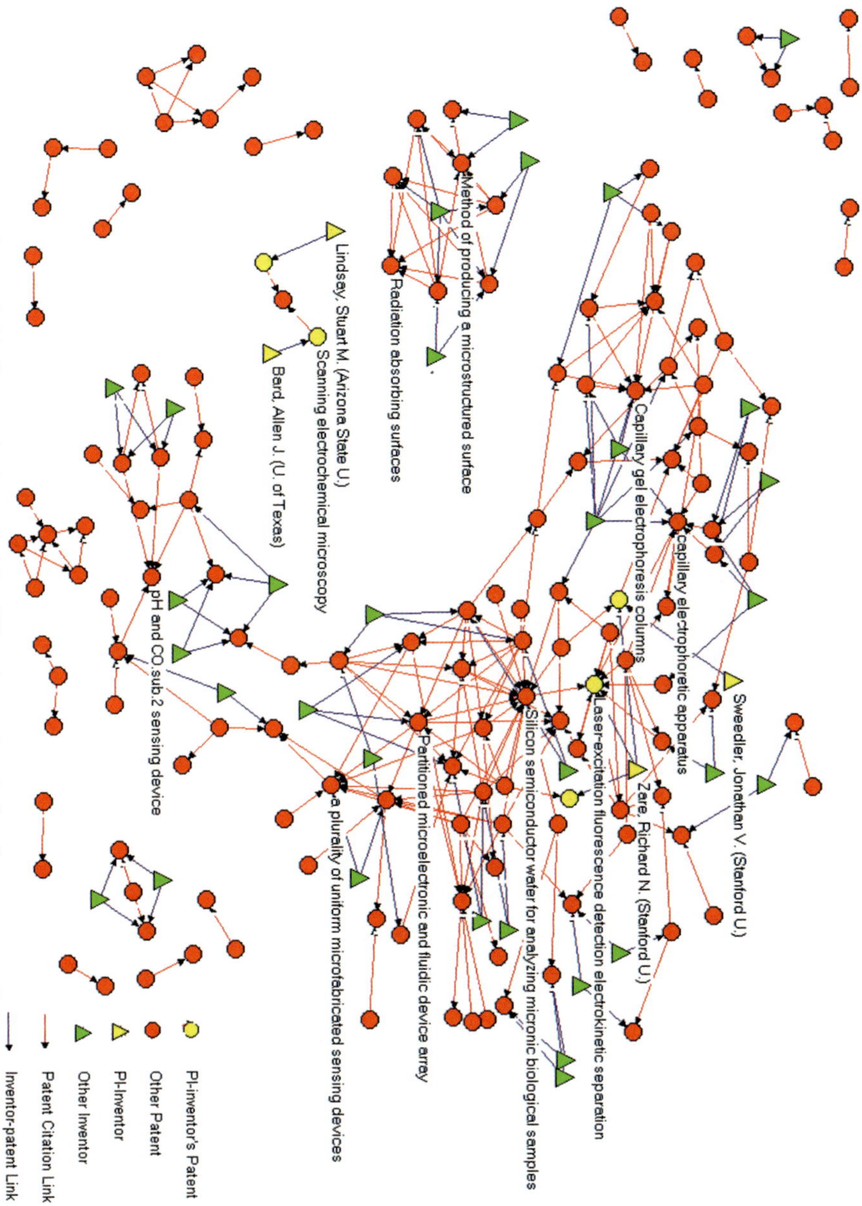

Figure 5-13. Patent citation network: "Chemistry: electrical and wave energy."

5. STATISTICAL HYPTHOSESIS TESTING GROUP COMPARISON

In order to determine the impact of NSF-funded PI-inventors (as a group), we compared their numbers of cites and Authority Scores with other benchmark comparison groups. We selected nine groups of inventors from the top assignee countries and institutions between 1976 and 2002:

1. NSF: PI-inventors and their patents (760 patents and 307 inventors)
2. IBM: Inventors and patents of IBM (2,106 patents and 2,756 inventors)
3. Top10: Inventors and patents of top 10 institutions (6,786 patents and 6,650 inventors)
4. UC: Inventors and patents of the University of California (the top academic institute in nanotechnology-related patenting, 767 patents and 894 inventors)
5. US: Inventors and patents of the United States (55,829 patents and 78,227 inventors)
6. EntireSet: Inventors and patents of the entire patent data set (77,605 patents and 108,378 inventors)
7. Japan: Inventors and patents of Japan (7,574 patents and 14,837 inventors)
8. European: Inventors and patents of European countries (4,046 patents and 9,560 inventors)
9. Others: Random inventors and patents of countries other than the U.S., Japan, and European countries (9,227 patents and 6,385 inventors)

We postulate the following hypotheses:

- H1: Patents associated with NSF-funded PI-inventors had higher numbers of cites than patents associated with other groups of inventors.
- H2: Patents associated with NSF-funded PI-inventors had higher Authority Scores than patents associated with other groups of inventors.
- H3: NSF-funded PI-inventors had higher numbers of cites than other groups of inventors.
- H4: NSF-funded PI-inventors had higher Authority Scores than other groups of inventors.

We conducted *Analysis of Variance* (ANOVA) tests (using the Minitab software, www.minitab.com) to validate the four hypotheses. The analysis results are shown in Figure 5-14. All four analyses had p-values of 0.000, indicating significant differences across different comparison groups.

Figure 5-14a shows that H1 was supported at the 95% confidence level. NSF-funded PI-inventors had patents with significantly larger numbers of cites (about 3 cites on average) compared to other groups (average about 1.5 cites for all patents), followed by the IBM and UC groups. US, EntireSet and Top10 were the next three groups. US patents had larger numbers of cites than patents of the EntireSet. Patents of Japan, European, and Others had smaller numbers of cites compared to other groups, while patents of the Others group had larger numbers of cites than the European group.

Figure 5-14b shows that H2 was also supported. The Authority Scores of NSF-funded PI-inventor patents were significantly larger (about 7.5 on average) than the scores of patents of other groups (less than 1 for all patents). Similar group ordering was observed as shown in Figure 5-14a, except that patents of the Japan group had relatively higher impact when measured by the Authority Score (comparable level with the UC group).

Figure 5-14c shows that H3 was supported. NSF-funded PI-inventors had a significantly larger number of cites (about 10 on average) than all other groups (average about 2 for all inventors). Inventors of Top10, IBM, and UC were in the second category, possessing significantly larger numbers of cites than the remaining groups.

Figure 5-14d shows that H4 was also supported. Only the UC inventors' Authority Score interval had overlap with that of the NSF-funded PI-inventors. IBM inventors had significantly larger Authority Scores than the Top10 inventors; while Figure 5-14c shows that Top10 inventors had larger numbers of cites. Similar to the pattern in Figures 5-14a and b, Japan inventors had relatively higher impact when measured by Authority Score than by number of cites. Inventors of the remaining groups had comparable Authority Scores.

Source	DF	SS	MS	F	P
Group	8	6879	860	44.19	0.000
Error	2.00E+05	3223869	19		
Total	2.00E+05	3230748			

95% CIs based on pooled StDev

Group	N	Mean	StDev	Lower	Upper
NSF	760	2.929	5.876	2.615	3.243
IBM	2106	2.102	6.364	1.914	2.290
Top10	6786	1.573	4.384	1.468	1.678
UC	767	2.033	4.744	1.721	2.345
US	56829	1.626	4.695	1.589	1.663
EntireSet	77605	1.489	4.374	1.458	1.520
Japan	7574	1.127	3.062	1.028	1.226
European	4046	0.903	3.04	0.767	1.039
Others	9227	1.186	3.593	1.096	1.276

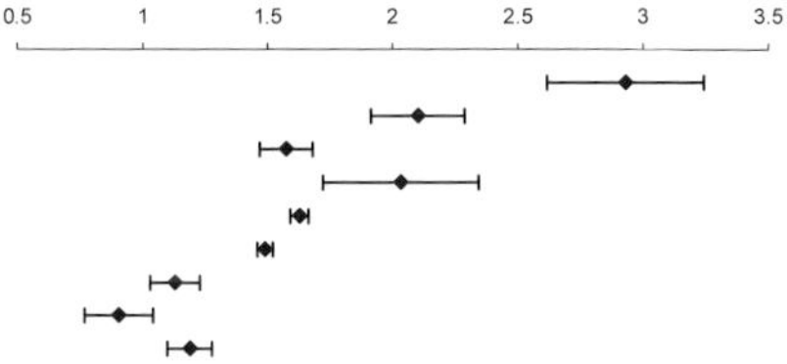

Pooled StDev = 4.411

Figure 5-14a. H1: Number of cites per patent

Source	DF	SS	MS	F	P
Group	8	79627	9953	38.48	0.000
Error	2.00E+05	42856058	259		
Total	2.00E+05	42935686			

Group	N	Mean	StDev	95% CIs based on pooled StDev Lower	Upper
NSF	760	7.32	45.37	6.177	8.463
IBM	2106	4.51	39.86	3.823	5.197
Top10	6786	1.25	16.96	0.867	1.633
UC	767	3.61	33.21	2.472	4.748
US	55829	0.96	14.73	0.827	1.093
EntireSet	77605	1	15.05	0.887	1.113
Japan	7574	2.37	24.63	2.008	2.732
European	4046	0.28	5.46	-0.215	0.775
Others	9227	0.46	7.47	0.132	0.788

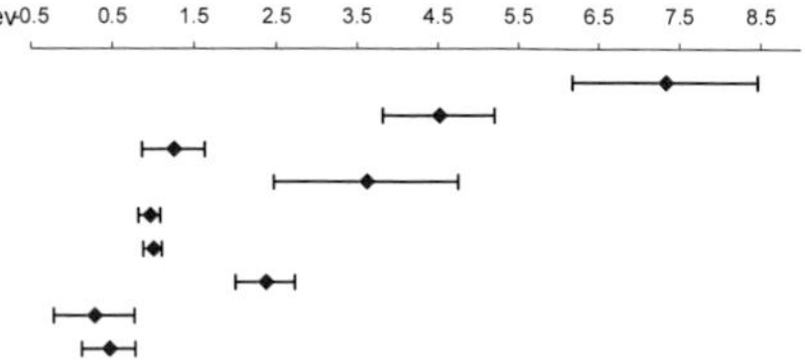

Pooled StDev = 16.08

Figure 5-14b. H2: Authority Score per patent

Source	DF	SS	MS	F	P
Group	8	210748	26344	192.43	0.000
Error	2.00E+05	31210767	137		
Total	2.00E+05	31421515			

Group	N	Mean	StDev	95% CIs based on pooled StDev Lower	Upper
NSF	307	10.04	23.6	8.731	11.349
IBM	2756	4.86	15.7	4.423	5.297
Top10	6650	5.63	22.38	5.349	5.911
UC	894	5.47	29.59	4.703	6.237
US	78227	2.54	12.29	2.458	2.622
EntireSet	108378	2.01	10.94	1.940	2.080
Japan	14837	0.8	7.89	0.612	0.988
European	9560	0.5	4.99	0.265	0.735
Others	6385	0.68	6.64	0.393	0.967

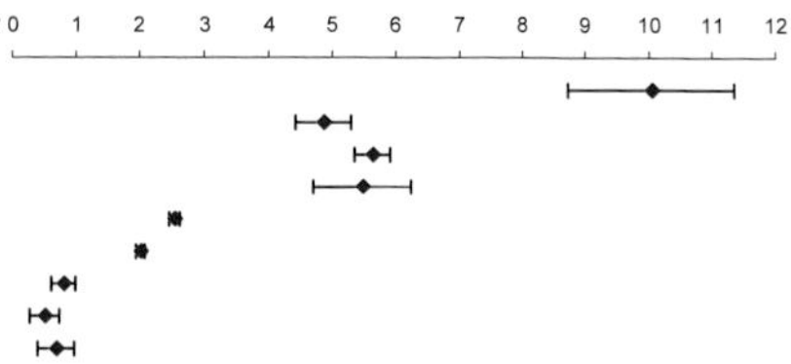

Pooled StDev = 11.70

Figure 5-14c. H3: Number of cites per inventor

Source	DF	SS	MS	F	P
Group	8	210748	26344	192.43	0.000
Error	2.00E+05	31210767	137		
Total	2.00E+05	31421515			

Group	N	Mean	StDev	95% CIs based on pooled StDev Lower	Upper
NSF	307	20.29	142.9	14.458	26.122
IBM	2756	9.94	82.43	7.993	11.887
Top10	6650	5.21	59.67	3.957	6.463
UC	894	17.28	309.84	13.862	20.698
US	78227	1.65	49.25	1.285	2.015
EntireSet	108378	1.81	47.18	1.500	2.120
Japan	14837	4.1	57.5	3.261	4.939
European	9560	0.77	22.64	-0.275	1.815
Others	6385	0.38	8.6	-0.899	1.659

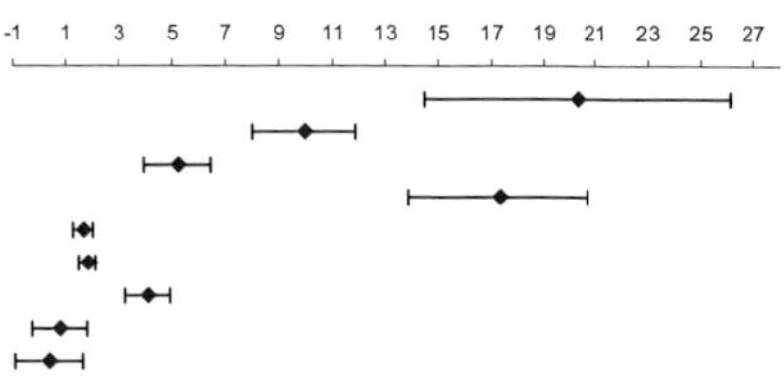

Pooled StDev = 52.14

Figure 5-14d. H4: Authority Score per inventor

Figure 5-14. Comparison of the impact of NSF-funded PI-inventors and other inventor groups.

In summary, our analysis shows that the NSF-funded PI-inventors had significantly higher impact on nanotechnology development than other comparison groups. We also observe that many PI-inventors are important players in the nanotechnology patenting activities as reflected in the patent citation data, although our results do not provide direct evidence of a causal relationship between NSF funding and the high impact of PI-inventors.

6. CONCLUSIONS

The NSF funding of nanotechnology research in the time period 1991-2002 and its impact on USPTO nanotechnology patents have been analyzed and correlated in this chapter. Using bibliometric, content map, critical inventor/patent and citation network analysis, we were able to identify general trends, key players, and evolution of technology topics in nanotechnology based on NSF awards and USPTO patents. The relative impact of NSF-funded researchers in the field of nanotechnology was compared with other top institution, country, and regional comparison groups. Among the key findings are:

- Nanotechnology R&D has grown rapidly in the past decade, as reflected in both NSF awards and USPTO patents.
- NSF-funded nanotechnology research and USPTO nanotechnology patents have substantial overlap in key technology topics such as materials and microscopes. However, NSF-funded research does not overlap much with drug or pharmaceutical patent topics, where industry funding is leading the development.
- NSF-funded researchers have significantly higher impact in nanotechnology development than other comparison groups as reflected in the patent citation data.

We believe the methodology presented in this chapter could help correlate publicly funded research with commercial R&D activities. The results have direct application for the evaluation of federal funding programs and R&D planning.

7. QUESTIONS FOR DISCUSSION

1. What are other major funding sources for nanotechnology research in the U.S.?
2. Which countries have experienced the largest increase in nanotechnology research funding in recent years?
3. What are other metrics that can be used to help evaluate the impact of federal funding on research and development?
4. Is there a direct correlation between federal funding and a country's industry leadership in technology in general?

Chapter 6

TOPOLOGICAL ANALYSIS OF PATENT CITATION NETWORKS: NANOTECHNOLOGY AT USPTO, 1976-2004

CHAPTER OVERVIEW

Patent citation is an indicator of technological development in individual institutions or countries. The patent citation networks provide additional insights into the technology landscape and knowledge transfer processes for a specific field. This chapter presents an analysis framework that applies three approaches to provide a systematic view of patent citation relationships. The three approaches – critical node analysis, core network analysis, and network topological analysis – facilitate the identification of key players/subfields and the knowledge transfer patterns among them. We applied our framework to nanotechnology and analyzed nanotechnology citation networks of patents, institutions, technology fields, and countries. The citation network analysis shows that the U.S. is the most important citation center in nanotechnology research. The *institution citation network* allows knowledge to transfer more quickly between institutions than a random network. The *country citation network* provides as efficient knowledge transfer capability as the random network. The *technology field citation network* and the *patent citation network* provide less efficient knowledge diffusion capability than the random network. The country citation network, the institution citation network, and the patent citation network of the nanotechnology field all show a tendency to form local citation clusters.

1. INTRODUCTION

Information on current technology outcomes and knowledge diffusion patterns is important in planning and managing technology development. In previous studies, bibliometric patent analysis is well accepted in the evaluation of the productivity and quality of industry research and development (R&D). Patent citation information has been used to assess the knowledge diffusion and transfer processes in research and development.

Typical performance measures based on patent citation counts (e.g., number of cites of a patent or an assignee) describe the "local" characteristics of knowledge conveyance. Leveraging recent scientific advances, a network view of the patent citation relations may provide us with a better understanding of the global characteristics of the knowledge diffusion process.

In this study, we propose a framework for the construction and analysis of patent citation networks. From the structured patent documents, we utilize the patent-to-patent citation information and important entities associated with each patent (i.e., country, institution, and technology field) to construct patent citation networks of different entities. *Critical node analysis* is applied to identify the key players/subfields of the field. The core network is visualized in order to identify the knowledge transfer patterns among the key players/subfields. We also apply *network topological modeling* techniques to analyze the global structure of the patent citation networks and characterize the global knowledge transfer process based on network topological measures.

2. RELEVANT LITERATURE

Patents represent critical technological advances for many industries. Patents and patent citations have frequently been used as indicators of research productivity and research impact (Narin, 1994). Lewison (Lewison, 1998) assessed different funding sources' impact on Gastroenterology research in the UK using patent analysis. Huang et al. (Huang et al., 2003a) explored research and development in the high-tech electronic companies of Taiwan based on patent information. In several large-scale longitudinal studies (Huang et al., 2003; 2004; 2005; 2006), Chen and his team at the University of Arizona analyzed the evolution and change of the international landscape of nanotechnology research and development using the complete nanotechnology patent data set collected from the United States Patent and Trademark Office (USPTO) database.

In addition to being used as an indicator of a patent's impact in the field, patent citation information can also be used to represent knowledge transfer between cited patents and citing patents (Karki, 1997; Oppenheim, 2000). Previous studies have investigated knowledge transfer patterns based on patent citations. For example, (Chakrabarti et al., 1993) analyzed the inter-organization patent citation patterns of defense-related research and development on the civilian sector. (Ong et al., 2004) studied the interactions between academia and industry by analyzing the paper-patent citations in the field of tissue engineering. (Verbeek et al., 2003) explored the geographic distribution of scientific research's impact on patents in the fields of biotechnology and information technology. (Singh, 2003) explored the impact of inventors' social distance on the knowledge flow within USPTO patents. While they all provided invaluable insight into the patent citation knowledge flow mechanisms, these knowledge diffusion studies were somewhat limited in terms of the focus on the citation patterns between individual entities. Recent advances in network analysis methodologies can greatly enhance our understanding of patents' knowledge flow by considering a network view of patent citations.

A patent citation network is often large and would be very difficult to analyze without recent advances in network topological analysis, which employs various statistical measures to characterize the topology of a network. Such analysis often uses the random graph model, the small-world model (Watts and Strogatz, 1998), and the scale-free model (Barabasi and Albert, 1999) to infer the governing principles of a network. The method has shown effectiveness in describing topological characteristics across a wide range of natural, social science, and technical networks (Albert and Barabasi, 2002). Topological analysis has also been incorporated into recent literature/patent citation network studies. For example, it was found that the literature citation network is a tree-like network and a scale-free network (Bilke and Peterson, 2001) with a power law degree distribution (Redner, 1998). The power law degree distribution phenomenon was also found in patent citation networks (Ong et al., 2004). Some recent studies have used the random graph model to study the knowledge diffusion process in a network. For example, the research of (Cowan and Jonard, 2004; Morone and Taylor, 2004) demonstrated that the knowledge diffusion process is efficient in a small-world network.

Network topological analysis can provide an in-depth understanding of the structure of large-scale networks. Combining patent citation network topological analysis and traditional bibliometric analysis, we may obtain a more comprehensive understanding of a technology field's knowledge diffusion status and characteristics. In this research, we integrate network topological analysis with bibliometric analysis to evaluate the knowledge transfer patterns in nanotechnology patent citation networks.

3. RESEARCH METHODOLOGY

The proposed research methodology consists of three steps: data acquisition, network construction, and network analysis (Figure 6-1).

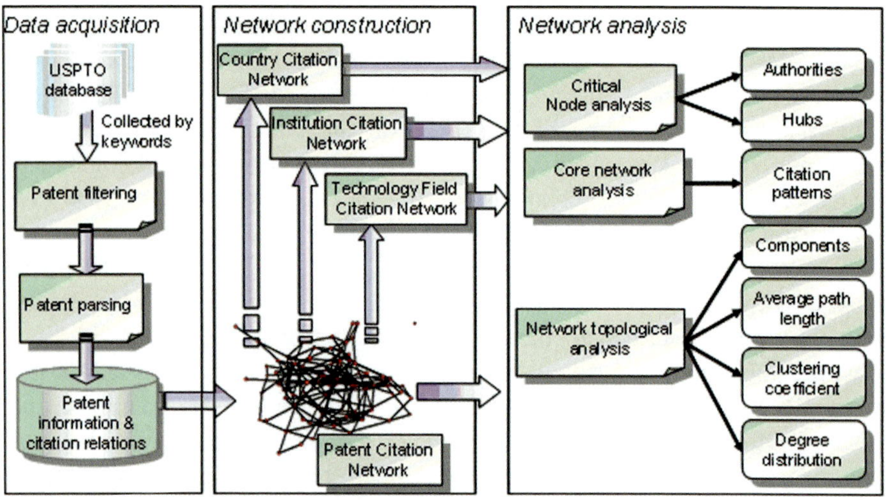

Figure 6-1. Patent citation network analysis framework.

3.1 Data Acquisition

During data acquisition, patents related to a specific scientific domain are collected from patent databases and parsed to obtain patent citation relations. A list of subject-specific keywords identified by domain experts is used to perform "full-text" search of patents (search is done on the "full-text" of patent titles, abstracts, claims, and descriptions). Besides "full-text" search, a more refined "title-claim" search (search is done on titles, abstracts, and claims) (Huang et al., 2003) can be performed. The "title-claim" search can provide more relevant (precise) results; while the "full-text" search can provide better coverage (recall). The collected patents are then parsed and the bibliographic and citation data are extracted.

3.2 Network Creation

From patent references, we can construct the complete patent citation relations. In this research, we are only concerned with the citations among patents in our extracted patent database; citation relations to patents outside the database were excluded.

From the parsed citation relations, we first constructed a patent-level citation network (referred to as *patent citation network* in this chapter). The nodes in the network are the patents in the database. The links indicate the citation relations between patents in the data set.

Patent citation relations can be aggregated into citation relations between different analytical units. In this research, we analyze the patent assignee country citation network (*country citation network*), patent assignee institution citation network (*institution citation network*), and the patent technology field citation network (*technology field citation network*). We use the United States Patent Classification (USPC) categories as a representation of the patent technology field. In these three networks, nodes represent the analytical units. Links represent citation relations between the patents that belong to any two analytical units (nodes). The weight of the link represents the number of citations between the nodes (analytical units). These three networks provide insight into the knowledge diffusion patterns at different analytical levels.

3.3 Network Analysis

In network analysis, we propose core network analysis, critical node analysis, and topological analysis on the four types of citation networks.

3.3.1 Core Network and Critical Node Analysis

In core network analysis, we study the core networks of the *country citation network*, *institution citation network*, and *technology field citation network*, which consist of the top citation relations according to link weight (number of citations). In our framework, we propose to use the top X links (determined by researchers and through experimentation) to create the core networks. We then visualize these networks using selected open source graph visualization software, e.g., *Graphviz*, developed by AT&T Labs (Gansner and North, 2000) (available at: http://www.research.att.com/sw/tools/graphviz/). Visualization of core networks allows us to identify the salient knowledge flow patterns among the analytical units.

In critical node analysis, we identify the hubs and authorities based on node degree in the citation networks.

- **Hub*:*** a node that cites many other nodes.
- **Authority*:*** a node that is cited by many other nodes.

Hubs and authorities indicate the important patents (within analytical units) in the patent citation network. The *country citation network*, *institution citation network,* and *technology field citation network* include self-citations, which should be removed when calculating the hubs and authorities. By analyzing the core networks and the critical nodes, we can identify the key nodes (countries, institutions, and technology fields) and find out how these key nodes directly interact with other nodes.

3.3.2 Topological Analysis

Network topological analysis employs various statistical measures to characterize the topology of the citation networks (Albert and Barabasi, 2002):

- ***Network size:*** We report the number of nodes and number of links. Network size shows the coverage of the field.
- ***Component size:*** A component is an isolated sub-network in a disconnected network. We report number of components (N_C), number of nodes of the giant component ($Node_C$), and number of links of the giant component ($Link_C$). The components represent independent groups in the field.
- ***Network diameter, D:*** The maximum value of the shortest path length between any pair of nodes in the network.
- ***Average path length, l:*** The average value of the shortest path length between any pair of nodes in the network. A short average path length means that knowledge will move to different parts of the graph more quickly.
- ***Clustering coefficient, C:*** A network's clustering coefficient, *C*, is the average of each node's clustering coefficient, *C'*. A node's clustering coefficient is the ratio of the number of edges between the node's neighbors to the number of possible edges between those neighbors (one node's neighbors are the nodes directly connected to it):

$$C' = \frac{\text{number of edges between the neighbors}}{\text{possible number of edges between the neighbors}}$$

The clustering coefficient of a patent citation network indicates the tendency for the patents/analytical units to form local clusters. Dense local clusters indicate that one node will receive more influence from its neighbors.

- ***Average degree, <k>:*** The average number of links that a node has to other nodes.
- ***Degree distribution, P(k):*** Degree distribution represents the probability that a selected node has exactly *k* links. Average degree and degree distribution indicate the ability of one patent/analytical unit to affect other patents.

$$P(k) = \frac{N(k)}{N}$$

N(k): the number of nodes with k links; *N*: number of nodes.
- ***In-degree and out-degree:*** In some networks, the links have directions. The in-degree of a node is the number of times a node is cited by other nodes. The out-degree of a node is the number of times a node cites other nodes. Specifically, in a citation network, the in-degree represents the number of times a node is cited by other nodes. The out-degree represents the number of times that a node cites other nodes.

From these topological measures, we can identify the citation network's global structure and infer the knowledge diffusion patterns in the citation network. Combining the topological analysis with the core network analysis and the critical node analysis may give us a better understanding of how knowledge was diffused from the core nodes to other parts of the citation network.

4. NETWORK ANALYSIS FOR NANOTECHNOLOGY

Nanotechnology's multidisciplinary characteristic and its widespread international development motivated several studies to assess its research and impact based on patent analysis. Chen and his team at the University of Arizona (Huang et al., 2003; 2004; 2005; 2006) studied the longitudinal research status of nanotechnology using content analysis and citation analysis approaches on the United States Patent and Trademark Office (USPTO) data. (Meyer, 2000; 2001) addressed the interaction between academia and industry in nanotechnology based on citation interactions between patents and papers. In this study, we apply our patent citation network analysis framework to assess the dynamics of the complex knowledge diffusion process in nanotechnology.

4.1 Data

In this research, we created a nanotechnology patent data set to test the proposed analysis framework. We used a list of nanotechnology-related keywords from our previous research (Huang et al., 2003; 2004; 2005). We collected nanotechnology patents that matched with these keywords in patent title, abstract, claim, and description ("full-text" search) from the USPTO database. After extracting patents from the USPTO database, we identified 78,609 patents published between 1976 and 2004. From this patent data set, we identified 22,219 assignees, 153,732 inventors, and 163 assignee countries. These patents cover 432 of 462 first-level United States Patent Classification (USPC) categories, which are used to represent technology fields.

4.2 Analysis Results

4.2.1 Country Citation Network

The *country citation network* represents the patent citation relationship at the assignee country level. Nodes in the *country citation network* are the assignee countries. In our testbed, the *country citation network* consists of 59 countries, 386 inter-country citation relations, and 37 self-citation relations. Table 6-1 shows the top assignee countries according to the number of patents published.

Table 6-1. Top nanotechnology patent assignee countries ("full-text" keyword search, 1976-2004).

Rank	Assignee Country	Number of Patents
1	United States	53,077
2	Japan	8,605
3	Federal Republic of Germany	2,651
4	France	2,354
5	Canada	1,161
6	United Kingdom	1,085
7	China (Taiwan)	547
8	Netherlands	546
9	Republic of Korea	535
10	Switzerland	534

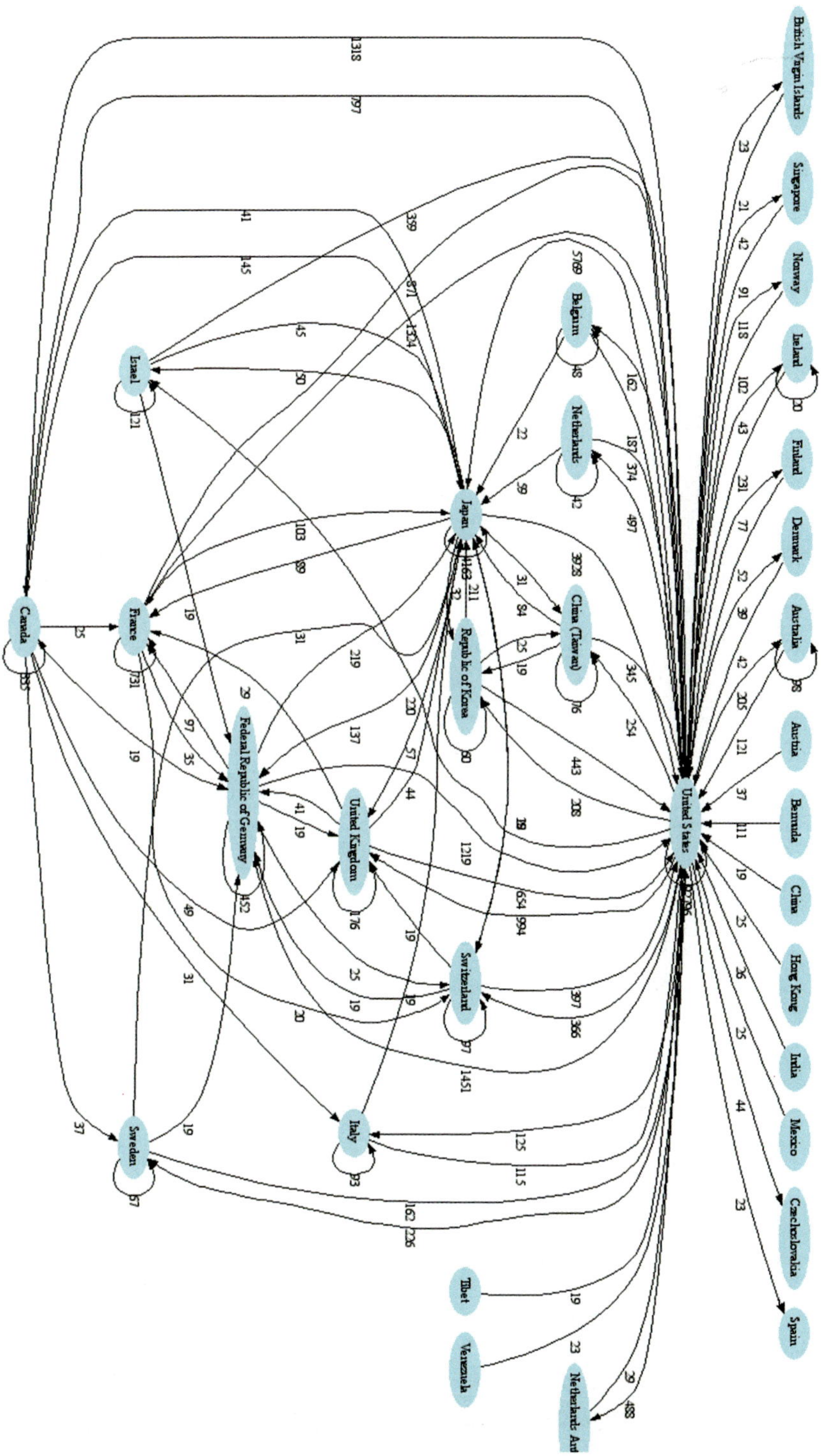

Figure 6-2. Country citation core network.

Core network and critical node analysis results

The core network of the nanotechnology patent *country citation network* between 1976 and 2004 is shown in Figure 6-2. It shows the top 100 country citation relations with the largest numbers of citations. Based on the analysis of the network, we observe that:

The U.S. dominated most of the citations and the U.S. patents heavily interact with patents of most other countries, especially Japan and Germany.

Japan, Germany, the United Kingdom, Switzerland, France, and Canada form secondary citation centers. These countries have many interactions with each other. Korea and Taiwan patents are heavily influenced by the U.S. and Japan patents.

Based on the top five authorities, hubs, and self-citation countries, as shown in Tables 6-2 to 6-4, we also observe that the U.S. and Japan played a very important role in nanotechnology research.

Table 6-2. Top 5 authorities (without self-citation) of the country citation network.

Rank	Country	Cited Countries	Number of Citations
1	United States	52	11,428
2	Japan	32	6,859
3	Federal Republic of Germany	24	1,818
4	Switzerland	24	489
5	France	23	1,655

Table 6-3. Top 5 hubs (without self-citation) of the country citation network.

Rank	Country	Citing Countries	Number of Citations
1	United States	38	13,611
2	Japan	23	4,442
3	Federal Republic of Germany	23	1,673
4	United Kingdom	23	857
5	France	19	1,092

Table 6-4. Top 5 self-citation countries of the country citation network.

Rank	Country	Number of Citations
1	United States	92,796
2	Japan	4,163
3	France	731
4	Federal Republic of Germany	452
5	Canada	335

The U.S. and Japan are the two largest citation centers and have the largest number of citations to other countries and self-citations. For example, 52 countries' patents cite U.S. patents 11,428 times. U.S. research directly affects 88.1% (52/59) of other countries' research in nanotechnology. Germany, France, the United Kingdom, and Canada form the group of secondary patent citation centers. Based on the number of citations, the U.S. has more self-citations than citations to other countries, which suggests that the knowledge from within the U.S. plays a more important role in its nanotechnology R&D than the knowledge from other countries.

Network topological analysis results

The topological measures of the network (Table 6-5) show that the *country citation network* only contains one connected component. There exists at least one path between every pair of nodes in the *country citation network.* In other words, every country in the network directly or indirectly affects the other countries through the published patents in nanotechnology R&D.

Table 6-5. Topological measures of the patent citation networks.

Networks	<k>	l	l_{rand}	C	C_{rand}	D	N_c	$Node_c$	$Link_c$
Country Citation Network	8.305	1.933	1.926	0.841	0.1432	4	1	59 (100%)	423 (100%)
Institution Citation Network	7.571	3.754	4.591	0.3342	0.0007	15	352	10,220 (93.95%)	39,770 (88.71%)
Technology Field Citation Network	58.317	2.007	1.472	0.7168	0.4907	6	3	395 (99.49%)	14,485 (99.98%)
Patent Citation Network	5.147	8.923	6.658	0.1781	9.41E-05	36	2,969	45,717 (83.53%)	133,769 (94.95%)

Notes: *<k>:* average degree; *l:* average path length; l_{rand}: average path length of a random network; *C*: clustering coefficient; C_{rand}: clustering coefficient of a random network; *D*: network diameter; N_C: number of components; $Node_C$: number of nodes in the largest component; $Link_C$: number of links in the largest component.

The average degree is the average number of countries that directly interact with one country. In the *country citation network*, the average degree of each node is 8.305, which is very high given that there are only 59 total nodes in the network. High average degree means that one country may affect and be affected by many other countries.

The *country citation network* has a small diameter (4) and a small average path length (1.933). The diameter and average path length indicate the number of links needed to propagate knowledge from one country to other countries. Such a small diameter and a small average path length show that one country's nanotechnology research and development results can be easily propagated to other countries. The average path length of the *country citation network* is close to that of a same-size random network (1.926). Thus, the knowledge diffusion process in the *country citation network* is as effective as that in a randomly connected citation network. In a random network, knowledge transfer is fast, especially at the beginning of the knowledge transfer process (Cowan and Jonard, 2004).

The *country citation network* has a larger clustering coefficient (0.8410) than that of the random network of the same size (0.1432). The large clustering coefficient means that the countries that have citation relations to a particular country may have a high tendency to interact with each other. Notice that in topological analysis the weight of the citation links (the citation frequency) is ignored and the network reveals the binary citation relationship (whether or not citation(s) occurred between the patents of the two counties). When the citation frequency is ignored and all citation links are included, the countries have a relatively high tendency to learn from each other and form local citation clusters in the *country citation network.*

From the topological analysis, we observe that the *country citation network* is a dense network with significant local clusters, which is hard to observe from the core network analysis and critical node network analysis. We also find that this *country citation network* is efficient from the knowledge diffusion perspective. Combining that with the results from core network analysis and critical node analysis, one can conclude that U.S. nanotechnology knowledge can be easily propagated to other countries.

4.2.2 Institution Citation Network

The *institution citation network* represents the patent citation relationships at the assignee institution level. Nodes in the *institution citation network* are the assignee institutions. The *institution citation network* of our testbed consists of 10,878 institutions, 42,950 inter-institution citation relations, and 1,878 self-citation relations. Table 6-6 shows the top assignee institutions according to the number of patents published. IBM, Minnesota Mining and Manufacturing Company, and Xerox Corporation are the top three nanotechnology assignees.

Table 6-6. Top nanotechnology patent assignee institutions.

Rank	Assignee Name	Number of Patents
1	International Business Machines Corporation	1,747
2	Minnesota Mining and Manufacturing Company	1,138
3	Xerox Corporation	1,130
4	The Regents of the University of California	972
5	Eastman Kodak Company	844
6	Micron Technology, Inc.	808
7	Motorola, Inc.	727
8	General Electric Company	670
9	NEC Corporation	634
10	Advanced Micro Devices, Inc.	615

Core network and critical node analysis results

We analyzed the core network of the nanotechnology patent *institution citation network* between 1976 and 2004. Self-citation is a very common phenomenon in the *institution citation network*, indicating that researchers tend to cite patents invented in their institutions. To show the citation relationships between institutions, we visualized the core network of the *institution citation network* that does not contain self-citations, as shown in Figure 6-3, and insets a, b, and c. Based on the analysis of the network, we observe that:

- IBM's patents heavily interact with patents of most of the other institutions, especially Hitachi, Canon, and Micron Technology, Inc. *(See inset a for details.)*
- The University of California (UofCA), Motorola, and Micron Technology are among the second group with active patent citations. *(See inset b.)*
- 3M Company, Xerox Corporation, the Dow Chemical Company, and Norton Company form a local citation cluster. *(See inset c.)*

From the hub and authority analyses (Tables 6-7 and 6-8), we observe that IBM and the University of California are the two largest citation centers in nanotechnology. For example, patents from 720 other institutions cite IBM patents 3,932 times. IBM patents also cite the patents from 445 institutions a total of 1,588 times. From the top self-citation institutions (Table 6-9), we observe that although IBM is the biggest nanotechnology patent assignee and the largest patent citation center in the network, it has a relatively small number of self-citations (998). The low self-citations indicate that IBM is less inclined to cite their own patents in comparison with other institutions.

The Xerox Corporation has a different pattern of self-citations. The Xerox Corporation has the largest number of self-citations (3,108), but it cites other institutions very few times (too few to be shown in the top 5 hubs). Considering its relatively large number of patents published (1,130) and relatively small number of citations from other institutions (885), we can infer that the Xerox Corporation is more self-dependent in research.

In the *institution citation network*, there is not a dominant institution like the U.S. in the *country citation network*. IBM, the University of California, and 3M Company all have strong influence in the field. These institutions tend to create different local clusters based on their tendency to cite other institutions' patents, which may relate to differences in research field, research approach, and so forth. Such local citation clusters are connected by several inter-institution citations.

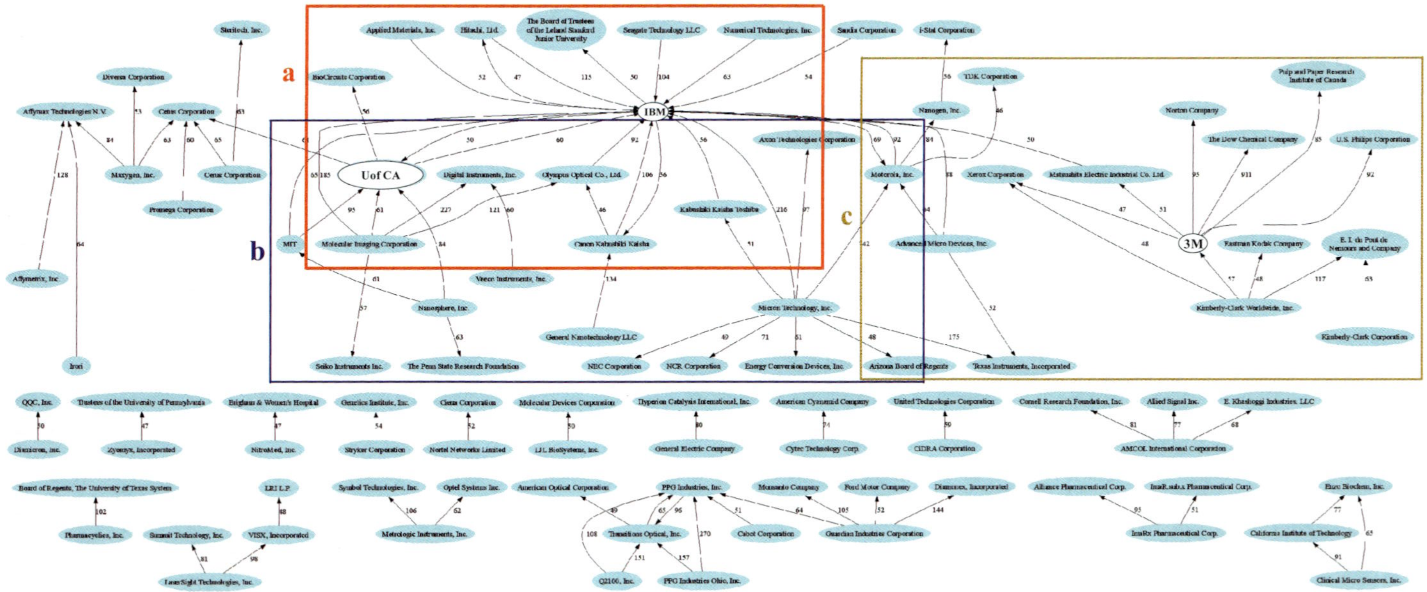

Figure 6-3 and insets a, b, and c. Institution citation core network

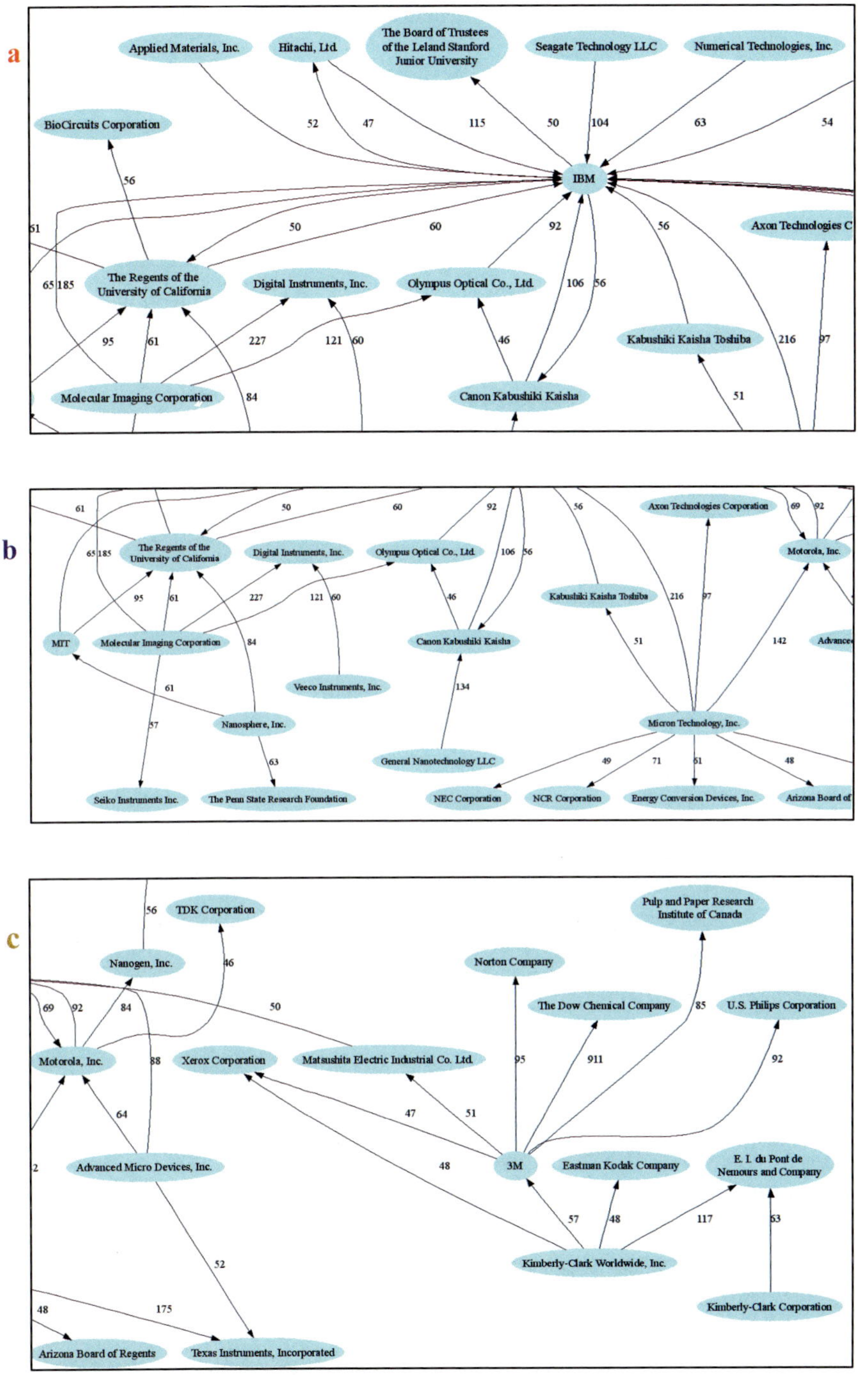
a
Applied Materials, Inc.
Hitachi, Ltd.
The Board of Trustees of the Leland Stanford Junior University
Seagate Technology LLC
Numerical Technologies, Inc.
BioCircuits Corporation
IBM
Axon Technologies C
The Regents of the University of California
Digital Instruments, Inc.
Olympus Optical Co., Ltd.
Kabushiki Kaisha Toshiba
Molecular Imaging Corporation
Canon Kabushiki Kaisha
b
Axon Technologies Corporation
The Regents of the University of California
Digital Instruments, Inc.
Olympus Optical Co., Ltd.
Motorola, Inc.
Kabushiki Kaisha Toshiba
MIT
Molecular Imaging Corporation
Canon Kabushiki Kaisha
Veeco Instruments, Inc.
Nanosphere, Inc.
Micron Technology, Inc.
General Nanotechnology LLC
Seiko Instruments Inc.
The Penn State Research Foundation
NEC Corporation
NCR Corporation
Energy Conversion Devices, Inc.
Arizona Board of
c
TDK Corporation
Pulp and Paper Research Institute of Canada
Nanogen, Inc.
Norton Company
The Dow Chemical Company
U.S. Philips Corporation
Motorola, Inc.
Xerox Corporation
Matsushita Electric Industrial Co. Ltd.
Advanced Micro Devices, Inc.
3M
Eastman Kodak Company
E. I. du Pont de Nemours and Company
Kimberly-Clark Worldwide, Inc.
Kimberly-Clark Corporation
Arizona Board of Regents
Texas Instruments, Incorporated

Table 6-7. Top 5 authorities (without self-citation) of the institution citation network.

Rank	Institution	Number of Cited Institutions	Number of Citations
1	International Business Machines Corporation	720	3,932
2	The Regents of the University of California	646	2,106
3	Minnesota Mining and Manufacturing Company	398	1,093
4	Massachusetts Institute of Technology	382	1,031
5	E. I. du Pont de Nemours and Company	337	972

Table 6-8. Top 5 hubs (without self-citation) of the institution citation network.

Rank	Institution	Number of Citing institutions	Number of Citations
1	The Regents of the University of California	520	1,553
2	International Business Machines Corporation	445	1,588
3	Minnesota Mining and Manufacturing Company	310	2,274
4	Massachusetts Institute of Technology	252	734
5	Motorola, Inc.	231	1,237

Table 6-9. Top 5 self-citation institutions of the institution citation network.

Rank	Institution	Number of Citations
1	Xerox Corporation	3,108
2	Minnesota Mining and Manufacturing Company	2,974
3	Micron Technology, Inc.	1,042
4	International Business Machines Corporation	998
5	Allergan, Inc.	653

Network topological analysis results

The topological measures of the network (Table 6-5) show that the *institution citation network* consists of 352 connected components. The giant component contains 10,220 (93.95%) institutions and 39,770 (88.71%) citation relations. In the *institution citation network*, most of the institutions directly or indirectly connect to each other. Thus, most institutions influence and are influenced by other institutions in nanotechnology.

The *institution citation network* has a much smaller average path length (3.754) than the random network of the same size (4.591). Compared with a random network, the institutions in the *institution citation network* are connected by fewer citation steps, indicatingthat knowledge can thus be transferred more easily between institutions. The current inter-institution patent citation relationships are much more efficient than randomly created inter-institution patent citation relationships.

The *institution citation network* has a much larger clustering coefficient (0.3342) than the random network (0.0007). The large clustering coefficient shows the institutions' high tendency to form local citation clusters. Institutions with similar interests and specializations may have dense citation relationships. It is also possible for collaborating institutions to create such local citation clusters. From the clustering coefficient measure, we can infer that intensive knowledge transfer among groups of peer institutions is a very common phenomenon.

The topological analysis is consistent with the core network analysis and critical node analysis results. The *institution citation network* shows the existence of major institution groups, such as IBM, the University of California, and 3M Company. The dense intra-group interactions and several inter-group interactions may be the reason for the short average path length for the entire *institution citation network*.

4.2.3 Technology Field Citation Network

The *technology field citation network* represents the patent citation relationships from the perspective of patent technology fields. The nanotechnology *technology field citation network* consists of 397 technology fields, 14,228 inter-technology field citation relations, and 259 self-citation relations. From Table 6-10, we observe that the technology field "*435 Chemistry: molecular biology and microbiology*" occupies the largest proportion of all the patents in nanotechnology.

Table 6-10. Top nanotechnology patent technology fields.

Rank	Technology Field		Number of Patents
1	435	Chemistry: molecular biology and microbiology	9,793
2	514	Drug, bio-affecting and body treating compositions	7,760
3	424	Drug, bio-affecting and body treating compositions	5,999
4	257	Active solid-state devices (e.g., transistors, solid-state diodes)	5,610
5	438	Semiconductor device manufacturing: process	5,387
6	428	Stock material or miscellaneous articles	5,101
7	536	Organic compounds – part of the class 532-570 series	4,729
8	530	Chemistry: natural resins or derivatives; peptides or proteins; lignins or reaction products thereof	4,655
9	250	Radiant energy	4,635
10	427	Coating processes	4,034

Core network and critical node analysis results

The core network of the *technology field citation network* between 1976 and 2004 is shown in Figure 6-4. The core network shows the top 100 technology field citations according to the number of citations. Base on the analysis of the network, we observe that:

- Technology fields "435 Chemistry: molecular biology and microbiology," "424 Drug, bio-affecting and body treating compositions," "436 Chemistry: analytical and immunological testing," "530 Chemistry: natural resins or derivatives; peptides or proteins; lignins or reaction products thereof," and "514 Drug, bio-affecting and body treating compositions" are closely related to each other.
- Technology fields "250 Radiant energy," "356 Optics: measuring and testing," and "073 Measuring and testing" form a local citation cluster. There are also some other local citation clusters, such as "430 Radiation imagery chemistry: process, composition, or product thereof," "428 Stock material or miscellaneous articles," and "427 Coating processes."

Analysis of the top authorities shows that the two citation clusters exhibit two different patterns. The technology fields "*428 Stock material or miscellaneous articles*," "*250 Radiant energy*," and "*427 Coating processes*" have been cited by the largest number of citing technology fields, indicating they may have a broader impact on the other technology fields. The technology fields "*435 Chemistry: molecular biology and microbiology*" (31,770 citations) and "*436 Chemistry: analytical and immunological testing*" (22,219 citations) (not shown in the table) have been cited by the largest number of patents from the other technology fields, indicating that they may have a stronger impact on the other technology fields.

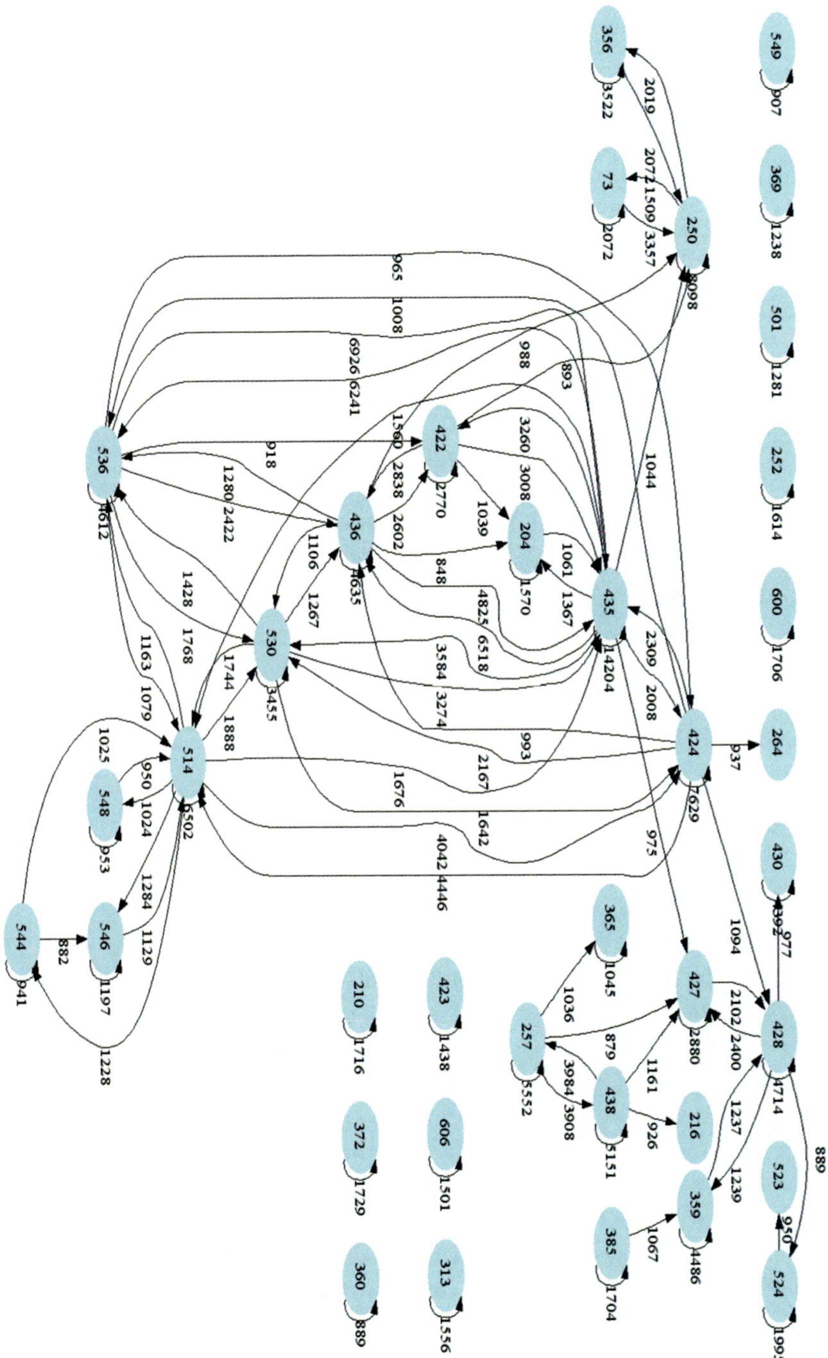

Figure 6-4. Technology field citation core network.

Table 6-11. Top 5 authorities (without self-citation) of the technology field citation network.

Rank	Technology Field		Number of Citing Technology Fields	Number of Citations
1	428	Stock material or miscellaneous articles	211	20,171
2	250	Radiant energy	208	16,871
3	427	Coating processes	201	18,941
4	359	Optics: systems (including communication) and elements	178	11,128
5	252	Compositions	174	9,462

Table 6-12. Top 5 hubs (without self-citation) of the technology field citation network.

Rank	Technology Field		Number of Cited Technology Fields	Number of Citations
1	428	Stock material or miscellaneous articles	198	20,416
2	427	Coating processes	193	16,411
3	250	Radiant energy	189	13,718
4	422	Chemical apparatus and process disinfecting, deodorizing, preserving, or sterilizing	179	14,983
5	359	Optics: systems (including communication) and elements	177	10,769

Table 6-13. Top 5 self-citation technology fields of the technology field citation network.

Rank	Technology Field		Number of Citations
1	435	Chemistry: molecular biology and microbiology	14,204
2	250	Radiant energy	8,098
3	424	Drug, bio-affecting and body treating compositions	7,629
4	514	Drug, bio-affecting and body treating compositions	6,502
5	257	Active solid-state devices (e.g., transistors, solid-state diodes)	5,552

Network topological analysis results

The topological measures of the *technology field citation network* (Table 6-5) identify three connected components. The largest component contains 395 (99.49%) technology fields and 14,485 (99.98%) citation relations. The other two components each contain one technology field. In nanotechnology-related research, the two separate technology fields only cite patents from their own field. The other technology fields have wider citation preferences and interact with each other through the citations in the network.

The *technology field citation network* has a very high average degree (58.317) with only 397 nodes. It is much higher than the average degree of the *country citation network* and the *institution citation network*. The high average degree shows the close relationship among the technology fields. Many of them interact with each other directly, indicating the interdisciplinary nature of nanotechnology.

The *technology field citation network* has a small average path length (2.007) and a small diameter (6). On average, the knowledge in one technology field can be transferred to the others through two steps of citation relations. However, the average path length is still large in comparison with the random network of the same size (1.472). This means that although knowledge could transfer quickly in the *technology field citation network*, it is slower than the knowledge transfer process in the random citation network.

The *technology field citation network* has a clustering coefficient (0.7168) which is similar in size to the random network (0.4907). Considering the high average degree, we can conclude that the *technology field citation network* is a dense network, but it does not show a distinct local cluster characteristic compared with the random network.

4.2.4 Patent Citation Network

The patent-level citation network (*patent citation network*) represents the citation relationships between patents. In our testbed, the *patent citation network* consists of 54,730 patents and 140,872 inter-patent citation relations.

Critical node analysis results

By analyzing the top five authorities, we found that:

- Kary Mullis's two patents on nucleic acid sequences have the highest number of citations among all the nanotechnology patents.
- Gerd Binnig (or Gerd Bennig) and Salvatore Pace have patents that have been cited more than 100 times.

By analyzing the top five hubs, we found that:

- Shaohua Xu's one patent and Wenhai Han's three patents all have more than 100 references, which are mostly in nanotechnology.
- Thomas Rueckes' patent has more than 200 references, but more than 50% of the references are from patents outside of nanotechnology.

Table 6-14. Top 5 authorities of the patent citation network.

Rank	PatentID	Title	Primary Inventor	Publication Date	No. of Citations
1	4683202	Process for amplifying nucleic acid sequences	Mullis, Kary B.	7/28/1987	418
2	4683195	Process for amplifying, detecting, and/or-cloning nucleic acid sequences	Mullis, Kary B.	7/28/1987	384
3	4343993	Scanning tunneling microscope	Binnig, Gerd	8/10/1982	136
4	4908112	Silicon semiconductor wafer for analyzing micronic biological samples	Pace, Salvatore J.	3/13/1990	128
5	4724318	Atomic force microscope and method for imaging surfaces with atomic resolution	Bennig, Gerd K.	2/9/1988	127

Table 6-15. Top 5 hubs of the patent citation network.

Rank	PatentID	Title	Primary Inventor	Publication Date	NSE Ref.	All Ref.
1	5874668	Atomic force microscope for biological specimens	Xu, Shaohua	10/24/1995	136	147
2	6134955	Magnetic modulation of force sensor for AC detection in an atomic force microscope	Han, Wenhai	1/11/1999	125	130
3	6835591	Methods of nanotube films and articles	Rueckes, Thomas	4/23/2002	121	245
4	5866805	Cantilevers for a magnetically driven atomic force microscope	Han, Wenhai	9/12/1996	120	125
5	5753814	Magnetically-oscillated probe microscope for operation in liquids	Han, Wenhai.	9/27/1996	115	119

Network topological analysis results

Topological measures

The topological measures of the network (Table 6-5) show that the *patent citation network* contains 2,969 connected components. The biggest component contains 45,717 (83.53%) patents and 133,769 (94.95%) relations.

The *patent citation network* has a much larger average path length (8.923) than the random network of the same size (6.658), which is different from most large scale networks (Albert and Barabasi, 2002). The *patent citation network* is not a small-world network. The knowledge transferring process in this network is not as efficient as that in a random network. Such a phenomenon may be caused by the nature of a citation network. In a random network, links may appear between any pair of nodes. Such shortcuts, with a path length of 1, will reduce the average path length. In a patent citation network, it seems that such shortcuts are rare. This may reflect the fact that industrial research and development in nanotechnology shows an interdisciplinary nature overall but each step of the innovation (revealed in one patent) can still be characterized by incremental development generally within the same or related disciplines. In the existing citation network, it is rare to see groundbreaking patents that join relatively distinct disciplines.

We observe that the *patent citation network* shows a very large clustering coefficient (0.1781) compared with the random network of the same size (9.41E-05), which indicates that nanotechnology patents have a very high tendency to form citation clusters.

Degree distribution and scale-free model

In this research, we studied the degree distribution of the *patent citation network.* Because the weight (number of citations) is not reflected in the degree distribution, we do not study the degree distributions of the *country citation network*, the *institution citation network*, and the *technology field citation network*, which are weighted networks.

Figure 6-5 shows the in-degree distribution and out-degree distribution of the *patent citation network.*

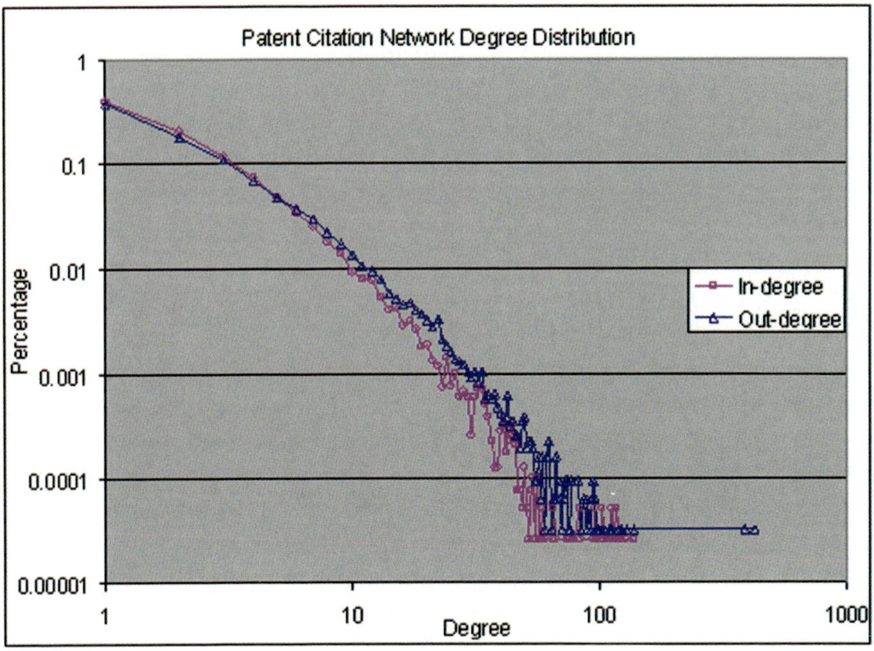

Figure 6-5. Patent citation network degree distribution.

The in-degree distribution shows the probability distribution of the number of citations one patent may receive. The out-degree distribution shows the probability distribution of the number of references one patent may have. In the log-log graph, the two degree distributions show the pattern of a straight line, which mean that they follow the power law distribution.

The power law distribution takes the form of P(*k*) ~ $k^{-\gamma}$, where P(k) is the probability that a node has exactly *k* links. The power law exponent γ and correlation coefficient r of the two degree distributions are shown in Table 6-16.

Table 6-16. Patent citation network degree distribution measures.

	power law exponent γ	*correlation coefficient r*
In-degree distribution	2.2925	0.6855
Out-degree distribution	2.1394	0.8528

The power law degree distribution shows that the *patent citation network* follows the scale-free model, which indicates that high degree (in-degree and out-degree) patents are less likely to exist. From the knowledge diffusion perspective, the scale-free characteristic indicates that the patent played different roles in the network. A few high out-degree patents represent critical far-reaching fundamental innovations that diffuse knowledge to many other patents. A few high in-degree patents integrate broad knowledge from many other patents and possibly combine existing innovations into a new technology. Most of the other (low in-degree and low out-degree) patents play a less important role in the network.

5. CONCLUSIONS

We identified the key entities, important knowledge transfer patterns, and the global knowledge transfer characteristics of nanotechnology based on a network analysis framework. The following important findings were identified:

In the *country citation network*, the U.S. is the most important citation center, which was cited 11,428 times by other countries' patents and cited the other countries' patents 13,611 times. Japan, Germany, the United Kingdom, and France are the secondary citation centers, which cited and were cited by other countries' patents more than 1,000 times.

In the *institution citation network*, self-citation is a very common phenomenon. We found that IBM, the largest authority of the institution citation network with 3,932 citations, has a small number of self-citations compared to the citations between IBM and other institutions. We also found that the Xerox Corporation most often cites the patents it invented.

Analyses on the *technology field citation network* and *patent citation network* help identify important citation centers in these networks.

By analyzing the topological structure of the four citation networks, we found that all four types of networks have a giant component that connects most of the nodes, indicating the wide interactions between the entities.

The four networks have different knowledge transfer efficiency as compared to a random network of the same size. According to the average path length measure, the *institution citation network* structure (average path length: 3.754) allows knowledge to transfer more quickly between institutions than a random network (average path length: 4.591). The *country citation network* (average path length: 1.933) provides as efficient knowledge transfer capability as the random network (average path length: 1.926). The *technology field citation network* structure (average path length: 2.007) and the *patent citation network* (average path length: 8.923) provide less efficient knowledge diffusion capability than the random network (the average path lengths are 1.472 and 6.658, respectively).

According to the clustering coefficient measure, the *country citation network* (clustering coefficient: 0.841), the *institution citation network* (clustering coefficient: 0.3342), and the *patent citation network* (clustering coefficient: 0.1781) all show a tendency to form local citation clusters, which indicates the intensive cooperation and knowledge exchange between these analytical units.

The degree distribution of the nanotechnology *patent citation network* shows that it is a scale-free network, indicating that a few high degree patents have more chance of being cited and play a more important role in the knowledge diffusion process.

6. QUESTIONS FOR DISCUSSION

1. What other topological network analysis techniques can be applied to patent citation network analysis?
2. Do nanotechnology patent citation networks follow the same principles as the small-world network or the scale-free network?
3. How can we apply selected graph partitioning techniques to identify subgroups of researchers and research topics that are of relevance to recent nanotechnology advancement?
4. How can we apply dynamic network analysis techniques to identify the hot topics that emerged in recent years?

Chapter 7

GOVERNMENT RESEARCH INVESTMENT AND NANOTECHNOLOGY INNOVATIONS: NSF FUNDING AND USPTO PATENT ANALYSIS, 2001-2004

CHAPTER OVERVIEW

Nanotechnology research has experienced rapid growth since 2001. It has also attracted significant government funding. Many countries have recognized nanotechnology as a critical research domain that promises to revolutionize a wide range of fields and applications. In this chapter, we present an analysis of the funding for nanotechnology at the National Science Foundation (NSF) and its relationship to technological innovation (patenting activities) in this field from 2001 to 2004. Using a combination of basic bibliometric and content analysis, we identify growth trends, research topics, the evolution in NSF funding, and commercial patenting activities recorded at the United States Patent and Trademark Office (USPTO) database. The patent citations are used to evaluate the impact of the NSF-funded research on nanotechnology development in comparison to other research funding organizations. The analysis shows that the NSF-funded researchers and patents authored by them have significantly higher impact based on patent citation measures in the four-year period than other comparison groups.

1. INTRODUCTION

Nanotechnology is an interdisciplinary field with wide applications in numerous economic sectors. In 2000, the United States announced the National Nanotechnology Initiative (NNI, www.nano.gov) (Roco et al., 2000). Since then, more than 65 countries have adopted national projects or programs. Increased attention is being given to stimulating nanotechnology innovation leading to economic benefits. In fiscal year 2007,the NNI investment was about $1.4 billion; NSF had the largest program with a budget over $390 million. Gaining an understanding of the effect of public research funding on nanotechnology innovation through the number of patents may contribute to shaping recommendations for future funding policies.

Scientific evaluation of the impact of public funding on research output and on the overall development of a scientific and engineering field is a difficult task and there is only sparse literature on this topic. The difficulty of associating research output with the general development of a field was documented by Adams and Griliches (Adams and Griliches, 1998). Most previous studies focused on the impact of public funding on research output based on scientific publications (Adams and Griliches, 1998; Arora and Gambardella, 1998; Narin, 1998; Payne and Siow, 2003). In our earlier research (also summarized in Chapter 5), we used patent documents and their citations in order to provide a more direct account of the impact of NSF funding on technological innovations between 1991 and 2002 (Huang et al., 2005).

In this chapter, we focus on nanotechnology research and development (R&D) during the time period of 2001 to 2004 using the patents in the USPTO database and the funded awards from the NSF database. This time period is important as it represents the first four years of the NNI program. After defining the NSF award and USPTO patent data sets used in the study, we assess the growth trends of the respective awards and patents in nanotechnology that reflect the funding activity and research productivity. The nanotechnology areas identified in patents and awards are presented in topic maps that reveal topic distributions and the evolution of funded research and patents in the field. Statistical analyses of patent citations are used to compare the patents and their inventors funded by NSF awards with those supported by other comparison groups.

2. NSF AWARD DATA AND USPTO PATENT DATA

The USPTO patent database provides full-text access to the patents filed with USPTO since 1976. From 2001 to 2004, USPTO issued 737,176 patents. We used the same nanotechnology keyword list given in Table 4-1 to collect nanotechnology-related patents by searching title, abstract, claim, and description of the patent documents ("full-text" search) (Huang et al., 2004; Huang et al., 2003). The resulting set of nanotechnology patents contains both the patents having nanotechnology products (identified in title, abstract and claims) and the patents using nanotechnology tools (typically described in the text). Our data set contains 32,180 nanotechnology patents ("full-text" search) issued during 2001-2004, invented by 53,597 authors from 9,470 assignees in 97 countries. These patents are associated with 386 (out of a total of 462) first-level United States Patent Classification categories. Using the same nanotechnology keyword list, we also collected nanotechnology-related patents by searching title, abstract, and claim of the patent documents ('title-claims" search). The result set contains 6,253 nanotechnology patents, invented by 11,449 authors from 2,325 assignees in 53 countries. These patents are associated with 298 (out of a total of 462) first-level United States Patent Classification categories. The "full-text" search results provide better coverage of the nanotechnology patents, and the "title-claims" search results provide better accuracy. In this chapter, we focused on the "full-text" search. We also report the "title-claims" search results as a comparison.

Table 7-1 lists the top 20 assignee countries according to the numbers of nanotechnology-related patents ("full-text" search) issued between 2001 and 2004. The United States produced the majority of the nanotechnology patents, followed by Japan, Germany, France, and Canada.

Table 7-1. Top assignee countries: Number of patents (2001-2004) identified by keyword "full-text" search.

Rank	Assignee Country	Number of Patents 2001-2004	Percentage of all Nano Patents (32,180)
1	United States	21,120	65.63%
2	Japan	3899	12.12%
3	Germany	1133	3.52%
4	France	883	2.74%
5	Canada	545	1.69%
6	Republic of Korea	394	1.22%
7	United Kingdom	375	1.17%
8	China (Taiwan)	354	1.10%
9	Netherlands	291	0.90%
10	Australia	288	0.89%
11	Switzerland	240	0.75%
12	Israel	194	0.60%
13	Sweden	160	0.50%
14	Italy	137	0.43%
15	Belgium	111	0.34%

Rank	Assignee Country	Number of Patents 2001-2004	Percentage of all Nano Patents (32,180)
16	Denmark	84	0.26%
17	Singapore	75	0.23%
18	Finland	73	0.23%
19	India	55	0.17%
20	Ireland	42	0.13%

Table 7-2 shows the top 20 assignee countries according to the numbers of nanotechnology-related patents by "title-claims" search between 2001 and 2004. The United States produced the majority of the nanotechnology patents, followed by Japan, Germany, France, and Korea.

Table 7-2. Top assignee countries: Number of patents (2001-2004) identified by keyword "title-claims" search.

Rank	Assignee Country	Number of Patents 2001-2004	Percentage of all Nano patents (6,253)
1	United States	4147	66.32%
2	Japan	501	8.01%
3	Federal Rep. of Germany	246	3.93%
4	France	192	3.07%
5	Republic of Korea	162	2.59%
6	China (Taiwan)	126	2.02%
7	Canada	91	1.46%
8	Netherlands	65	1.04%
9	United Kingdom	62	0.99%
10	Switzerland	43	0.69%
11	Israel	42	0.67%
12	Australia	35	0.56%
13	Belgium	33	0.53%
14	Italy	30	0.48%
15	Singapore	25	0.40%
16	Sweden	24	0.38%
17	India	19	0.30%
18	Ireland	16	0.26%
19	China	10	0.16%
20	Denmark	9	0.14%

Table 7-3 shows the top 20 assignee institutions according to the numbers of nanotechnology-related patents published between 2001 and 2004. The top five assignees are International Business Machines Corporation (IBM), Micron Technology Inc., Advanced Micro Devices Inc., the Regents of the University of California, and Minnesota Mining and Manufacturing Company (3M).

Table 7-3. Top assignee institutions: Number of patents by "full-text" search (2001-2004).

Rank	Assignee Name	Number of Patents
1	International Business Machines Corporation	730
2	Micron Technology, Inc.	620
3	Advanced Micro Devices, Inc.	519
4	The Regents of the University of California	461
5	Minnesota Mining and Manufacturing Company	396
6	Xerox Corporation	356
7	Intel Corporation	268
8	General Electric Company	239
9	NEC Corporation	234
10	Motorola, Inc.	227
11	Eastman Kodak Company	219
12	Canon Kabushiki Kaisha	181
13	Corning Incorporated	165
14	Applied Materials, Inc.	162
15	Hewlett-Packard Development Company, L.P.	162
16	Hitachi, Ltd.	154
17	Massachusetts Institute of Technology	144
18	Silverbrook Research PTY LTD	143
19	Kabushiki Kaisha Toshiba	142
20	Fuji Photo Film Co., Ltd.	141

We used the first-level United States Patent Classification categories (available at: http://www.uspto.gov/go/classification/selectnumwithtitle.htm) as representations of the patents' technology fields. In this ontology, some categories have identical names; however, the detailed specifications of such categories are different. We used the category name and the U.S. Patent Classification ID number to label each technology field. Table 7-4 presents the top 20 technology fields according to the numbers of the nanotechnology-related patents issued between 2001 and 2004. Technology fields "435 Chemistry: molecular biology and microbiology," "438 Semiconductor device manufacturing: process," "257 Active solid-state devices," "514 Drug, bio-affecting and body treating compositions," and "424 Drug, bio-affecting and body treating compositions" are at the top of the list.

Table 7-4. Top U.S. patent technology fields: Number of patents by "full-text" search (2001-2004).

Rank	Technology Field	Number of Patents
1	435: Chemistry: molecular biology and microbiology	4191
2	438: Semiconductor device manufacturing: process	3249
3	257: Active solid-state devices (e.g., transistors, solid-state diodes)	3030
4	514: Drug, bio-affecting and body treating compositions	2923
5	424: Drug, bio-affecting and body treating compositions	2549
6	536: Organic compounds -- part of the class 532-570 series	2251

Rank	Technology Field	Number of Patents
7	428: Stock material or miscellaneous articles	2236
8	530: Chemistry: natural resins or derivatives; peptides or proteins; lignins or reaction products thereof	1776
9	427: Coating processes	1542
10	250: Radiant energy	1468
11	359: Optics: systems (including communication) and elements	1299
12	430: Radiation imagery chemistry: process, composition, or product thereof	1284
13	436: Chemistry: analytical and immunological testing	1151
14	356: Optics: measuring and testing	1004
15	385: Optical waveguides	889
16	422: Chemical apparatus and process disinfecting, deodorizing, preserving, or sterilizing	849
17	524: Synthetic resins or natural rubbers -- part of the class 520 series	793
18	204: Chemistry: electrical and wave energy	706
19	252: Compositions	655
20	313: Electric lamp and discharge devices	616

2.1 Award Data

The NSF funds science and engineering through research and education awards (grants, contracts, fellowships, and cooperative agreements). It accounts for about 20% of federal support to academic institutions for basic research (NSF, http://www.nsf.gov/home/grants.htm). In 2005, more than 5% of the NSF budget was dedicated to support nanotechnology research.

The nanotechnology-related NSF awards data set was provided by the NSF. The data set was created by matching the award title and abstract ("title-abstract" search) in the NSF's award database using a nanotechnology keyword list. Although this keyword list is slightly different from the one used for patent collection (Table 7-5), domain experts believe that they have similar coverage. To ensure the accuracy and coverage of our nanotechnology award data set, the large awards (over $500,000) were manually checked to ensure their relevance to nanotechnology. We also added awards that came from nanotechnology program solicitations that were missed by the keyword search.

We identified 3,891 nanotechnology-related awards out of the total 43,855 awards issued by the NSF between 2001 and 2004. These awards have 5,957 investigators and were made by 44 NSF divisions and 712 NSF programs. For both the patent and award collections, the majority of the documents were obtained by searching the keyword 'nano*' (91% of the patent collection and 88% of the award collection).

Table 7-5. Keyword lists used for patent and award collection.

Patent Collection Keywords	Award Collection Keywords
atomic force microscope	atomic force microscop*
atomic force microscopic	afm
atomic force microscopy	atom model*, atom* simulat*
atomic-force-microscope	
atomic-force-microscopy	
atomistic simulation	
biomotor	
molecular device	
molecular electronics	
molecular modeling	molec mod*, molec* simulat*
molecular motor	molec motor
molecular sensor	
molecular simulation	
nano*	nano*
quantum computing	nems
quantum dot*	quantum dot
quantum effect*	
scanning tunneling microscope	scanning tunnel*
scanning tunneling microscopic	stm
scanning tunneling microscopy	
scanning-tunneling-microscope	
scanning-tunneling-microscopy	
self assembled	self assembl*
self assembling	
self assembly	
selfassembl*	selfassemble*
self-assembled	
self-assembling	
self-assembly	

Tables 7-6 and 7-7 present the top 20 NSF divisions and programs, respectively, according to the numbers of nanotechnology-related awards issued during 2001-2004. The Division of Materials Research (DMR) was dominant with more than a quarter of the total number of nanotechnology awards, followed by the Division of Design, Manufacture and Industrial Innovation (DMI); Division of Chemistry (CHE); Division of Chemical and Transportation Systems (CTS); and the Division of Electrical and Communication Systems (ECS). The top 5 NSF Programs funding nanotechnology research were: Major Research Instrumentation (Program 1189); Electronics, Photonics, and Device Technologies (Program 1517); Small Business Phase I (Program 5371); Condensed Matter Physics (Program 1710); and Polymers (Program 1773).

Table 7-6. Top NSF divisions funding nanotechnology research: Number of awards (2001-2004).

Rank	Division	Number of Awards
1	DMR Division of Materials Research	997
2	DMI Div of Design, Manufac & Industrial Innov	505
3	CHE Division of Chemistry	466
4	CTS Division of Chemical & Transport Systems	408
5	ECS Division of Electrical & Communications Systems	381
6	CMS Division of Civil & Mechanical Systems	217
7	CCF Division of Computing and Communications Foundations	150
8	OISE/INT Office of International Science and Engineering	143
9	DMS Division of Mathematical Sciences	127
10	BES Division of Bioengineering & Environmental Systems	111
11	EEC Division of Engineering Education & Centers	104
12	CCR DIV OF Computer-Communications Research	93
13	PHY Division of Physics	77
14	MCB Division of Molecular & Cellular Biosciences	59
15	EAR Division of Earth Sciences	57
16	DBI Division of Biological Infrastructure	47
17	DUE Division of Undergraduate Education	46
18	EIA Division of Experimental & Integ Activit	30
19	EPS Office of Exper Prog to Stim Comp Rsch	20
20	DGE Division of Graduate Education	17

Table 7-7. Top NSF programs funding nanotechnology research: Number of awards (2001-2004).

Rank	Program		Number of Awards
1	1189	MAJOR RESEARCH INSTRUMENTATION	235
2	1517	ELECT, PHOTONICS, & DEVICE TEC	218
3	5371	SMALL BUSINESS PHASE I	202
4	1710	CONDENSED MATTER PHYSICS	164
5	1773	POLYMERS	104
6	4710	DES AUTO FOR MICRO & NANO SYS	100
7	1750	INSTRUMENT FOR MATERIALS RSRCH	97
8	1414	INTERFAC TRANS,& THERMODYN PRO	81
9	1676	NANOSCALE: EXPLORATORY RSRCH	80
10	1775	ELECTRONIC MATERIALS	78
11	1972	ELECTROCHEMISTRY & SURFACE CHE	75
12	1762	SOLID-STATE CHEMISTRY	62
13	1771	METALS	60
14	1788	NANOMANUFACTURING	54
15	1765	MATERIALS THEORY	53
16	1630	MECHANICS & STRUCTURE OF MATER	51
17	1415	PARTICULATE & MULTIPHASE PROCSS	50
18	1765	CONDENSED MATTER & MAT THEORY	49
19	5373	SMALL BUSINESS PHASE II	49
20	1674	NANOSCALE: INTRDISCPL RESRCH T	39

We observe that the dominant NSF divisions are related to material science, chemistry, and device design. This finding is consistent with the related technology areas reflected in the patent data. However, the dominant fields related to biology and pharmaceutical research in nanotechnology patents were not reflected in major NSF divisions. This difference may reflect the NSF's mission and respective decision to differentiate its funding scope in nanotechnology from other federal funding agencies such as the National Institutes of Health (NIH).

2.2 Linking NSF Award and USPTO Patent Data

The key linkage we rely on is the set of inventors who received funding from the NSF and filed nanotechnology patents (Huang et al., 2004). Specifically, we have identified the nanotechnology patent inventors who were also the principal investigators of nanotechnology-related NSF awards in past years. We refer to them as "PI-inventors" in the rest of the chapter. To provide a more complete assessment of the PI-inventors, we used the nanotechnology patent data from 1976 to 2004 and the NSF award data from 1991 to 2004 for inventor-investigator mapping. We identified 705 PI-inventors using name and institution matching in the award and patent data sets. Among these PI-inventors, there were 524 PI-inventors associated with 839 nanotechnology patents and 747 NSF awards in the time period from 2001 to 2004. In this study, we consider these patents to be the research contributions for the NSF-funded research conducted by the PI-inventors.

3. TREND ANALYSIS

Tables 7-8 and 7-9 show the nanotechnology-related USPTO patents and NSF awards publication trends between 1991 and 2004. New awards refer to the awards started in the respective year. Active awards refer to all awards receiving NSF funding in the respective year.

Table 7-8. Trend analysis: Numbers and percentages of nanotechnology-related USPTO patents (1991-2004).

Year	Number of Nano Patents for "full-text" Search	Number of Nano Patents for "title-claims" Search	Number of All Patents	Nano Patents Percentage for "full-text" Search	Nano Patents Percentage for "title-claims" Search
1991	1,864	290	107,259	1.74%	0.27%
1992	2,088	371	108,156	1.93%	0.34%
1993	2,303	346	110,540	2.08%	0.31%
1994	2,032	343	114,564	1.77%	0.30%
1995	2,734	452	114,764	2.38%	0.39%

Year	Number of Nano Patents for "full-text" Search	Number of Nano Patents for "title-claims" Search	Number of All Patents	Nano Patents Percentage for "full-text" Search	Nano Patents Percentage for "title-claims" Search
1996	2,719	471	122,953	2.21%	0.38%
1997	3,906	598	125,884	3.10%	0.48%
1998	4,989	752	166,801	2.99%	0.45%
1999	5,559	860	170,265	3.26%	0.51%
2000	5,884	952	176,350	3.34%	0.54%
2001	**7,015**	**1,219**	**184,172**	**3.81%**	**0.66%**
2002	**7,733**	**1,382**	**179,764**	**4.30%**	**0.77%**
2003	**8,630**	**1,670**	**187,147**	**4.61%**	**0.89%**
2004	**8,802**	**1,982**	**181,443**	**4.85%**	**1.09%**

Table 7-9. Trend analysis: Numbers and percentages of nanotechnology-related NSF awards (1991-2004).

Year	Number of Nano New Awards	Number of All New Awards	New Nano Awards Percentage	Number of Nano Active Awards	Number of All Active Awards	Active Nano Awards Percentage
1991	153	10,645	1.44%	222	27002	0.82%
1992	176	10,832	1.62%	383	32985	1.16%
1993	235	9,748	2.41%	542	35409	1.53%
1994	273	10,429	2.62%	736	37083	1.98%
1995	309	9,843	3.14%	941	37822	2.49%
1996	325	9,575	3.39%	1,083	37775	2.87%
1997	365	10,256	3.56%	1,226	37219	3.29%
1998	453	9,919	4.57%	1,390	37266	3.73%
1999	525	9,645	5.44%	1,586	37342	4.25%
2000	716	10,478	6.83%	1,905	38377	4.96%
2001	**823**	**10,485**	**7.85%**	**2,367**	**39365**	**6.01%**
2002	**910**	**10,923**	**8.33%**	**2,911**	**40693**	**7.15%**
2003	**1,255**	**11,592**	**10.83%**	**3,596**	**42924**	**8.38%**
2004	**1,172**	**10,893**	**10.76%**	**3,694**	**44209**	**8.36%**

Figure 7-1 shows the numbers of patents for "full-text" search and "title-claims" search, as well as the numbers of new awards and active awards between 1991 and 2004 reported in Tables 7-8 and 7-9. An almost identical number of nanotechnology patents (by both search methods) were published by USPTO between 2001 and 2004 (four years) and between 1991 and 2000 (ten years). More nanotechnology-related NSF awards were made between 2001 and 2004 than between 1991 and 2000.

Figure 7-2 shows the proportion of nanotechnology patents/awards as compared to the total number of patents/awards, respectively. We observe that the nanotechnology patents/awards percentage is increasing (see Table 7-7 and Figure 7-2). After 2003, more than ten percent of the new NSF awards were related to nanotechnology.

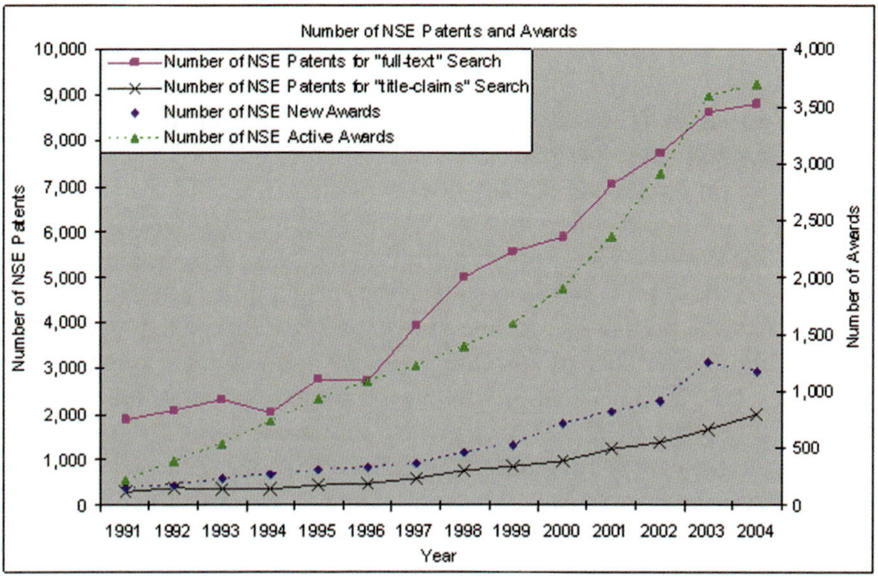

Figure 7-1. Trend analysis: Numbers of nanotechnology-related USPTO patents and NSF awards (1991-2004).

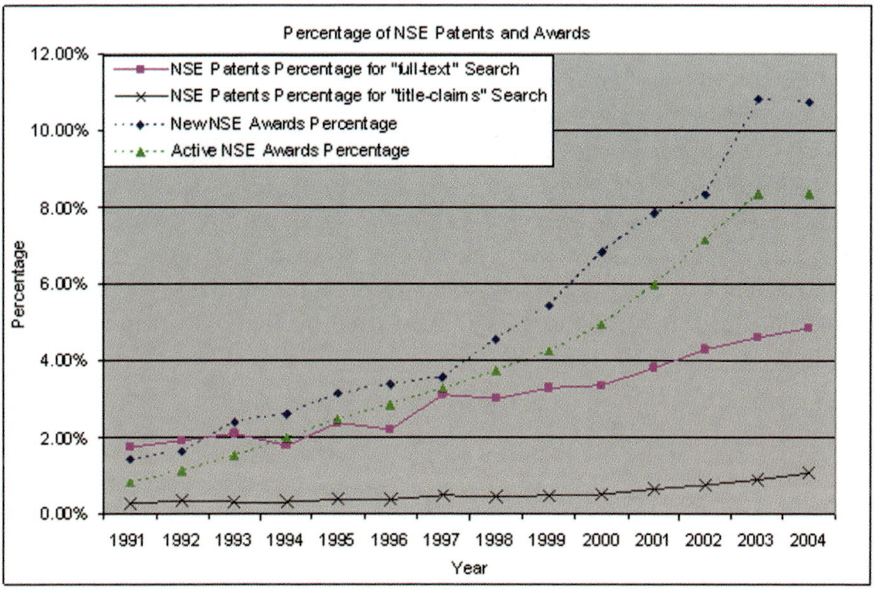

Figure 7-2. Trend analysis: Percentages of nanotechnology-related USTPO patents and NSF awards (1991-2004).

4. CONTENT MAP ANALYSIS

We prepared content maps for both the nanotechnology NSF awards and USPTO patents in order to identify the major nanotechnology research topics in different time periods. The nanotechnology award/patent content maps for the time intervals of 2001-2002 (Figures 7-3 and 7-4) and 2003-2004 (Figures 7-5 and 7-6) visualize the changes of topic areas from the previous time period. The growth rate of a topic area was computed as the ratio between the number of documents in the current time period and that of the previous time period. A baseline growth rate was computed as the ratio between the total number of documents in the current time period (2001-2002 and 2003-2004) and that of the previous time period. A topic region with similar growth rate to the base growth rate was assigned a green color that is consistent with the region color in the 2001-2002 content maps. The higher (lower) the growth rate of a topic region is, the warmer (colder) the assigned color. The red color indicates new research topics.

4.1 Content Map Analysis for 2001-2002

Figures 7-3 and 7-4 present nanotechnology award and patent content maps from 2001 to 2002 respectively.

Figure 7-3 shows that in 2001-2002, NSF-funded nanotechnology awards cover 26 topics, which are concentrated in "microscope"-related topics, "quantum"-related topics and "molecular"-related topics.

Figure 7-4 shows that nanotechnology patents cover broader technology topics (37 topics in total) including "optical fibers," "thin films," "nucleic acids," "electromagnetic radiation," "semiconductor substrates," "semiconductor devices," and "pharmaceutical compositions." As the topic labels are noun phrases extracted using the Arizona Noun Phraser, their capitalization varies. However, phrases with capitalization as well as morphological and inflectional variations were treated as the same phrase for the patent/award topic area representation used by the self-organizing map algorithm.

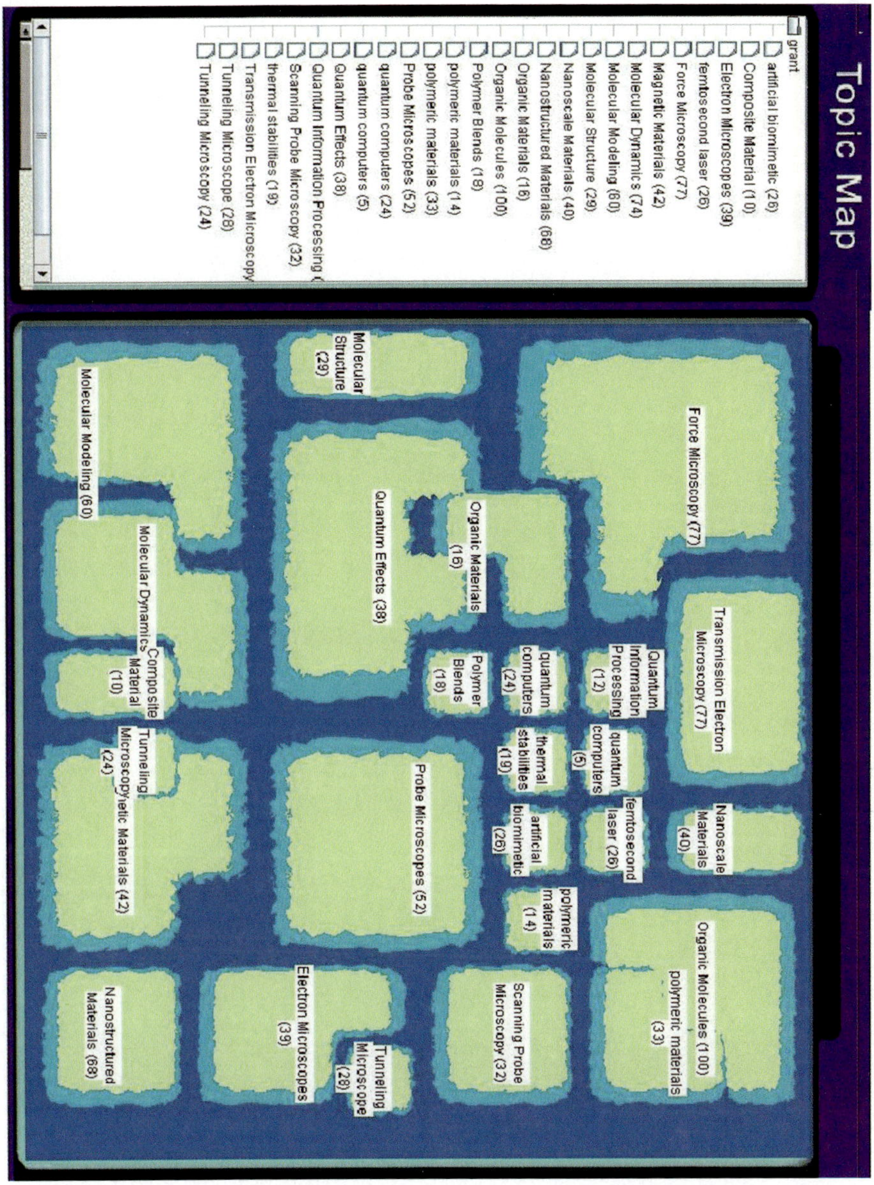

Figure 7-3. Nanotechnology NSF award content map (2001-2002).

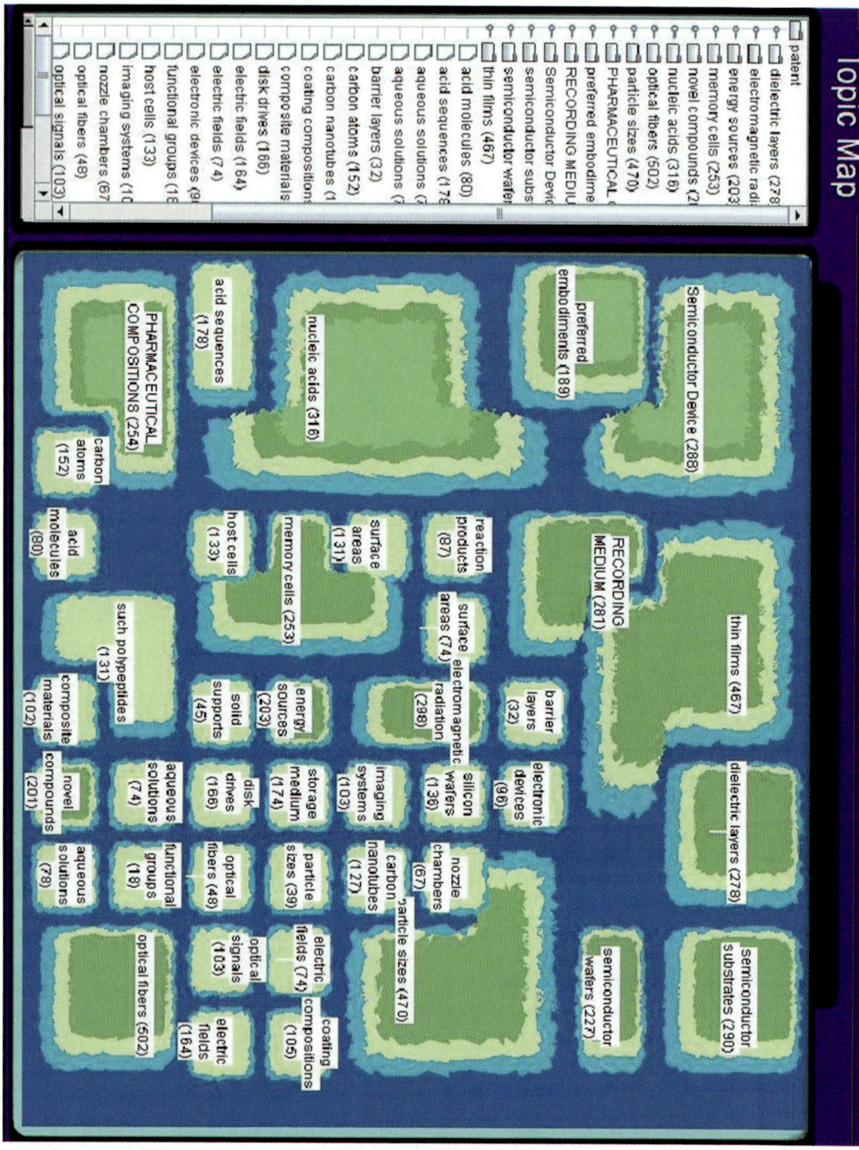

Figure 7-4. Nanotechnology USPTO patent content map (2001-2002).

4.2 Content Map Analysis for 2003-2004

Figure 7-5 presents the award content map for the years 2003 and 2004. With the color scheme described previously, Figure 7-5 displays many new award topics (shown as red regions), for instance, "condensed matter physics," "electronic devices," "optical properties," and "liquid crystal." The "microscope"-related, "quantum"-related and "molecular"-related topics are still active. The detailed topic region growth rates and base growth rate are presented in Table 7-10.

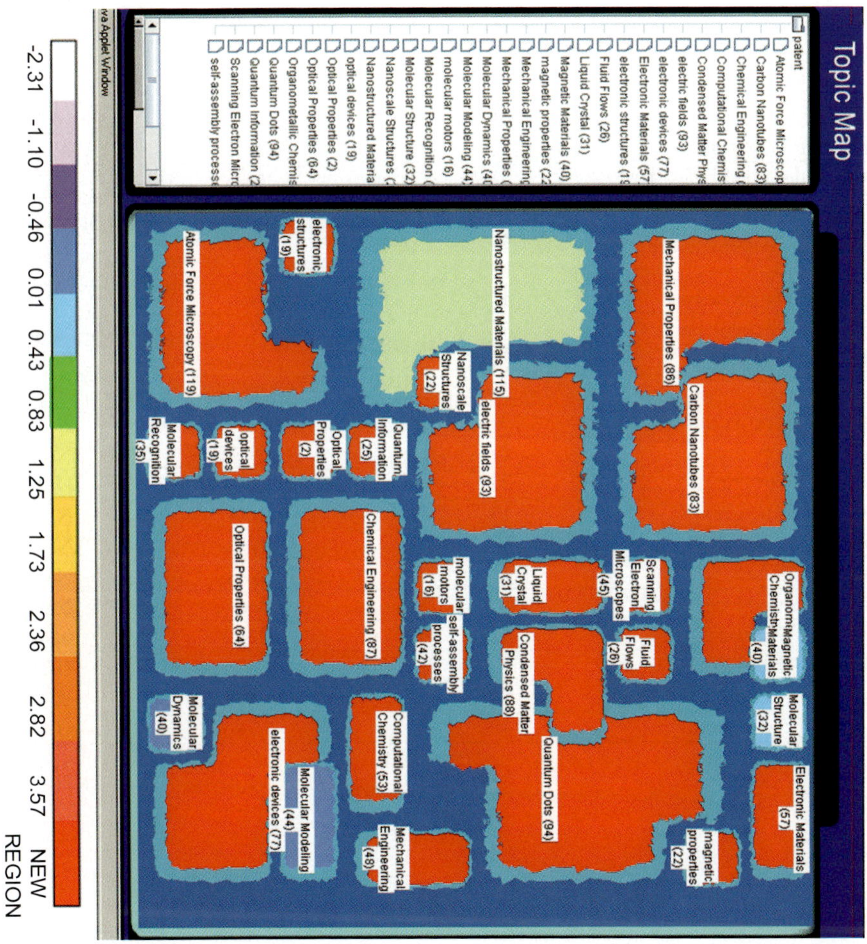

Figure 7-5. Nanotechnology award content map (2003-2004): The color scale shows the rate of increase of the number of awards in the respective topic.

Table 7-10. Nanotechnology award changes from (2001-2002) to (2003-2004).

Region Label	# of Grants in the region (2003-2004)	# of Grants in the region (2001-2002)	Growth Rate
Atomic Force Microscopy	119	N/A	N/A
Quantum Dots	94	N/A	N/A
electric fields	93	N/A	N/A
Condensed Matter Physics	88	N/A	N/A
Chemical Engineering	87	N/A	N/A
Mechanical Properties	86	N/A	N/A
Carbon Nanotubes	83	N/A	N/A
Organometallic Chemistry	80	N/A	N/A
electronic devices	77	N/A	N/A
Optical Properties	66	N/A	N/A
Electronic Materials	57	N/A	N/A
Computational Chemistry	53	N/A	N/A
Mechanical Engineering	48	N/A	N/A
Scanning Electron Microscopes	45	N/A	N/A
self-assembly processes	42	N/A	N/A
Molecular Recognition	35	N/A	N/A
Liquid Crystal	31	N/A	N/A
Fluid Flows	26	N/A	N/A
Quantum Information	25	N/A	N/A
magnetic properties	22	N/A	N/A
Nanoscale Structures	22	N/A	N/A
electronic structures	19	N/A	N/A
optical devices	19	N/A	N/A
molecular motors	16	N/A	N/A
Nanostructured Materials	115	68	0.69
Molecular Structure	32	29	0.10
Magnetic Materials	40	42	-0.05
Molecular Modeling	44	60	-0.27
Molecular Dynamics	40	74	-0.46

Region Label	# of Grants in the region (2003-2004)	# of Grants in the region (2001-2002)	Growth Rate
Composite Material	N/A	10	N/A
Quantum Information Processing	N/A	12	N/A
Organic Materials	N/A	16	N/A
Polymer Blends	N/A	18	N/A
thermal stabilities	N/A	19	N/A
artificial biomimetic	N/A	26	N/A
femtosecond laser	N/A	26	N/A
quantum computers	N/A	29	N/A
Scanning Probe Microscopy	N/A	32	N/A
Quantum Effects	N/A	38	N/A
Electron Microscopes	N/A	39	N/A
Nanoscale Materials	N/A	40	N/A
polymeric materials	N/A	47	N/A
Probe Microscopes	N/A	52	N/A
Tunneling Microscopy	N/A	52	N/A
Force Microscopy	N/A	77	N/A
Transmission Electron Microscopy	N/A	77	N/A
Organic Molecules	N/A	100	N/A
Baseline Growth Rate			*0.631*

Figure 7-6 presents the colored patent content map of the same time period. The dominant patent topics were “pharmaceutical compositions,” “semiconductor devices,” “optical fibers,” “nucleic acids,” “carbon nanotubes,” and “dielectric layers.” New patent topics include “metal oxides,” “magnetic layers,” “conductive materials,” “laser beams,” “amino acid sequences,” “refractive indexes,” and “silicon substrates.” Table 7-11 presents the detailed growth rate information. Similar to the topic areas in the award content map, many patent topic areas were new or had a higher growth rate than the base rate, indicating that the patents issued in 2003-2004 were dominated by new and growing topic areas.

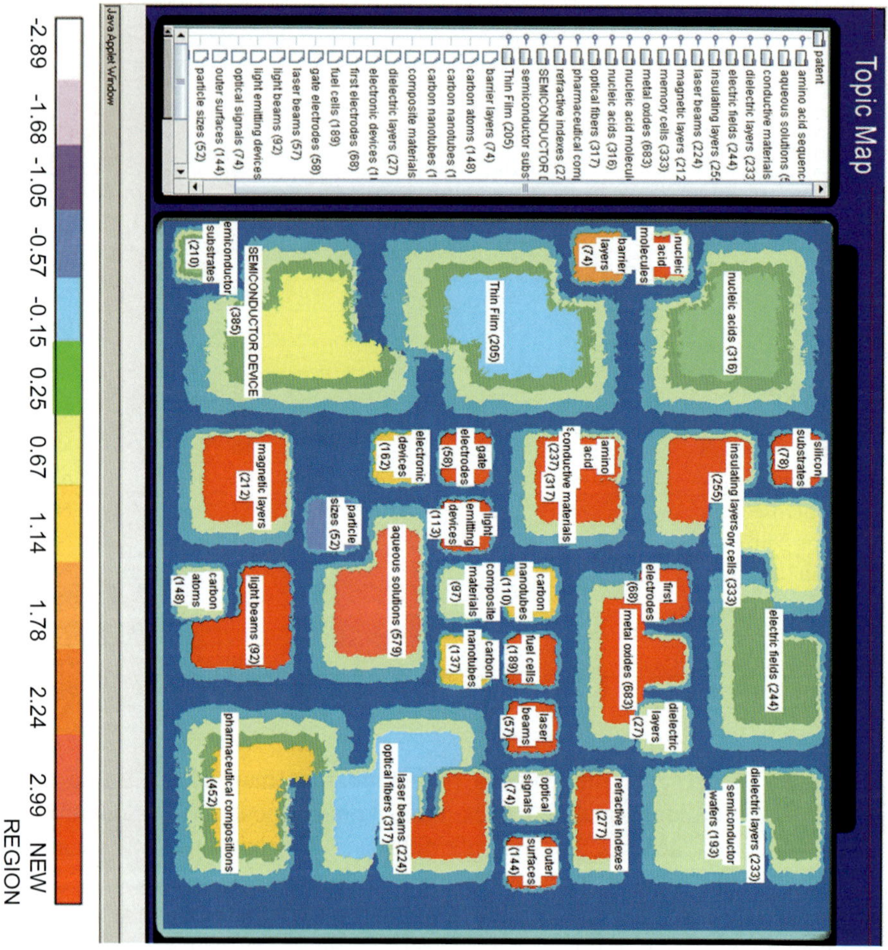

Figure 7-6. Nanotechnology patent content map (2003-2004).

Table 7-11. Nanotechnology patent changes from (2001-2002) to (2003-2004).

Region Label	# of Patents in the region (2003 - 2004)*	# of Patents in the region (2001 - 2002)*	Growth Rate
metal oxides	683	N/A	N/A
conductive materials	317	N/A	N/A
laser beams	281	N/A	N/A
refractive indexes	277	N/A	N/A
insulating layers	255	N/A	N/A
amino acid sequences	237	N/A	N/A
nucleic acid molecules	220	N/A	N/A
magnetic layers	212	N/A	N/A
fuel cells	189	N/A	N/A

Region Label	# of Patents in the region (2003 - 2004)*	# of Patents in the region (2001 - 2002)*	Growth Rate
outer surfaces	144	N/A	N/A
light emitting devices	113	N/A	N/A
light beams	92	N/A	N/A
silicon substrates	78	N/A	N/A
first electrodes	68	N/A	N/A
gate electrodes	58	N/A	N/A
aqueous solutions	579	178	2.25
barrier layers	74	32	1.31
carbon nanotubes	247	127	0.94
pharmaceutical compositions	452	254	0.78
electronic devices	162	96	0.69
semiconductor device	385	288	0.34
memory cells	333	253	0.32
electric fields	244	238	0.03
nucleic acids	316	316	0.00
carbon atoms	148	152	-0.03
composite materials	97	102	-0.05
dielectric layers	260	278	-0.06
semiconductor wafers	193	227	-0.15
semiconductor substrates	210	290	-0.28
optical signals	74	103	-0.28
optical fibers	317	550	-0.42
Thin Film	205	467	-0.56
particle sizes	52	509	-0.90
functional groups	N/A	18	N/A
solid supports	N/A	45	N/A
nozzle chambers	N/A	67	N/A
acid molecules	N/A	80	N/A
reaction products	N/A	87	N/A
imaging systems	N/A	103	N/A
coating compositions	N/A	105	N/A
such polypeptides	N/A	131	N/A
host cells	N/A	133	N/A
silicon wafers	N/A	136	N/A
disk drives	N/A	166	N/A
storage medium	N/A	174	N/A
acid sequences	N/A	178	N/A
preferred embodiments	N/A	189	N/A
novel compounds	N/A	201	N/A
energy sources	N/A	203	N/A
surface areas	N/A	205	N/A
recording medium	N/A	281	N/A
electromagnetic radiation	N/A	298	N/A
Baseline Growth Rate			*0.046*

We observe that among all 33 topics of the patents published in 2003-2004, 15 were new topics. Twenty-four of the 29 topics of the awards issued in 2003-2004 were new topics. The baseline growth rate of awards issued in 2003-2004 was 0.631, which is significantly higher than that of patents published in 2003-2004 (0.046). The topics in NSF awards changed at a faster rate than the topics in USPTO patents. This could indicate that topic changes of new awards were "upstream" of the topic changes in patents.

5. PATENT AND INVENTOR IMPACT ANALYSIS

In this section we analyze the patent and inventor impact based on their citation in patents. Changes over time are evaluated. We also compare different patent and inventor groups to assess the impact of different funding sources.

5.1 Measures

We use the number of times an inventor/patent is cited by others (number of cites) to measure its impact. In USPTO patent documents, the inventors are required by law to cite all important prior works on the pertinent topic. These citations eventually determine the scope of coverage of that patent. Although the debates on the value and quality of the patent citations are still ongoing, the number of citations received by a patent/inventor has been the core measure in evaluating the impact and quality of patents/inventors and the technology impact of institutions and countries. In our study, the number of cites for a patent is the number of later patents from which this patent receives citations. The number of cites for an inventor is the sum of the number of cites of all his/her patents. In our previous research, we also used a network structure-based Authority Score measure to evaluate patent/inventor impacts (Huang et al., 2003). As the 2001-2004 citation network is relatively sparse, such a network-based measure is not suitable in this case.

5.2 Patent Citation Growth

To compare the impact of recent nanotechnology research results with previous years, we report the citation growth of the nanotechnology patents. Figure 7-7 shows the average number of cites for the patents published in 1991-2004 as a function of the number of years after the date of publication. In this figure, we have included the 1991-2000 patent data for comparison purposes. We can observe that there are steady increases in the average numbers of cites after the patents' publication. In this graph, the slopes of the lines indicate the speed of citation accumulation. Generally, the citation accumulation speed of recently published patents is faster than that of the older patents, and the patents published in 2001-2004 follow this pattern. The more recent patents clearly have greater impact on future patents. Figure 7-8 details the average number of cites for the patents published between 2001 and 2004. In these four years, the patents have similar citation accumulation trends.

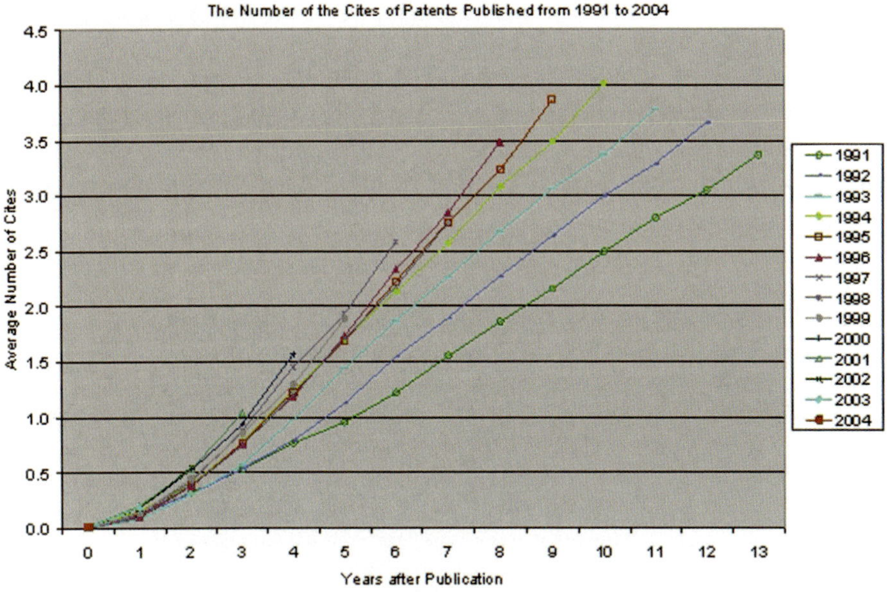

Figure 7-7. Average number of cites for nanotechnology patents published between 1991 and 2004.

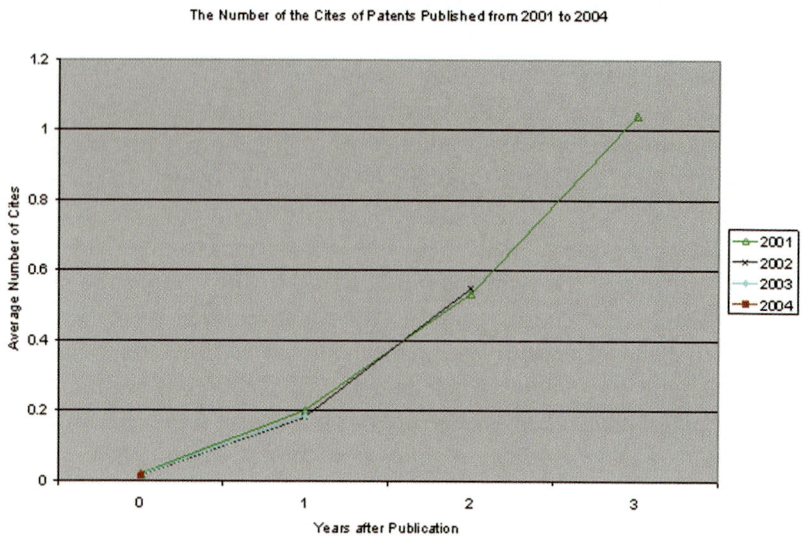

Figure 7-8. Average number of cites for patents published between 2001 and 2004.

The percentage of NSF patents cited by the nanotechnology patents in each year is shown in Figure 7-9. One can observe that there is an increasing trend of the percentage of NSF-supported patents in all the patents cited by nanotechnology patents, from 0.71% in 2000 to about 1.30% in 2004.

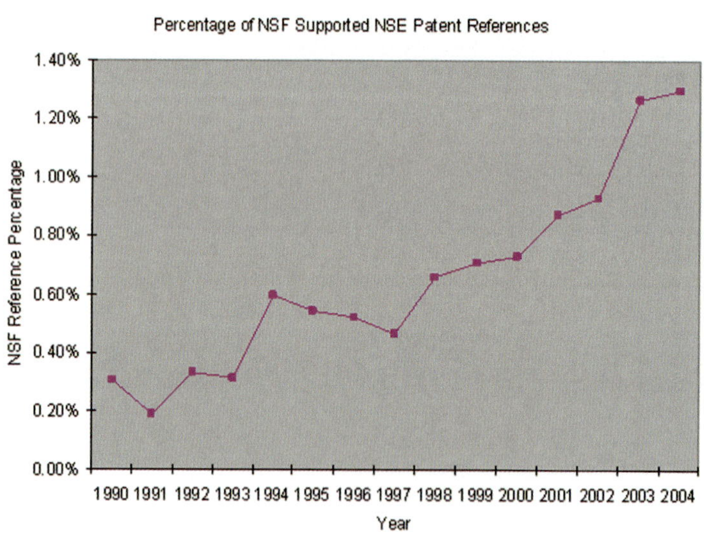

Figure 7-9. Percentage of NSF-supported patents by PI-inventors cited by all nanotechnology patents.

5.3 Comparison of the Impact of Different Groups

In order to compare the impact of publicly funded NSF research with the impact of other kinds of funding resources, we formed nine comparison groups (see explanations in Chapter 5):

- NSF: PI-inventors and their patents
- IBM: Inventors and patents of IBM
- Top10: Inventors and patents of top 10 institutions
- UC: Inventors and patents of the University of California
- US: Inventors and patents of the United States
- EntireSet: Inventors and patents of the entire patent data set
- Japan: Inventors and patents of Japan
- European: Inventors and patents of European countries
- Others: Random inventors and patents of countries other than the U.S., Japan, and European countries

Among these nine groups, only the NSF group represents a public funding source. The IBM group, which is the biggest assignee institution, represents the top commercial funding source. The UC group, which is the biggest academic assignee institution, represents the top funding source through academic channels. The U.S., Japan, and European groups are the main countries/regions in nanotechnology. The Top10 group represents the main research institutions in nanotechnology.

To assess in detail the status of the critical patents/inventors in recent years, we compared different groups' patents/inventors over three time periods:

- Patents published from 2001 to 2004
- Patents published from 2001 to 2002
- Patents published from 2003 to 2004

For the patents published between 2001 and 2002, the number of patent citations can be calculated in two ways:

(1) *restricted citation*: only the citations from patents issued in the same two years (2001-2002) are counted;

(2) *extended citation*: citations from all patents issued between 2001 and 2004 that cited 2001-2002 patents are counted.

One notes that the Top 10 assignee institutions, based on the number of cites, changes over time as illustrated in Table 7-12.

Table 7-12. Top 10 assignee institutions after the number of patent citations.

Rank	Assignee Name	Number of Patents
	2001-2004	
1	International Business Machines Corporation	730
2	Micron Technology, Inc.	620
3	Advanced Micro Devices, Inc.	519
4	The Regents of the University of California	461
5	Minnesota Mining and Manufacturing Company	396
6	Xerox Corporation	356
7	Intel Corporation	268
8	General Electric Company	239
9	NEC Corporation	234
10	Motorola, Inc.	227
	2001-2002	
1	International Business Machines Corporation	381
2	Micron Technology, Inc.	285
3	Advanced Micro Devices, Inc.	235
4	The Regents of the University of California	212
5	Minnesota Mining and Manufacturing Company	204
6	Xerox Corporation	191
7	NEC Corporation	149
8	Motorola, Inc.	116
9	Canon Kabushiki Kaisha	115
10	Silverbrook Research PTY LTD	115
	2003-2004	
1	International Business Machines Corporation	349
2	Micron Technology, Inc.	335
3	Advanced Micro Devices, Inc.	284
4	The Regents of the University of California	249
5	Intel Corporation	196
6	Minnesota Mining and Manufacturing Company	192
7	Xerox Corporation	165
8	Hewlett-Packard Development Company, L.P.	162
9	General Electric Company	138
10	Eastman Kodak Company	133

5.4 Statistical Analysis

We conducted a series of statistical hypothesis tests to compare the nine groups' impact in four settings: 2001-2004, 2001-2002 restricted citation, 2001-2002 extended citation, and 2003-2004.

(1) Hypotheses on the impact of patents (Figure 7-10):

a) Patents associated with PI-inventors had a higher number of cites measure than patents associated with other groups of inventors during 2001-2004.

b) Patents associated with PI-inventors had a higher number of cites measure than patents associated with other groups of inventors during 2001-2002, where we restricted the citing patents to those published between 2001 and 2002.

c) Patents associated with PI-inventors had a higher number of cites measure than patents associated with other groups of inventors during 2001-2002, including all citing patents published between 2001 and 2004.

d) Patents associated with PI-inventors had a higher number of cites measure than patents associated with other groups of inventors during 2003-2004, where we restrict the citing patents to those published between 2003 and 2004.

(2) Hypotheses on the impact of inventors (Figure 7-11):

a) PI-inventors had a higher number of cites measure than other groups of inventors during 2001-2004.

b) PI-inventors had a higher number of cites measure than other groups of inventors during 2001-2002, if we restricted the citing patents to those published between 2001 and 2002.

c) PI-inventors had a higher number of cites measure than other groups of inventors during 2001-2002, including all citing patents published between 2001 and 2004.

d) PI-inventors had a higher number of cites measure than other groups of inventors during 2003-2004.

We conducted ANOVA tests according to our eight hypotheses and analysis results are presented in Figures 7-10 and 7-11.

Figure 7-10a shows that the respective hypothesis was supported at the 95% level. During 2001-2004, NSF-funded PI-inventors' patents have a significantly larger number of cites measure (average about 1) than patents in other comparison groups such as IBM (0.56), Top10 (0.57), UC (0.64), and US (0.46). These four groups' average number of cites are significantly larger than the EntireSet average (0.41), while the Japan (0.35), European (0.22) and Other (0.34) groups have a smaller-than-average number of cites.

Figure 7-10b shows that the respective hypothesis was not supported at the 95% level. With the citations restricted to 2001-2002, there is no significant difference in the number of cites measure for the comparison groups NSF (0.17), Top10, and UC. These groups' numbers of cites are significantly larger than the European group and Others group.

Figure 7-10c shows that the hypothesis was supported at the 95% level. During 2001-2002, with the extended count of number of cites, NSF-funded

PI-inventors' patents have a significantly larger number of cites than patents in other comparison groups, followed by the groups IBM, Top10, UC, US, and EntireSet. The Japan and Others groups had a significantly smaller number of cites than these five groups and a significantly larger number of cites than the European group.

Figure 7-10d shows that the hypothesis was not supported at the 95% level. During 2003-2004, there is no significant difference in the number of cites measure of the comparison groups NSF, IBM, Top10, and UC. But the NSF group's number of cites is significantly larger than the US, EntireSet, Japan, Others, and European groups.

				95% CIs based pooled StDev	
Group	N	Mean	StDev	Lower	Upper
NSF	839	0.996	2.96	0.903	1.089
IBM	730	0.562	1.455	0.463	0.661
Top10	4049	0.571	1.692	0.529	0.613
UC	464	0.649	2.445	0.524	0.774
US	21120	0.465	1.502	0.447	0.483
Entire Set	32180	0.412	1.37	0.397	0.427
Japan	3889	0.35	1.165	0.307	0.393
European	3585	0.219	0.741	0.174	0.264
Others	3682	0.353	1.201	0.309	0.397

Pooled StDev = 1.428

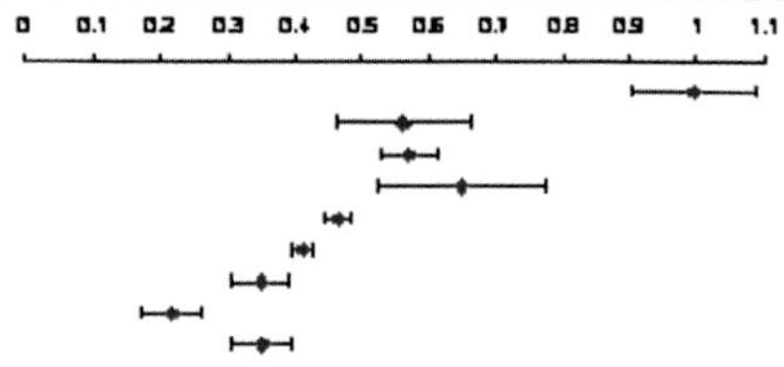

Figure 7-10a. Number of cites per patent, 2001-2004.

				95% CIs based pooled StDev	
Group	N	Mean	StDev	Lower	Upper
NSF	311	0.2026	0.6228	0.153	0.252
IBM	381	0.0997	0.4707	0.055	0.144
Top10	2003	0.1658	0.6254	0.146	0.185
UC	215	0.1628	0.8842	0.103	0.222
US	9760	0.1206	0.4948	0.112	0.129
Entire Set	14748	0.104	0.4456	0.097	0.111
Japan	1658	0.0965	0.3762	0.075	0.118
European	1594	0.0471	0.237	0.025	0.069
Others	1736	0.0703	0.3441	0.049	0.091

Pooled StDev = 0.4649

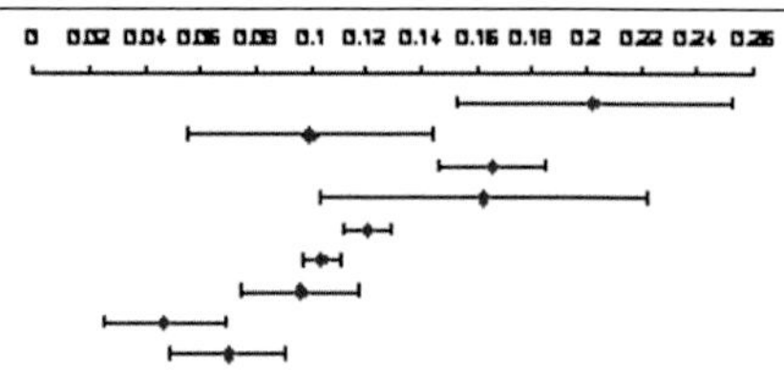

Figure 7-10b. Number of cites per patent, 2001-2002 with "restricted citations" in 2001-2002.

				95% CIs based pooled StDev	
Group	N	Mean	StDev	Lower	Upper
NSF	311	2.389	4.459	2.178	2.600
IBM	381	0.963	1.845	0.772	1.154
Top10	2003	1.06	2.293	0.977	1.143
UC	215	1.307	3.452	1.053	1.561
US	9760	0.881	2.08	0.843	0.919
Entire Set	14748	0.782	1.902	0.751	0.813
Japan	1658	0.698	1.86	0.606	0.790
European	1594	0.417	0.996	0.324	0.510
Others	1736	0.645	1.635	0.556	0.734

Pooled StDev = 1.979

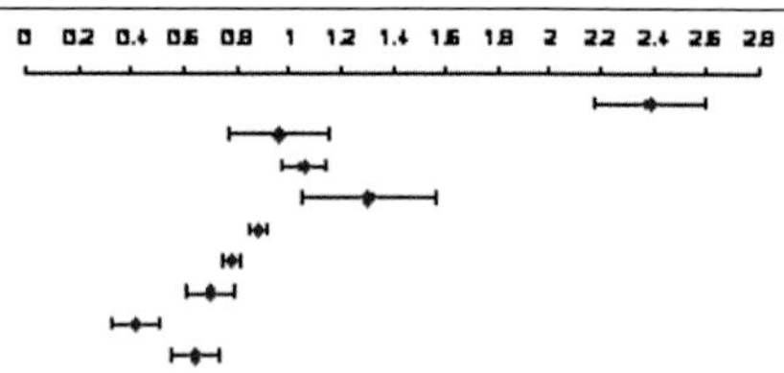

Figure 7-10c. Number of cites per patent, 2001-2002 with "extended citations" in 2001-2004.

				95% CIs based pooled StDev	
Group	N	Mean	StDev	Lower	Upper
NSF	528	0.1761	0.647	0.139	0.213
IBM	349	0.1232	0.5353	0.077	0.169
Top10	2202	0.1281	0.5229	0.110	0.146
UC	249	0.0803	0.4231	0.026	0.135
US	11360	0.1085	0.4529	0.100	0.117
Entire Set	17432	0.0989	0.4369	0.092	0.105
Japan	2241	0.0924	0.4096	0.074	0.110
European	1991	0.0608	0.3708	0.042	0.080
Others	1946	0.092	0.4481	0.073	0.111

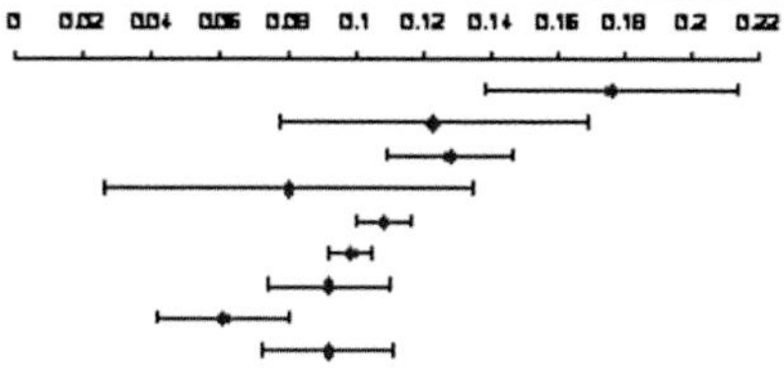

Pooled StDev = 0.4474

Figure 7-10d. Number of cites per patent, 2003-2004 with "restricted citations" in 2003-3004.

Figure 7-10. ANOVA results for hypothesis testing for citations per patent.

The hypothesis in Figure 7-11a was supported at the 95% level. During 2001-2004, NSF-funded PI-inventors have a significantly higher number of cites measure than inventors in other comparison groups, followed by the groups IBM, Top10, and UC. The US group had a significantly smaller number of cites than these three groups and a significantly larger number of cites than the EntireSet group. The Japan, Others, and European groups had a significantly smaller number of cites compared to other groups.

The hypothesis in Figure 7-11b was not supported at the 95% level. During 2001-2002, with the restricted count of number of cites, there is no significant difference in the number of cites measure of the comparison groups NSF, IBM, Top10, and UC. But these five groups' numbers of cites are significantly larger than the US, EntireSet, Japan, European, and Others groups.

The hypothesis in Figure 7-11c was supported at the 95% level. During 2001-2002, with the non-restricted count of number of cites, NSF-funded PI-inventors have significantly higher number of cites measures than inventors in other comparison groups, followed by the groups IBM, Top10, and UC. The US group had a significantly smaller number of cites than these three groups and a significantly larger number of cites than the EntireSet group. The Japan, Others, and European groups had a significantly smaller number of cites compared to other groups.

The hypothesis in Figure 7-11d was supported at the 95% level. During 2003-2004, NSF-funded PI-inventors have significantly higher number of cites measures than inventors in other comparison groups, followed by the Top10 group. The IBM, UC, US, EntireSet, Japan, and Others groups had a significantly smaller number of cites than the NSF and Top10 groups. The European group had a significantly smaller number of cites compared to other groups.

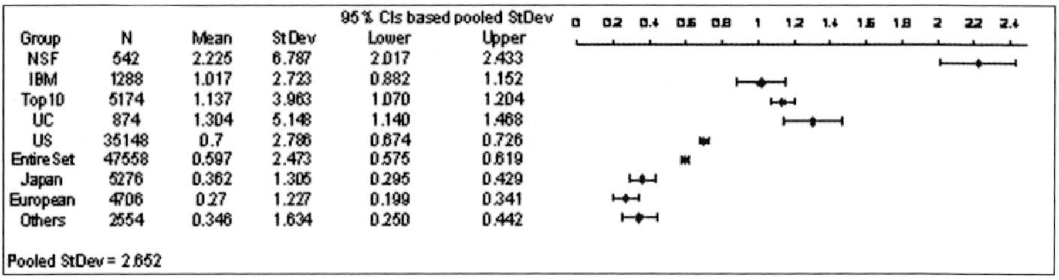

Group	N	Mean	St Dev	Lower	Upper
NSF	542	2.225	6.787	2.017	2.433
IBM	1288	1.017	2.723	0.882	1.152
Top10	5174	1.137	3.963	1.070	1.204
UC	874	1.304	5.148	1.140	1.468
US	35148	0.7	2.786	0.674	0.726
Entire Set	47558	0.597	2.473	0.575	0.619
Japan	5276	0.362	1.305	0.295	0.429
European	4706	0.27	1.227	0.199	0.341
Others	2554	0.346	1.634	0.250	0.442

Pooled StDev = 2.652

Figure 7-11a. Number of cites per inventor, 2001-2004 with citations in 2001-2004.

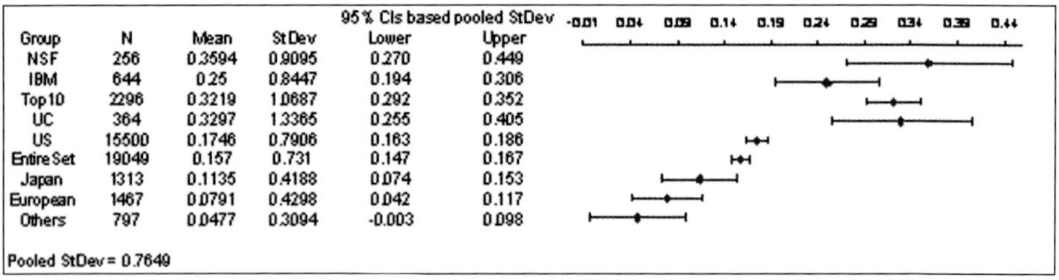

Group	N	Mean	St Dev	Lower	Upper
NSF	256	0.3594	0.9095	0.270	0.449
IBM	644	0.25	0.8447	0.194	0.306
Top10	2296	0.3219	1.0687	0.292	0.352
UC	364	0.3297	1.3365	0.255	0.405
US	15500	0.1746	0.7906	0.163	0.186
Entire Set	19049	0.157	0.731	0.147	0.167
Japan	1313	0.1135	0.4188	0.074	0.153
European	1467	0.0791	0.4298	0.042	0.117
Others	797	0.0477	0.3094	-0.003	0.098

Pooled StDev = 0.7649

Figure 7-11b. Number of cites per inventor, 2001-2002 with "restricted citations" in 2001-2002.

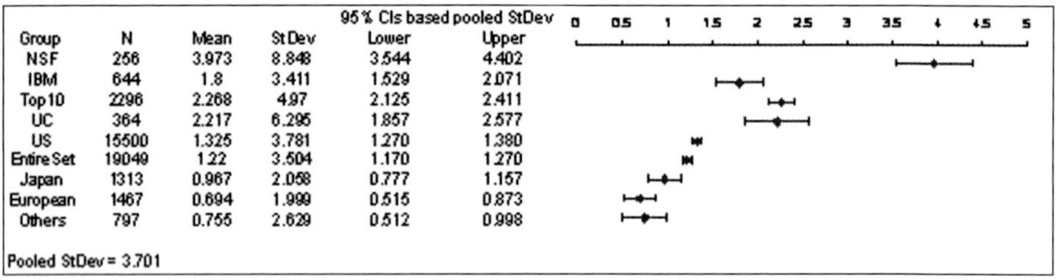

Group	N	Mean	St Dev	Lower	Upper
NSF	256	3.973	8.848	3.544	4.402
IBM	644	1.8	3.411	1.529	2.071
Top10	2296	2.268	4.97	2.125	2.411
UC	364	2.217	6.295	1.857	2.577
US	15500	1.325	3.781	1.270	1.380
Entire Set	19049	1.22	3.504	1.170	1.270
Japan	1313	0.967	2.058	0.777	1.157
European	1467	0.694	1.999	0.515	0.873
Others	797	0.755	2.629	0.512	0.998

Pooled StDev = 3.701

Figure 7-11c. Number of cites per inventor, 2001-2002 with citations in 2001-2004.

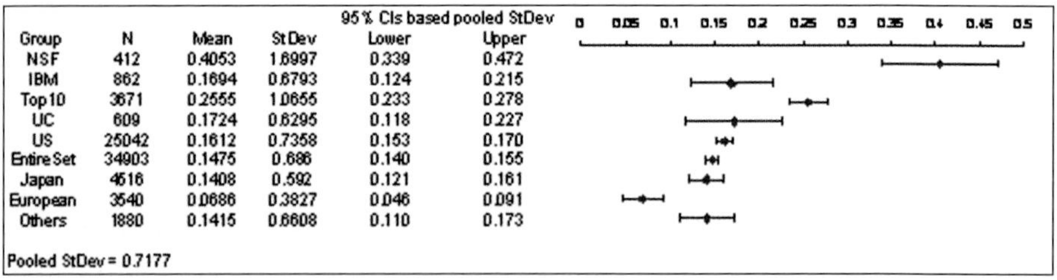

Group	N	Mean	St Dev	Lower	Upper
NSF	412	0.4053	1.6997	0.339	0.472
IBM	862	0.1694	0.6793	0.124	0.215
Top10	3671	0.2555	1.0655	0.233	0.278
UC	609	0.1724	0.6295	0.118	0.227
US	25042	0.1612	0.7358	0.153	0.170
Entire Set	34903	0.1475	0.686	0.140	0.155
Japan	4616	0.1408	0.592	0.121	0.161
European	3540	0.0686	0.3827	0.046	0.091
Others	1880	0.1415	0.6608	0.110	0.173

Pooled StDev = 0.7177

Figure 7-11d. Number of cites per inventor, 2003-2004 with "restricted citations" in 2003-2004.

Figure 7-11. ANOVA results for hypothesis testing for citations per inventor.

Table 7-13 summarizes the hypotheses testing results. As hypotheses 1a and 2a are supported, we can conclude that NSF-funded PI-inventors and their patents have higher impact than the inventors and their patents in other groups in 2001-2004. As hypotheses 1b and 2b are not supported, for patents published in 2001-2002 only considering the citing patents published in 2001-2002, PI-inventors do not have significantly higher impact than other groups. As hypotheses 1c and 2c are supported, we can conclude that PI-inventors and their patents are more influential than the inventors and their patents in other groups in 2001-2002 as time passes, considering the impact throughout 2001-2004. Only considering the patent citations in a two-year period may be too short to differentiate different patents' impacts. In the test on the third time period, 1d is not supported, which is consistent with the 2001-2002 results. But 2d is supported, which indicates that PI-inventors had stronger influence on nanotechnology patents in 2003-2004. Significantly larger numbers of cites for PI-inventors than other groups were observed even within the two-year time period. We observe that it is difficult to differentiate the influence level of inventor groups in a short time period such as 1-2 years. As the life of a patent increases, the NSF PI inventors' impact generally increases over other groups. This may be explained by the long-term fundamental research usually sponsored by the NSF.

Table 7-13. Hypotheses testing results.

Hypotheses	Patent		Inventor	
2001-2004	1a	YES	2a	YES
2001-2002 (restricted citation)	1b	NO	2b	NO
2001-2002 (extended citation)	1c	YES	2c	YES
2003-2004	1d	NO	2d	YES

6. CONCLUSIONS

The NSF support of nanotechnology research awards in 2001-2004 and its impact on technological innovation, as described by the USPTO patent data set, have been evaluated in this chapter using trend analysis, topic map analysis, and inventor/patent impact analysis. Key findings are:

The rate of increase in the number of NSF awards and USPTO patents in nanotechnology in 2001-2004 is higher than the respective increases in all technology fields. The percentage of nanotechnology patents has increased from 3.8% in 2001 to 4.9% in 2004 ("full-text" search) and from about 0.66% to about 1.09% ("title-claims" search). The corresponding percentage of NSF active awards with full or partial nanotechnology contents increased from 6.0% in 2001 to 8.4% in 2004.

The nanotechnology topics in both patents and NSF awards changed significantly in only four years. NSF awards covered a smaller number of topics than the USPTO patents in nanotechnology (26 versus 37 in 2001-2002 and 29 versus 33 in 2003-2004). NSF awards covered the topics related to microscopy, quantum effects, and molecular study. NSF awards covered fewer biology-related and pharmaceutical-related topics than USPTO patents did.

From 2001 to 2004, the NSF's PI-inventors and their patents had more impact than the inventors and patents in other groups based on the "number of cites" measure, suggesting the importance of fundamental longer-term research. In the shorter term (1-2 years), it is generally difficult to differentiate the impact level of inventor groups.

7. QUESTIONS FOR DISCUSSION

1. What impacts do other federal funding sources have on nanotechnology research?
2. What additional resources, in addition to patents, can be used to quantify science and technology impacts?
3. How can we measure the success of different funding programs within the NSF or other funding agencies?
4. How can we measure the success of research performed by individual principal investigators (PIs)?

Chapter 8

ACADEMIC LITERATURE CITATION IN PATENTS: A LONGITUDINAL STUDY OF USPTO PATENTS, 1976-2004

CHAPTER OVERVIEW

Academic nanotechnology research provides a foundation for nanotechnology innovation reported in patents. About 60% of the nanotechnology-related patents identified by "full-text" keyword searching between 1976 and 2004 at the United States Patent and Trademark Office (USPTO) have an average of 18 academic citations. In this research, we evaluated the most cited academic journals, individual researchers, and research articles in the nanotechnology area over a 29-year period. The most influential articles were cited about 90 times on the average, while the most influential author was cited more than 700 times by other nanotechnology-related patents. Thirteen mainstream journals accounted for about 20% of all citations. *Science*, *Nature,* and *Proceedings of the National Academy of Sciences (PNAS)* have consistently been the top three most cited journals. There are also influential specialty journals, represented by *Biosystems* and *Origin of Life*, which have very few articles cited but have exceptionally high frequencies of cites. The number of academic citations per year from the ten most cited journals has increased by over 15 times in the time period 1990-1999 as compared to 1976-1989, and again over 3 times in the time period 2000-2004 as compared to 1990-1999. This is an indication of the increased influence of academic research on the nanotechnology-related patents.

1. INTRODUCTION

Recent science and engineering developments in nanotechnology promise fundamental changes in a wide range of industries that will lead to applications such as new materials and nanomanufacturing, detecting and treating diseases, monitoring and protecting the environment, converting and storing energy, and building complex structures for electronic circuits or airplanes. Better understanding of the impact of nanotechnology academic research on technological innovation may help to better plan and justify the research and development (R&D) investments in this area.

Most patent offices require citations to general literature as well as to previously issued patents, and previous studies have focused primarily on the citations among patents. Patent documents contain a large number of citations to academic research articles (such as those listed in the "Other References" Section of USPTO patents).

Citation analysis of academic articles is well recognized since the introduction of the Science Citation Index (SCI) by Eugene Garfield (1964). Citation indices for journals, authors, and articles have analyzed in order to assess their impact and contribution to specific scientific disciplines (Garfield, 1972; 1978a; 1978b). Recent studies (Kostoff, 2006a; 2006b) have used citation analysis to identify high-impact nanotechnology-related articles based on the citation information of nanotechnology-related articles available in the Thomson SCI database.

Patent citations to academic research articles reported in this chapter may provide data for better understanding the connection between academic research and industrial innovation. We extracted nanotechnology patents issued between 1976 and 2004 from the U.S. Patent and Trademark Office (USPTO) database using selected keywords validated by domain scientists (see Table 4-1). An automatic citation parsing tool was adapted to obtain structured citation entries containing author, journal, and publication year information. The impact of academic journals, individual researchers, and research articles on patents issued between 1976 and 2004 was evaluated using this database. A major challenge for large-scale analysis of the patent-to-article citations is the lack of a uniform reporting standard in patent documents, resulting in noisy, unstructured, and often incomplete citation entries. We report in detail how we addressed these challenges in this chapter.

2. PATENT CITATIONS TO ACADEMIC LITERATURE

A total of 78,609 nanotechnology-related patents from the USPTO database were identified for 1976-2004 using "full-text" nanotechnology keyword search (title, abstract, claims, and specifications). A large subset -- 49,567 patents of the 78,609 nanotechnology patents -- contains entries in the "Other References" section. In this subset of patents, an average of 17.9 articles was cited per patent, with a total of 886,212 citations.

The patents searched by "full-text" include either nanotechnology products and processes usually referred to in the title, abstract and claims, or use of nanotechnology knowledge and tools usually referred to in the patent document specifications. Only a fraction of these patents claim new nanotechnology products and processes exploiting direct control and restructuring of matter at the nanoscale (those identified by "title-claims" keyword search). "Full-text" search may favor a more comprehensive coverage of several related nanotechnology fields. The reported results in this chapter are based on "full-text" search, except where it is specified otherwise.

For comparison purposes, we present in Table 8-1 the number of patents matching the reference keyword list by searching using both approaches. The "full-text" and "title-claims" searches respectively lead to 8,802 and 1,982 nanotechnology patents issued by the USPTO in 2004.

The numbers of nanotechnology patents recorded at USPTO between 1976 and 2004 from all countries and from the United States are shown in Figure 8-1. There is a good correlation between the "full-text" search (Figure 8-1a) and the "title-claims" search (Figure 8-1b). Since 1990, the number of patents extracted by "full-text" search has been 5 to 7 times larger than the "title-claims" search results. USPTO patents from all fields are plotted at a different scale (shown on the right Y-axis) for reference. The figures show that the nanotechnology patents grew significantly faster than the USPTO patents as a whole, especially after 1997. The number of nanotechnology patents (Figure 8-1a and b) increased about 70% between 2000 and 2004. This is significantly larger than the approximately 3% increase for patents in all fields (USPTO, 2004). In 2004, the nanotechnology patents represent 4.9% of the USPTO patents for all fields when using "full-text" search and 1.1% when using "title-claims" search.

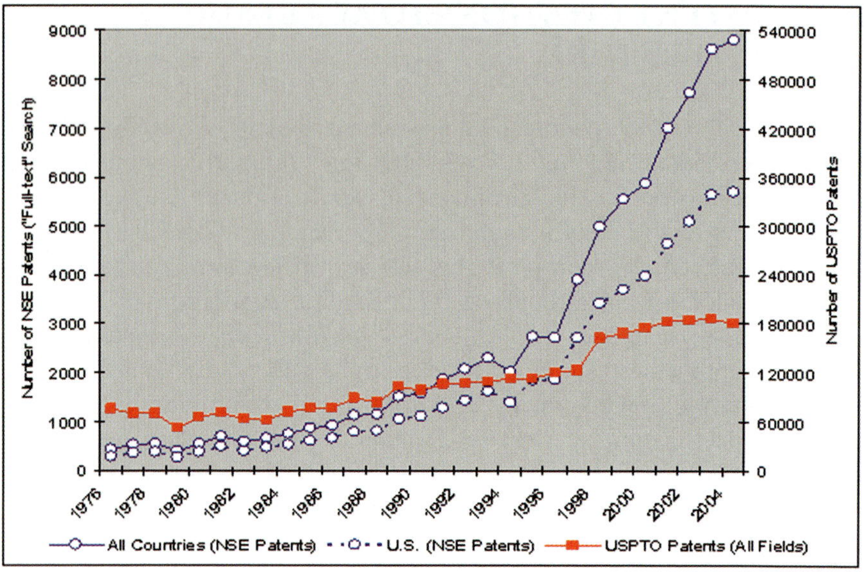

Figure 8-1a. Nanotechnology patents assigned to the U.S. as compared to all countries: "Full-text" nanotechnology keyword search.

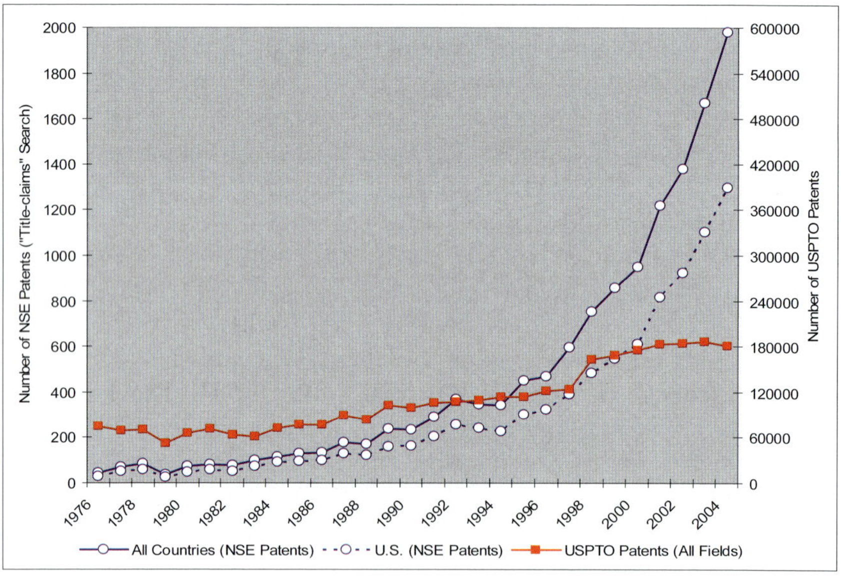

Figure 8-1b. Nanotechnology patents assigned to the U.S. as compared to all countries: "Title-claims" nanotechnology keyword search.

Table 8-1. Nanotechnology patents assigned to the U.S. as compared to all countries by "Full-text" and "Title-claims" nanotechnology keyword search.

Year	All Countries Nanotech Patents		U.S. Nanotech Patents		USPTO Patents (All Fields)
	Full-Text	Title-claims	Full-Text	Title-claims	
1976	429	43	291	30	75,521
1977	538	70	363	53	69,886
1978	557	84	372	58	70,604
1979	401	39	272	26	52,498
1980	556	76	386	50	66,219
1981	700	83	486	61	71,114
1982	579	77	396	51	63,307
1983	655	100	466	73	62,016
1984	754	115	535	93	72,681
1985	866	131	607	97	77,273
1986	927	136	663	100	77,041
1987	1130	178	794	132	89,598
1988	1154	171	822	124	84,439
1989	1521	237	1059	162	102,690
1990	1584	235	1113	164	99,220
1991	1864	290	1295	204	106,840
1992	2088	371	1445	256	107,511
1993	2303	346	1632	244	109,890
1994	2032	343	1402	227	113,704
1995	2734	452	1853	302	113,955
1996	2719	471	1866	325	121,805
1997	3906	598	2714	393	124,147
1998	4989	752	3430	486	163,206
1999	5559	860	3710	548	169,145
2000	5884	952	3985	612	176,083
2001	7015	1219	4656	818	184,047
2002	7733	1382	5104	926	184,425
2003	8630	1670	5654	1103	187,050
2004	8802	1982	5706	1300	181,322

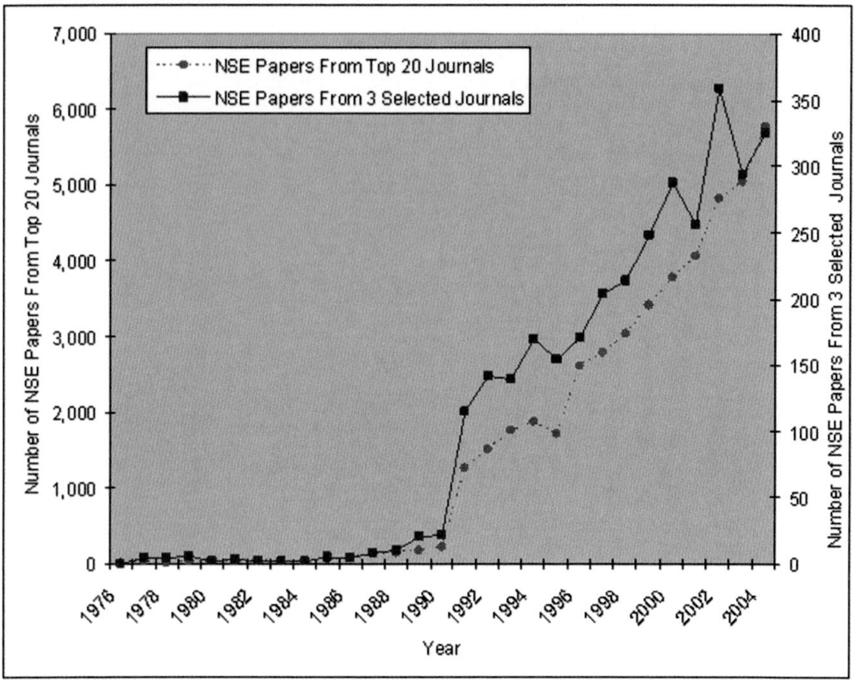

Figure 8-2. Nanotechnology papers from Top 20 journals and 3 selected journals. (The top 20 journals are based on the number of citations they received in 2004 from Science Citation Index (SCI). The 3 selected journals are *Science*, *Nature,* and *PNAS.*)

Table 8-2. Nanotechnology papers from Top 20 journals and 3 selected journals.*

Year	Nanotech Papers from Top 20 Journals			Nanotech Papers from 3 Selected Journals		
	Number of Nano Papers	Number of All Papers	Nanotech Paper Percentage	Number of Nano Papers	Number of All Papers	Nanotech Paper Percentage
1976	15	15,592	0%	0	3,692	0.00%
1977	23	16,549	0%	4	4,056	0.10%
1978	17	16,266	0%	4	3,998	0.10%
1979	25	16,938	0%	6	3,954	0.15%
1980	26	15,969	0%	2	2,490	0.08%
1981	36	16,560	0%	3	2,382	0.13%
1982	42	16,883	0%	2	2,299	0.09%
1983	30	17,135	0%	2	2,203	0.09%
1984	46	17,613	0%	2	2,055	0.10%
1985	59	19,947	0%	5	3,814	0.13%
1986	94	21,166	0%	4	3,973	0.10%
1987	128	21,392	1%	8	3,907	0.20%
1988	151	22,652	1%	10	3,964	0.25%

Year	Nanotech Papers from Top 20 Journals			Nanotech Papers from 3 Selected Journals		
	Number of Nano Papers	Number of All Papers	Nanotech Paper Percentage	Number of Nano Papers	Number of All Papers	Nanotech Paper Percentage
1989	176	23,669	1%	20	3,928	0.51%
1990	227	24,184	1%	22	3,985	0.55%
1991	1,269	25,465	5%	115	4,225	2.72%
1992	1,508	26,367	6%	142	4,475	3.17%
1993	1,754	27,844	6%	139	4,398	3.16%
1994	1,882	28,873	7%	169	4,515	3.74%
1995	1,713	29,502	6%	154	4,400	3.50%
1996	2,616	34,014	8%	170	4,702	3.62%
1997	2,791	34,016	8%	204	4,560	4.47%
1998	3,031	35,070	9%	214	4,683	4.57%
1999	3,419	35,705	10%	248	4,494	5.52%
2000	3,790	36,725	10%	288	4,624	6.23%
2001	4,070	37,724	11%	256	4,545	5.63%
2002	4,834	39,059	12%	359	4,823	7.44%
2003	5,048	38,859	13%	294	4,537	6.48%
2004	5,776	40,852	14%	325	4,866	6.68%

* The top 20 journals are based on the number of citations they received in 2004 from Science Citation Index (SCI). The 3 selected journals are *Science*, *Nature,* and *PNAS*.

Table 8-2 and Figure 8-2 show the increase in the number of nanotechnology papers in two groups of journals: (a) in the top twenty most cited journals (as ranked in 2004), and (b) in three leading journals (*Science, Nature*, and *Proceedings of the National Academy of Science*). Figure 8-2 shows that the nanotechnology papers grew significantly faster after 1991, about 6 years earlier than a similar change in the rate of increase of the nanotechnology patents. One may note that the NSF issued the first call for proposals on nanoparticles in 1990. The number of nanotechnology papers (Table 8-2) for the top twenty journals increased about 52% between 2000 and 2004 (about the same rate increase as for the nanotechnology patents in the same time interval, Table 8-1). The nanotechnology papers represent about 14% of all fields in 2004.

2.1 Patent-to-Article Text Parsing Challenges

The citations to academic articles in patent documents are not regulated by USPTO reporting standards. Many entries are arbitrarily composed and do not follow the common citation formats used in research articles. Many entries do not even contain complete information regarding the author(s), journal, and publication year. Figure 8-3 shows examples of different citation formats used in nanotechnology patents. Generally only the first author is reported for multiple-author articles.

Mountain TIBTECH 18:119-128 2000. cited by examiner.
EST database Accession No. Al307166, q179c06.x1 Soares_NhHMPu_S1 Homo sapiens cDNA clone IMAGE:1878538 3'. available Jan. 29, 1999. cited by examiner.
Pennica et al., 1998, PNAS USA 95:14717-14722. cited by examiner.
Esmon, The FASEB J., 1995; 9-946-955. cited by other.
Dillman, J. Clin-Oncol. 1994; 12:1497-1515. cited by other.
Blast Results A1-A20, Gen Bank. cited by other.
Pollack, Jonathan et al., "Microarray Analysis Reveals a Major Direct Role of DNA Copy Number Alteration in the Transcriptional Program of Human Breast Tumors," PNAS, 99:(20) 12963-12968 (2002). cited by other.

Figure 8-3. Example of articles cited in USPTO patents.

We adopted the well-tested open source toolkit ParaCite (Jewell, 2004) to parse the patent citation entries into several basic fields: author name, journal, and publication year. ParaCite was developed for EPrints and OpCit projects (Hitchcock et al., 2002). The toolkit was reported to achieve 79% accuracy for parsing about 2,500 citations (Huang, Hung et al., 2004). ParaCite employs a template matching mechanism. A citation entry is compared with a collection of 235 high-level, structured citation templates (e.g., [_AUTHORS_, _YEAR_, _JOURNAL_]). The best matching template is then used to divide the citation entry into subfields. For example, the template [_AUTHORS_, _YEAR_, _JOURNAL_] is the best match for the citation entry "*Pennica et al., 1998, PNAS USA 95:14717-14722. cited by examiner.*"

From the 886,212 raw citations, author names were successfully extracted for 317,445 citations (35.8%), journal names for 675,810 citations (76.3%), and publication year for 763,582 (86.2%). We manually checked the parsing results of sample citations. We found that the parsing results for author names and publication year are highly accurate, while there was some noise in the parsing results of journal names. We also found that most noise came from journal names that appeared fewer than 5 times. So for our analysis, we only included the parsed journal names which appear more than 5 times in our data set. Eventually 573,793 (64.7%) citations among the 886,212 raw citations had the journal names correctly parsed. Our analysis is inevitably biased toward those patents with well-formatted citations to academic articles.

Figure 8-4a shows the growth pattern of the total number of patent-to-article citations and the average number of citations per patent from 1976 to

2004. We observe that the average number of citations increased steadily from 1976 to 1993 and experienced a drastic increase from 1994 to 2002, reaching about 25 citations per patent in 2002. This demonstrates the significantly increased impact of academic research on industrial nanotechnology innovation, which may be due to increased funding and research productivity in the nanotechnology field during this time. We also observe a significant drop of average number of citations in 2003, back to about 20 citations per patent, followed by an increase in 2004 to more than 21 citations per patent. This drop was associated with a decrease in the total number of citations, from 130,848 in 2002 to 104,619 in 2003, while the number of patents continued to increase from 5,135 to 5,227. Further study is needed to identify the cause of this sudden drop in the average number of citations in patents issued in 2003. Figure 8-4b shows the growth pattern of nanotechnology-related patents and nanotechnology patents with citations to academic articles in the "Other References" section. We found that the number of nanotechnology-related patents with citations constantly increased over the entire time period, and the rate of growth significantly increased after 1994.

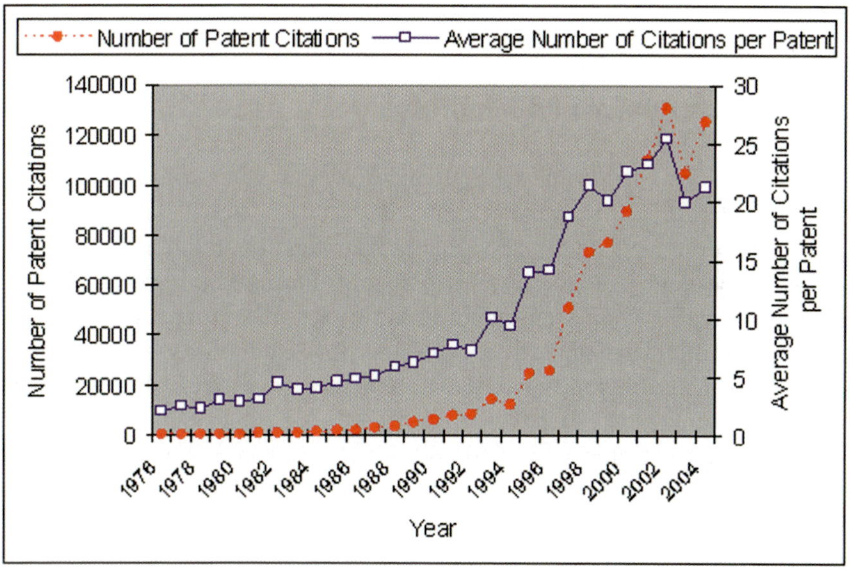

Figure 8-4a. The growth pattern of nanotechnology patent ("full-text" search) and article citations in patents: number of citations and average number of citations.

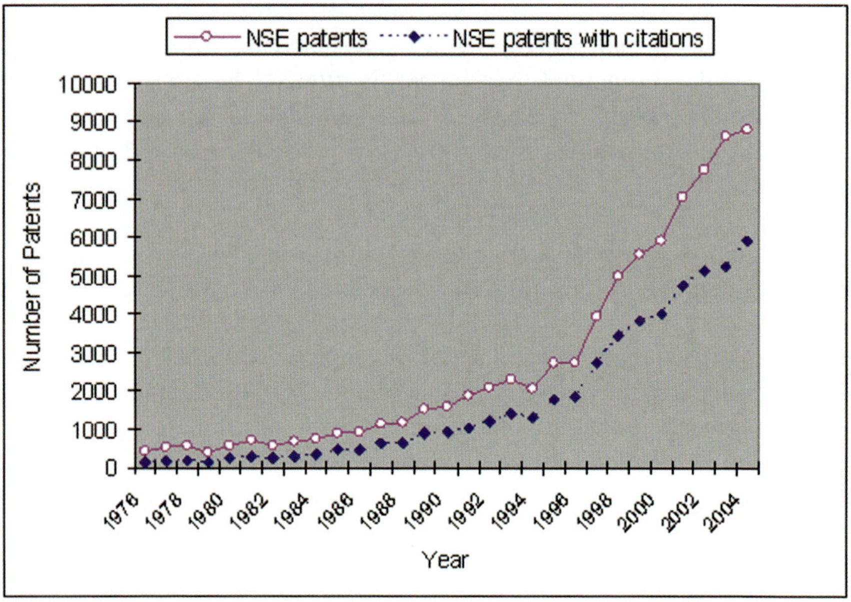

Figure 8-4b. The growth pattern of nanotechnology patent ("full-text" search) and article citations in patents: numbers of patents and patents with citations.

2.2 Technology Field Analysis

Table 8-3 shows the number of patent and citation statistics for the top 10 technology fields in the nanotechnology area. The technology fields correspond to the first-level U.S. Patent Classification categories (available at: http://www.uspto.gov/go/classification/selectnum withtitle.htm). We used the category name and its assigned classification ID number to label each technology field. The top 10 nanotechnology technology fields can be categorized into two groups based on the average number of citations per patent:

(a) technology fields 435, 514, 424, 536, 530 and 436 have average numbers of citations ranging from 27 to 39 (highlighted in bold face in Table 8-3). These technology fields are mostly in the chemical and pharmaceutical industries, with the academic foundation in biology and chemistry.

(b) technology fields 257, 438, 428, and 250 have relatively smaller average numbers of citations, ranging from 11 to 13. These technology fields are generally related to material and semiconductor industries, with the academic foundation in physics and electrical engineering.

The groupings based on citation statistics coincide with the groupings based on general scientific disciplines and industries. Our results demonstrate significant differences in the impact of academic research in these two groups of technology fields. Potential explanations for this disparity may relate to research productivity in these disciplines and the relationship between academic research and industrial innovation in these diverse fields.

Table 8-3. Statistics of nanotechnology patent citations in the top 10 technology fields (1976-2004).

Tech. field ID	Technology field name	No. of patents	No. of citations	Average number of citations per patent
435	Chemistry: molecular biology and microbiology	8846	311146	35.17
514	Drug, bio-affecting and body treating compositions	6411	177964	27.76
424	Drug, bio-affecting and body treating compositions	4821	159245	33.03
536	Organic compounds -- part of the class 532-570 series	4386	169411	38.63
530	Chemistry: natural resins or derivatives; peptides or proteins; lignins or reaction products thereof	4337	150212	34.64
257	Active solid-state devices (e.g., transistors, solid-state diodes)	3649	48561	13.31
438	Semiconductor device manufacturing: process	3335	37388	11.21
428	Stock material or miscellaneous articles	2885	35231	12.21
436	Chemistry: analytical and immunological testing	2861	80325	28.08
250	Radiant energy	2560	31868	12.45

In order to assess the longitudinal change in the impact of academic research, we performed analysis for three time periods between 1976 and 2004: (1) 1976-1989; (2) 1990-1999; and (3) 2000-2004.

These intervals roughly correspond to significant time periods in public funding activities in nanotechnology in the U.S. In 1990, the first NSF program solicitation focusing on nanoparticle synthesis and processes was issued. Then, in 1994-1995, NSF announced the National Nanotechnology User Network and program solicitation on instrumentation for nanotechnology, followed by the Functional Nanostructures initiative (1998). The last time period, 2000-2004, corresponds to the beginning of the National Nanotechnology Initiative (NNI, www.nano.gov).

Table 8-4 shows that the average number of literature citations per chemical/pharmaceutical patents (highlighted in bold face) was about four times larger during 1990-1999 than during 1976-1989. For the same time periods, the average number of citations was two to three times larger for material/semiconductor fields. During 2000-2004, the average number of citations per patent continued to increase in all fields at comparable rates. The number of citations per patent roughly doubled as compared to 1990-1999. The average number of citations for these top 10 fields in each year is also shown in Figure 8-5.

Table 8-4. Statistics of top 10 technology fields at different time periods.

Technology field ID	Average number of citations per patent		
	1976 - 1989	1990-1999	2000-2004
435	8.04	435	8.04
514	6.27	514	6.27
424	8.13	424	8.13
536	8.54	536	8.54
530	8.65	530	8.65
257	3.91	257	3.91
438	4.35	438	4.35
428	4.64	428	4.64
436	6.90	436	6.90
250	3.64	250	3.64

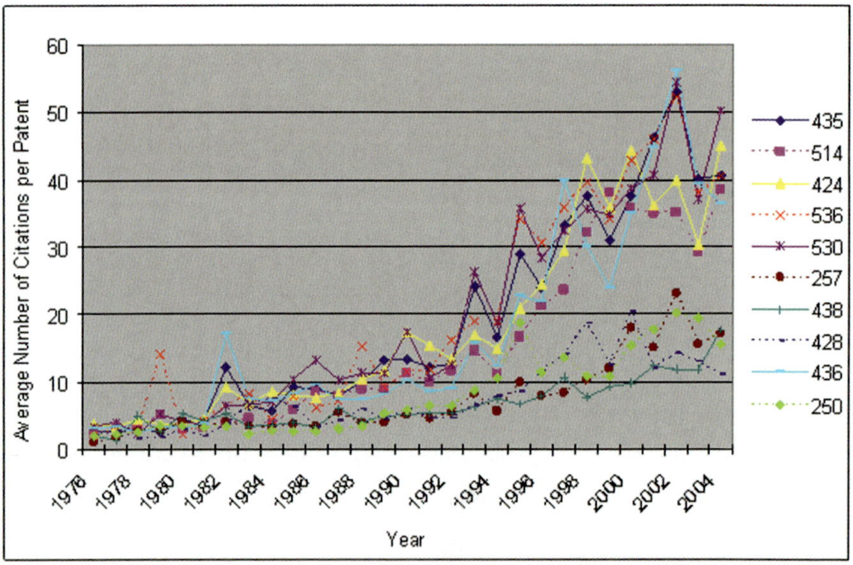

Figure 8-5. Average number of citations per patent for top 10 nanotechnology fields.

In addition to yearly fluctuations, it appears that in 2003 several top fields experienced a decrease in the average number of citations, which is closely related to the unusual drop in 2003 reported in Figure 8-4.

3. PATENT CITATIONS TO SIGNIFICANT RESEARCH ARTICLES, INDIVIDUALS, AND JOURNALS

Table 8-5. Top 26 most cited articles by nanotechnology patents from 1976 to 2004 (identified by the author name, journal, and publication year).

Rank	First initial	Last name	Journal	Year	Times cited
1	S	Guarini	EXPERIMENTIA	1983	91
2	I	Caramazza	THROMBOSIS RESEARCH	1991	89
3	A	Dal Pozzo	THROMBOSIS RESEARCH	1989	89
4	S.W	Fox	ARCHIVES OF BIOCHEMISTRY AND BIOPHYSICS	1960	89
5	S	Gurrieri	THERMOCHIMICA ACTA	1973	89
6	B	Heinz	BIOSYSTEMS	1981	89
7	G	Hennon	BIOCHIMIE	1975	89
8	Y	Ishima	BIOSYSTEMS	1981	89
9	W.W	Mcalhaney	BIOSYSTEMS	1976	89
10	T	Nakashima	BIOSYSTEMS	1981	89
11	J	Raymond	JOURNAL OF THE AMERICAN CHEMICAL SOCIETY	1994	89
12	J.W	Ryan	BIOSYSTEMS	1973	89
13	M.A	Saunders	BIOSYSTEMS	1974	89
14	S.W	Fox	NATURWISSENSCHAFTEN	1980	88
15	S	Guarini	PHARMACOLOGICAL RESEARCH COMMUNICATION	1985	88
16	J.R	Jungck	NATURWISSENSCHAFTEN	1973	88
17	C.J	Murphy	SCIENCE	1993	88
18	T	Nakashima	J.MOL.EVOL.	1980	88
19	S.W	Fox	BIOSYSTEMS	1976	87
20	S.W	Fox	ORIGINS OF LIFE	1974	87
21	L.L	Hsu	BIOSYSTEMS	1976	87
22	K	Matsuno	BIOSYSTEMS	1984	87
23	P	Melius	BIOORGANIC CHEMISTRY	1975	87
24	V.J.A	Novak	ORIGINS OF LIFE	1984	87
25	A.T	Przybylski	BIOSYSTEMS	1985	87
26	D.L	Rohlfing	SCIENCE	1970	87

Table 8-5 lists the top 26 research articles (eight articles are tied at 19–26 with 87 citations) most frequently cited by nanotechnology patents from 1976-2004. The most influential academic research papers in the nanotechnology field were cited by about 90 patents.

The top 20 authors most frequently cited by nanotechnology patents are listed in Table 8-6. We manually identified the institutions of these authors. The most frequently cited author was cited by more than 700 patents. Among these most cited authors, twelve are from the U.S., three from Japan, two from the U.K., and one each from Germany, Taiwan, and Italy, respectively. Half of them work in biology- and chemistry-related areas. This is consistent with our findings above that nanotechnology patents in biology- and chemistry-related technology fields have a larger average number of citations per patent. We also found that 12 of these top 20 authors produced 33 of the 100 most cited articles.

Table 8-6. Top 20 most cited authors by nanotechnology patents (1976-2004).

Rank	First initial	Last name	Institution	Times cited
1	SW	Fox	Department of Histopathology, St. George's Hospital Medical School, UK	707
2	DL	Rohlfing	Department of Biology, University of South Carolina, USA	435
3	J	Raymond	Department of Chemistry, University of Massachusetts, USA	389
4	H	Watanabe	RIKEN Genomic Sciences Centre, Japan	350
5	G	Binnig	Physics, IBM Zurich Research Laboratory, Switzerland	330
6	P	Melius	Department of Chemical Engineering, Auburn University, USA	266
7	K	Matsuno	Department of Biological Science and Technology, Tokyo University of Science, Japan	260
8	AT	Przybylski	NFCR Laboratory of the Institute for Molecular and Cellular Evolution, University of Miami, USA	252
9	S	Wolf	Genetics Institute, USA	240
10	G	Krampitz	Abteilung für Biochemie, Institut für Anatomie, Physiologie und Hygiene der Hasutiere, Universität Bonn, Germany	214
11	CR	Lowe	Institute of Biotechnology, University of Cambridge, UK	199
12	P	Boolchand	Department of ECECS, University of Cincinnati, USA	188
13	SM	Sze	National Nano Device Lab, Taiwan	187
14	S	Guarini	Department of Biomedical Sciences, University of Modena and Reggio Emilia, Italy	182
15	A	Heller	Department of Chemical Engineering, University of Texas at Austin, USA	179
16	T	Nakashima	Second Department of Surgery, Kagawa Medical University, Japan	179
17	X	Ma	Department of Mechanical Engineering, Texas A&M University, USA	179
18	LL	Hsu	Department of Pediatrics, Emory University, USA	176
19	J	Folkman	Department of Surgery, Children's Hospital, USA	159
20	Y	Degani	Department of Biological Chemistry, University of California at Los Angeles, USA	146

Figure 8-6 shows the journal distribution as a function of their nanotechnology patent-to-article citations. All journals were ranked by the number of citations they received from nanotechnology patents. We plotted the number of citations received by each journal (left panel) and the cumulative percentage of citations by adding individual journals according to their rankings (right panel). Our results show that nanotechnology patent-to-article citations follow a power law distribution with the exponent of −0.92. The top 13 journals accounted for about 20% of all citations; the top 172 journals accounted for about 50%; and about 1,700 journals accounted for 80%. This predominance of a small core group of journals in citations was also found in academic article citation analysis studies in various disciplines (Garfield et al., 1973; Wood, 1966).

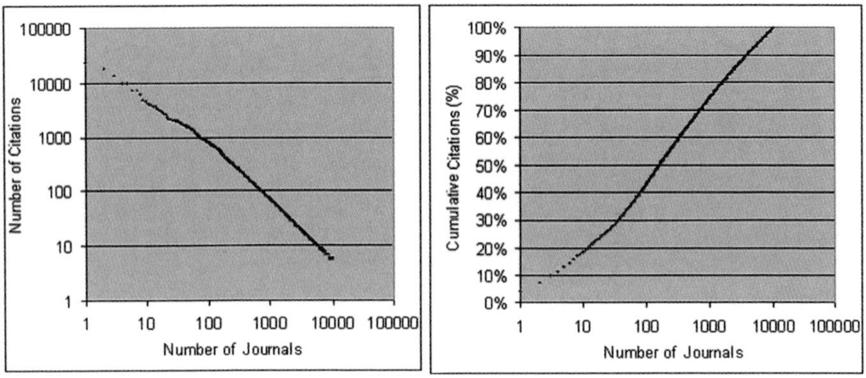

Figure 8-6. Journal distribution of nanotechnology patent citations (1976-2004).

Table 8-7 shows the top ten most cited journals for the entire time period (1976-2004) as well as the three time intervals identified earlier: 1976-1989, 1990-1999, and 2000-2004.

One notes that the top journals have remained at about the same position from 1976 to 2004 while their rankings slightly varied. *PNAS* and *Science* have stayed in the top three during all intervals, while *Nature* was among the top three from 1990 to 2004. The number of citations for these top 10 journals was constantly growing. Moreover, there was a sharp increase of citations between the first two time intervals, 1976-1989 and 1990-1999. These patterns are consistent with the growth pattern of nanotechnology patent citations presented in Figure 8-4. Figure 8-7 presents the number of citations from nanotechnology patents to the top 10 journals in each year from 1976 to 2004. We observe that the numbers of citations to *Science, PNAS, Nature, Journal of Biological Chemistry*, and *Cell* all dropped in 2003, which is consistent with our observation in Figure 8-4. In addition, the number of citations to *Science* and *Journal of the American Chemical Society* drastically decreased in 2004.

We also studied the journal citation distribution for individual technology fields and found that these top journals ranked similarly for different types of fields (chemical/pharmaceutical or material/semiconductor). The number of academic citations per year from the ten most cited journals has increased by over 15 times in the interval 1990-1999 as compared to 1976-1989, and again over 3 times in the interval 2000-2004 as compared to 1990-1999. This is an indication of the increased influence of academic research on technological innovation.

Table 8-7. Top 10 most cited journals by nanotechnology patents in different time periods (1976-2004).

	1976 - 2004		2000 - 2004		1990-1999		1976-1989	
Rank	Journal	Times cited	Journal	Times cited	Journal	Times cited	Journal	Times cited
1	PNAS	24400	SCIENCE	14935	SCIENCE	8865	PNAS	772
2	SCIENCE	21424	PNAS	13197	PNAS	7455	SCIENCE	600
3	NATURE	18678	NATURE	11376	NATURE	6768	JOURNAL OF BIOLOGICAL CHEMISTRY	549
4	JOURNAL OF BIOLOGICAL CHEMISTRY	15671	JOURNAL OF BIOLOGICAL CHEMISTRY	9386	JOURNAL OF BIOLOGICAL CHEMISTRY	5736	NATURE	534
5	JOURNAL OF THE AMERICAN CHEMICAL SOC	12882	JOURNAL OF THE AMERICAN CHEMICAL SOC	8226	BIOCHEMI-STRY	4428	BIOCHEMI-STRY	405
6	APPLIED PHYSICS LETTERS	11561	APPLIED PHYSICS LETTERS	7401	JOURNAL OF THE AMERI-CAN CHEMI-CAL SOC	4399	NUCLEIC ACIDS RESEARCH	402
7	NUCLEIC ACIDS RESEARCH	11351	BIOCHEMI-STRY	6728	APPLIED PHYSICS LETTERS	3742	JOURNAL OF THE AMERICAN CHEMICAL SOC	257
8	BIOCHEMI-STRY	10481	NUCLEIC ACIDS RESEARCH	6562	NUCLEIC ACIDS RESEARCH	3517	CELL	234
9	CELL	7650	CELL	4594	CELL	2777	APPLIED PHYSICS LETTERS	208
10	CHEMICAL ABSTRACTS	6738	CHEMICAL ABSTRACTS	4079	CHEMICAL ABSTRACTS	2511	CHEMI-CAL ABSTRACTS	148

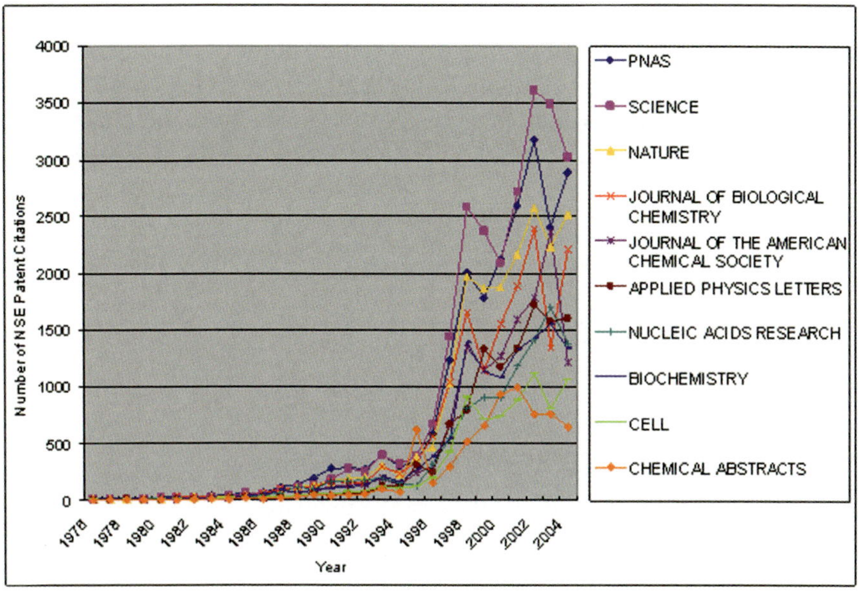

Figure 8-7. Yearly nanotechnology patent citation distribution of the top 10 most cited journals (1976-2004).

Table 8-8 shows the rank of journals where the top 100 articles most frequently cited by nanotechnology patents were published. This ranking turns out to be quite different from the overall ranking shown in Table 8-7. In particular, *Biosystems* was ranked first with 22 articles among the top 100 articles but was ranked 23rd in the overall ranking. *Origin of Life* was ranked second with seven top articles (tied with the overall second-place journal *Science*), but had an overall ranking of only 146. This finding contradicts the intuition that highly cited journals like *Science*, *PNAS,* and *Nature* would have a larger number of highly cited articles than others. Detailed analysis in Table 8-9 shows that *Biosystems* and *Origin of Life* only had 39 and 13 articles, respectively, ever cited by any nanotechnology patents, but about half of these articles were among the top 100 most cited articles, resulting in drastically larger average numbers of citations per article than other top journals. Based on our analysis, we believe there are two types of high-impact journals in the nanotechnology area:

(a) Broad-coverage, prestigious journals such as *Science*, *Nature,* and *PNAS* usually include a wide range of topics. Nanotechnology patents cited a large number of articles from these journals. However, on average each article was cited only about three times.

(b) Specialty journals such as *Biosystems* and *Origins of Life* usually focus on relatively narrow domains. A very small number of articles in these journals were cited heavily by nanotechnology patents. They were cited on average about 50 times, which is significantly greater than the citations per article in broad-coverage journals.

Table 8-8. Top 13 journals of the top 100 most cited articles by nanotechnology patents (1976-2004).

Rank	Journal	Number of articles among the top 100
1	BIOSYSTEMS	22
2	ORIGINS OF LIFE	7
3	SCIENCE	7
4	J.AM.CHEM.SOC.	6
5	ARCHIVES OF BIOCHEMISTRY AND BIOPHYSICS	3
6	J.MED.CHEM.	3
7	BIOCHIM.BIOPHYS.ACTA	2
8	CHEMISTRY AND INDUSTRY	2
9	IEEE TRANSACTIONS ON ELECTRON DEVICES	2
10	J.PHYS.CHEM.	2
11	NATURE	2
12	PROC.NATL.ACAD.SCI.USA	2
13	THROMBOSIS RESEARCH	2

Table 8-9. A comparison of the journals with the top 100 most cited articles.

Journal	The Number of Top 100 Most Cited Articles	Total Number of Citations	Total Number of Articles being Cited	Average Number of Citations per Article
Biosystems	22	1942	39	49.795
Origins of Life	7	590	13	45.385
Science	7	7359	2217	3.319
J.Am.Chem.Soc.	6	3510	1255	2.797
Nature	2	5714	2673	2.138
PNAS	2	4461	2121	2.103

4. CONCLUSIONS

In our research patent citations to academic articles were used to evaluate the impact of the scientific literature on technological innovation in nanotechnology from 1976-2004. The number of patents and article citations in patent documents has increased faster for the nanotechnology area than for all areas considered. Two distinct types of technology fields were identified in the nanotechnology area based on citation patterns: (a) the chemical/pharmaceutical fields with average patents citing about 40 articles;

and (b) the material/semiconductor fields with average patents citing about 10 articles.

The impact of articles, authors, and journals on the nanotechnology patents was investigated using selected automatic citation parsing and analysis tools. To the best of our knowledge, this is the first study that investigates the patent-to-literature connection for a major technical field over a long time period.

We found that the top articles most cited by the nanotechnology patents have about 90 citations each in the time interval 1976-2004. The most cited author, S. W. Fox from the Department of Histopathology at St. George's Hospital Medical School, London, U.K., was cited more than 700 times. The number of patent citations to journals follows a power law distribution. Thirteen journals accounted for about 20% of all citations. *Science*, *Nature,* and *PNAS* have consistently been the top three most cited journals, with an article being cited on average three times from 1976 to 2004. A few specialty journals, mainly represented by *Biosystems* and *Origin of Life*, had very few articles cited by nanotechnology patents but those cited averaged about 50 citations.

Research article citations by patents provide valuable information for assessing the impact of academic research on industrial innovation, especially for an emerging application such as nanotechnology that heavily relies on academic research. The findings on the difference in citation patterns for different nanotechnology subfields and the impact of articles, authors, and journals provide an overall picture of the longitudinal impact of academic nanotechnology research. Such analysis may be useful in providing insights to basic research funding policy and directions. The proposed analysis framework can be extended to other technology areas.

5. QUESTIONS FOR DISCUSSION

1. What are the most influential journals in various subfields of nanotechnology?
2. Which journals have become most influential in the past decade in nanotechnology and patenting?
3. What are some ways to improve the parsing accuracy of journal and author names?
4. What are the most influential and prestigious nanotechnology conferences and meetings?

Chapter 9

WORLDWIDE NANOTECHNOLOGY DEVELOPMENT: A COMPARATIVE STUDY OF USPTO, EPO, AND JPO PATENTS, 1976-2004

CHAPTER OVERVIEW

To assess the worldwide development of nanotechnology, we conducted bibliographic analysis, content map analysis, and citation network analysis on multiple patent offices' nanotechnology patents at three analytical unit levels (country, institution, and technology field). We compared the nanotechnology patents identified by keywords from the United States Patent Office (USPTO), European Patent Office (EPO), and Japan Patent Office (JPO). The numbers of nanotechnology patents published in USPTO and EPO have continued to increase exponentially since 1980, while those published in JPO stabilized after 1993. Institutions and individuals located in the same region as a repository's patent office have a higher contribution to the nanotechnology patent publication in that repository, which illustrates the "home advantage" effect. Bibliographic analysis on USPTO and EPO patents shows that researchers in the United States and Japan published larger numbers of patents than other countries, and that their patents were more frequently cited by other patents. Nanotechnology patents covered physics research topics in all three repositories. However, USPTO showed the broadest coverage in biomedical and electronics areas.

1. INTRODUCTION

Nanotechnology has been identified as a critical indicator of a country's technological competence. In 2000, the United States announced the National Nanotechnology Initiative (NNI, www.nano.gov) (Roco, Williams, and Alivisatos, 2000). Since then, more than 65 countries have adopted national projects or programs partially stimulated by the establishment of the NNI. Comprehensive assessments of nanotechnology development are not only necessary for research policy decisions but are also of interest to academic and industry communities.

Patent analysis has been used to assess innovation, research and development of a technology field (Karki, 1997; Narin, 1994; Oppenheim, 2000). Patent publication is managed by specialized offices in different countries. In general, inventors can file their patents in one or more countries. However, domestic applicants tend to file more patents with their home country patent office than do foreign applicants (European Commission, 1997). This "home advantage" effect influences the composition of the patents in the patent repositories. In previous research, Ganguli (1998) observed the home advantage effect in patents filed in USPTO, EPO, and JPO during 1995 and 1996. Criscuolo (2005) determined that patents filed by multinational enterprises in EPO and USPTO showed strong home advantage effects in every technology area. The existence of the home advantage effect indicates that individual patent office repositories may not provide comprehensive coverage of all patents in a technology domain.

Patent offices have different policies and examination procedures, which may also affect the patents filed in their repositories. For example, the USPTO's Duty of Disclosure, Candor, and Good Faith rule requires applicants to disclose and cite all prior related work of which they are aware. However, the EPO has no such requirement. Most EPO patent citations were added by examiners. For this reason, USPTO patents usually have more citations per patent than do EPO patents (Bacchiocchi and Montobbio, 2004). Also, a larger proportion of USPTO patent applications are granted than are EPO applications (Quillen, Webster, and Eichmann, 2002).

The inventors' preferences and the patent offices' policies affect both patent contents and the repositories' coverage. Hence, it is necessary to study the patents filed in multiple patent offices to obtain a comprehensive view of a technology area's development status. In other domains, patent analysis studies have combined data from different patent offices. For example, to determine the contribution of Italian professors to patents owned by science-based technological companies, Balconi et al. (2004) studied both USPTO and EPO patents. Lukach and Plasmans (2001) also examined both repositories in their study of inter-firm and intra-firm knowledge diffusion patterns using patents published by Belgian companies. However, in the nanotechnology field, few studies have employed multiple repositories to reveal its research and development status. Previous nanotechnology patent analyses may be biased by the characteristics of individual databases.

Our previous research showed that the United States, Japan, and European countries played an important role in worldwide nanotechnology research. Thus, we focus on the (English) nanotechnology patents at the USPTO, EPO, and JPO, which may cover a large cross-section of the nanotechnology research efforts in the world. Although several other countries have their own patent offices and accept patent filings in local languages, we do not consider them in this comparative research. We analyze the English patents documented in the three repositories using three types of analysis techniques: bibliographic analysis, content map analysis, and citation network analysis.

2. METHODS

Our research methodology contains three steps: data acquisition, patent parsing, and research status analysis (Figure 9-1).

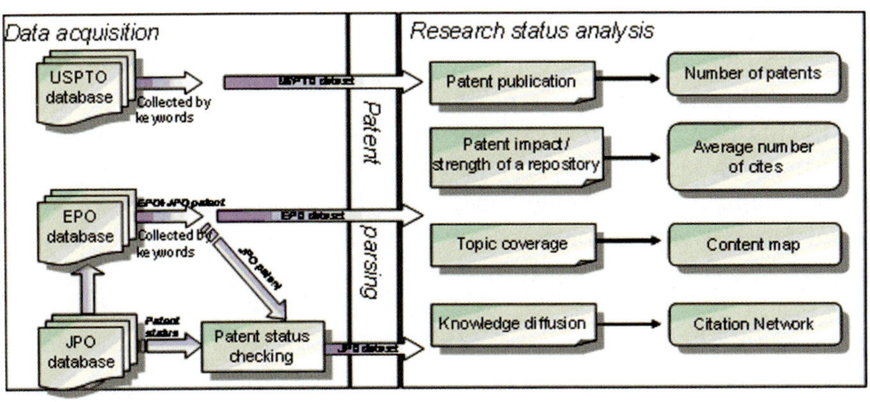

Figure 9-1. Research design.

2.1 Data Acquisition

The nanotechnology patents were identified in the USPTO, EPO, and JPO databases using the list of nanotechnology keywords given in Table 9-1 (same as those in Table 4-1) (Huang et al., 2003; Huang et al., 2004). Each of the three databases provides an online search interface. The USPTO database provides the full-text of the patents issued since 1976, which can be searched using almost all of a patent's data fields. The EPO repository, esp@cenet, documents EPO patents issued since 1978 together with patent applications from the patent offices of more than 70 countries. This system supports search on title, abstract, and some of the bibliographic data. The JPO official patent database, Patent Abstracts of Japan (PAJ), contains the patents issued since 1976.

System limitations make the PAJ database difficult to search. However, because its patents and patent applications are also stored in esp@cenet, we were able to retrieve JPO data from esp@cenet, check the retrieved patents' status (whether application or registered patent) through PAJ, and remove the ongoing applications from the data set.

In our previous research, we retrieved nanotechnology patents from the USPTO database by searching the nanotechnology-related keyword list in each patent's title, abstract, and claims ("title-claims" search) as well as in all patent data fields ("full-text" search) (Huang et al., 2004; Huang et al., 2003). Because of the search function limitations of esp@cenet, we can only collect nanotechnology patents in EPO and JPO by searching the keyword list in patent title and abstract ("title-abstract" search). To be comparable with the patents retrieved from these two databases, in this research we also collected the USPTO data set using "title-abstract" search.

Table 9-1 shows the number of patents retrieved in the three databases using different approaches. "Title-abstract" search provides significantly fewer search results than our previous methods ("title-claims" and "full-text"). However, the numbers of patents retrieved by different keywords are proportional to each other across the three search methods. After inspecting the data, domain experts believe that using the "title-abstract" search in the three databases is appropriate and can provide consistent results for making comparisons across the three repositories.

Table 9-1. Nanotechnology keywords and retrieved patents.

	USPTO (1976-2004)			EPO (1978-2004)	JPO (1976-2004)
	Full-text search	Title-claims search	Title-abstract search	Title-abstract search	Title-abstract search
atomic force microscope	2,309	375	241	69	62
atomic force microscopic	53	2	2	2	1
atomic force microscopy	1,679	103	68	21	0
atomic-force-microscope	6	0	0	0	0
atomic-force-microscopy	3	0	0	0	0

	USPTO (1976-2004)			EPO (1978-2004)	JPO (1976-2004)
	Full-text search	Title-claims search	Title-abstract search	Title-abstract search	Title-abstract search
atomistic simulation	5	0	0	0	0
biomotor	6	1	0	0	0
scanning tunneling microscope	1,097	205	145	47	79
scanning tunneling microscopic	21	1	0	0	1
scanning tunneling microscopy	809	50	28	8	0
scanning-tunneling-microscope	24	0	0	0	0
scanning-tunneling-microscopy	1	1	0	0	0
molecular device	164	13	6	4	3
molecular electronics	284	3	3	3	3
molecular modeling	1,787	37	26	2	1
molecular motor	88	3	0	3	0
molecular sensor	31	7	0	2	1
molecular simulation	43	2	2	1	1
nano*	72,762	12,220	4,497	2,024	635
quantum computing	66	25	19	4	1
quantum dot*	609	185	117	49	81
quantum effect*	563	58	36	16	65
selfassembl*	23	4	2	0	0
self assembly	1,802	192	121	33	0
self assembled	1,682	297	170	31	1
self assembling	877	158	111	48	5
self-assembly	1,625	173	108	0	5
self-assembled	1,587	277	158	0	0
self-assembling	807	147	102	0	0
Total	**90,813**	**14,539**	**5,962**	**2,367**	**945**
Unique Total	**78,609**	**13,463**	**5,363**	**2,328**	**923**

2.2 Patent Parsing

The nanotechnology patents retrieved from USPTO, EPO, and JPO are in free-text format. We first parse the free-text data into a structured format. Table 9-2 lists the data fields used in this research (the "applicant" in EPO and JPO is the same as "assignee" in USPTO). Because JPO patents do not contain assignee country and patent citation information, we are unable to study the publication trend of assignee country and country group and the citation networks of the JPO patents.

We use the classification codes assigned by each patent office to represent the patent's technology field. USPTO supports United States Patent Classification (USPC) and International Patent Classification (IPC) codes. EPO and JPO support European Patent Classification (EPC) and IPC codes. To compare the patents in the three databases, we adopt IPC as a representation of patent technology fields.

IPC is a five-level hierarchical ontology that contains eight first-level categories ("section"), 120 second-level categories ("class"), and 631 third-level categories ("subclass"). In this research we use the subclass IPC categories to represent technology fields in order to compare these results to those of our previous studies using the equivalent 462 class categories in USPC.

Table 9-2. Data fields of USPTO, EPO, and JPO patents.

	USPTO	EPO	JPO
Patent ID	√	√	√
Publication date	√	√	√
Inventor name	√	√	√
Assignee (applicant) institution name	√	√	√
Assignee (applicant) country	√	√	N/A
Patent classification code	IPC, USPC	IPC, EPC	IPC, EPC
Patent citation information	√	√	N/A
Title	√	√	√
Abstract	√	√	√
Claim	√	√	√
Description	√	√	√

2.3 Analysis of Research Status

We analyze the patents at three analytical levels: country (country group), assignee institution, and technology field (represented by subclass IPC categories). We assess the nanotechnology field's research status from four perspectives. First, we identify the number of patent publications by country (and country group), assignee institution, and technology field. Secondly, we use the average number of citations per patent to assess the degree to which different countries, assignee institutions, and technology fields have influenced and/or dominated the nanotechnology field in the three repositories. Thirdly, we generate the content maps of the three repositories' patents in different time intervals and compare which nanotechnology topics dominated each time period. Lastly, we analyze the citation network created by each country, assignee institution, and technology field to explore the knowledge diffusion of the invention process (as noted above, citation analysis could not be performed for JPO patents). Table 9-3 lists the types of analysis we were able to conduct in all three repositories.

Table 9-3. Research status analysis of USPTO, EPO, and JPO.

	USPTO	EPO	JPO
Patent publication trend			
Number of patents by country in each year	√	√	No
Number of patents by country group in each year	√	√	No
Number of patents by assignee institution in each year	√	√	√
Number of patents by technology field in each year	√	√	√
Patent impact			
Average number of cites by country	√	√	No
Average number of cites by assignee institution	√	√	No
Average number of cites by technology field	√	√	No
Topic coverage			
Content map analysis	√	√	√
Knowledge diffusion			
Country citation network analysis	√	√	No
Assignee institution citation network analysis	√	√	No
Technology field network analysis	√	√	No

3. DATA DESCRIPTION

USPTO: The USPTO database has more than 6.5 million patents with 3,500 to 4,000 newly granted patents each week. In May 2005 we conducted a "title-abstract" search in the USPTO database and collected 5,363 USPTO nanotechnology patents published between 1976 and 2004 (U.S. patents granted before 1976 do not have full-text access). These patents were submitted by 2,196 assignee institutions, 8,405 inventors, and 46 countries.

EPO: The EPO database has more than 1.5 million patents with more than 1,000 newly granted patents each week. The "title-abstract" search we conducted in esp@cenet in May 2005 identified 2,328 nanotechnology patents published between 1978 and 2004 (the EPO database does not document its patents before 1978). These patents were submitted by 1,168 assignee institutions, 5,400 inventors, and 43 countries.

JPO: The JPO database has more than 1.7 million patents with 2,000 to 3,000 newly granted patents each week. We retrieved the patents using "title-abstract" search in esp@cenet and checked their publication status in PAJ. We collected 923 JPO registered patents submitted by 348 assignee institutions and 1,729 inventors (JPO patents do not contain country information) between 1976 and 2004. Table 9-4 and Figure 9-2 present the numbers of nanotechnology patents published in the three repositories each year.

From the log scale graph of Figure 9-2, we observe that the numbers of nanotechnology patents in USPTO and EPO show an exponential growth pattern. JPO patents show a rapid increase before 1993 and a stable number of nanotechnology patent publications after that (about 50 to 86 patents per year).

Table 9-4. Number of nanotechnology patents collected by "title-abstract" search in USPTO (1976-2004), EPO (1978-2004), and JPO (1976-2004) in each year.

Year	Nanotechnology Patents		
	USPTO	EPO	JPO
1976	9	0	9
1977	18	0	18
1978	23	0	23
1979	11	0	11
1980	15	3	15
1981	25	3	25
1982	25	5	25
1983	24	6	24
1984	25	7	25
1985	33	7	33
1986	27	9	27
1987	46	10	46
1988	39	23	39
1989	65	24	65
1990	57	47	57
1991	85	35	85
1992	121	28	121
1993	123	42	123
1994	128	67	128
1995	160	73	160
1996	205	71	205
1997	238	93	238
1998	297	112	297
1999	367	125	367
2000	422	141	422
2001	524	235	524
2002	582	308	582
2003	739	364	78
2004	930	478	50

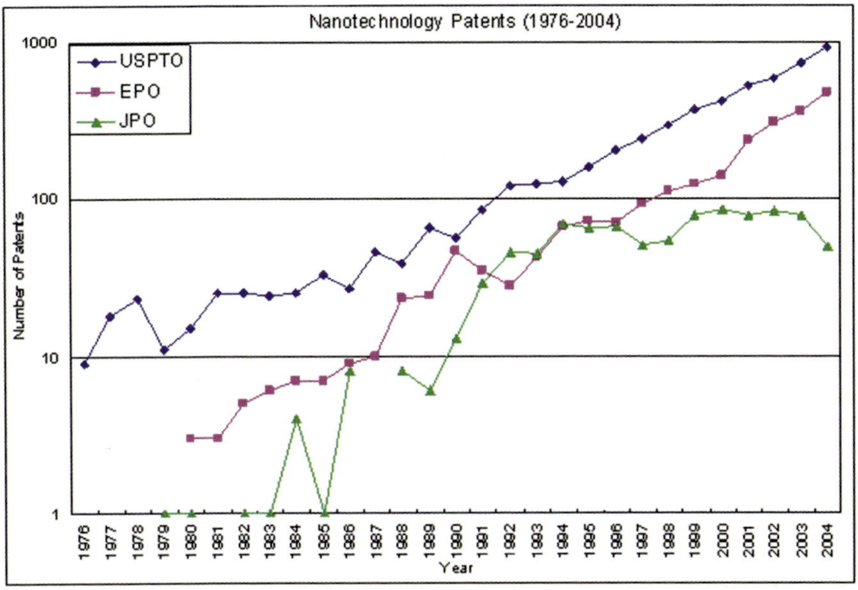

Figure 9-2. Number of nanotechnology patents collected by "title-abstract" search in USPTO (1976-2004), EPO (1978-2004), and JPO (1976-2004) in each year (log scale).

4. BASIC BIBLIOGRAPHIC ANALYSIS

We use the number of patents published by each country (and country group), assignee institution, and technology field to assess their productivity in nanotechnology. We analyze the impact using the average number of citations the patents generated. We counted a patent's number of citations only in the collected patent data set (using "title-abstract" search). We calculated the average number of citations for each country (country group), assignee institution, and technology field.

4.1 Country Analysis

4.1.1 Top Country Analysis

Tables 9-5 and 9-6 show the top 20 assignee countries which have the most nanotechnology patent publications in USPTO and EPO. The JPO patents do not contain assignee country information. We observe that the top 20 assignee countries and their rankings are very similar in both repositories, with the United States publishing the most patents in both databases. Although their ranks are reversed in USPTO and EPO, Japan and Germany have the most patent publications after the United States, followed by France and Republic of Korea. Canada and China (Taiwan) have much higher ranks and numbers of patents in the USPTO than in EPO; which may indicate that the inventors in these two countries prefer to file patents in USPTO. On the other hand, inventors in Switzerland prefer to file patents in EPO.

Table 9-5. Top 20 nanotechnology patent assignee countries in USPTO ("title-abstract" search, 1976-2004).

Rank	Assignee Country	Number of Patents
1	United States	3,450
2	Japan	517
3	Germany	204
4	France	156
5	Republic of Korea	131
6	Canada	104
7	China (Taiwan)	71
8	United Kingdom	60
9	Netherlands	54
10	Switzerland	41
11	Australia	34
12	Belgium	24
13	Israel	21
14	Italy	20
15	Ireland	20
16	Sweden	18
17	India	14
18	Spain	12
19	China	9
20	Singapore	8

Table 9-6. Top 20 nanotechnology patent assignee countries in EPO ("title-abstract" search, 1978-2004).

Rank	Assignee Country	Number of Patents
1	United States	925
2	Germany	343
3	Japan	323
4	France	201
5	Republic of Korea	98
6	Switzerland	77
7	United Kingdom	72
8	Netherlands	51
9	Belgium	42
10	Italy	34
11	Canada	26
12	Ireland	26
13	Spain	22
14	Israel	20
15	Sweden	18
16	Australia	14

Rank	Assignee Country	Number of Patents
17	Austria	13
18	China	9
19	Finland	7
20	India	7

Figures 9-3 and 9-4 show the yearly patent publication trends of the top 20 assignee countries in the USPTO and EPO in log scale. Many of the top assignee countries had an increasing trend of nanotechnology patent publication in both repositories. The United States filed more nanotechnology patents than other countries in almost every year, and the number of U.S. patents grew exponentially in both repositories. In the USPTO database, Japanese patents had a rapid growth before 1994 and then slowed. In the EPO database, Japanese patents remained relatively steady between 1989 and 2000. After 2000, the number of Japanese patents in EPO grew rapidly. Although the number of German patents in the USPTO was continuously increasing, the number published in the EPO remained stable after 2000. While the number of patents from France was consistently increasing in the EPO, the yearly publication of French patents in the USPTO decreased after 2002. The differences in these countries' publication patterns in the two repositories may indicate the inventors' change of interests in the two repositories.

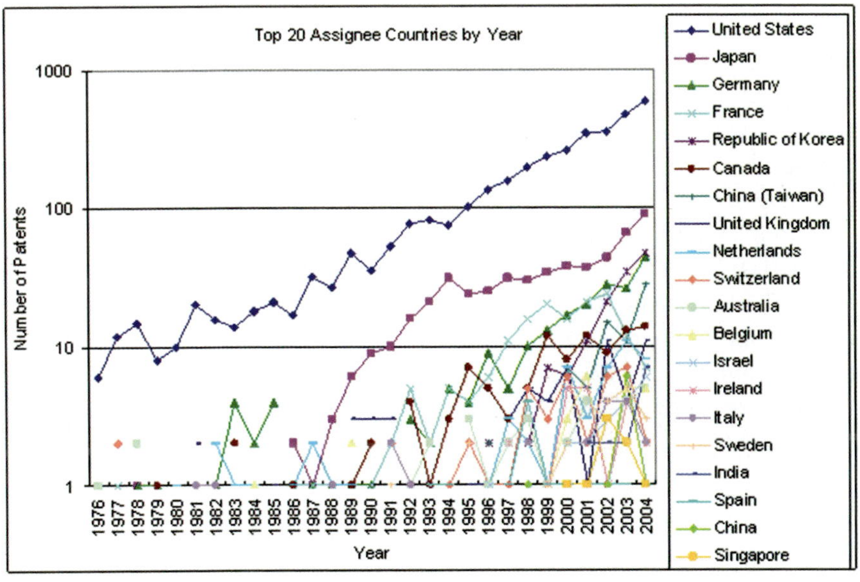

Figure 9-3. Top 20 nanotechnology assignee countries in USPTO and their patents by year ("title-abstract" search, 1976-2004) (log scale).

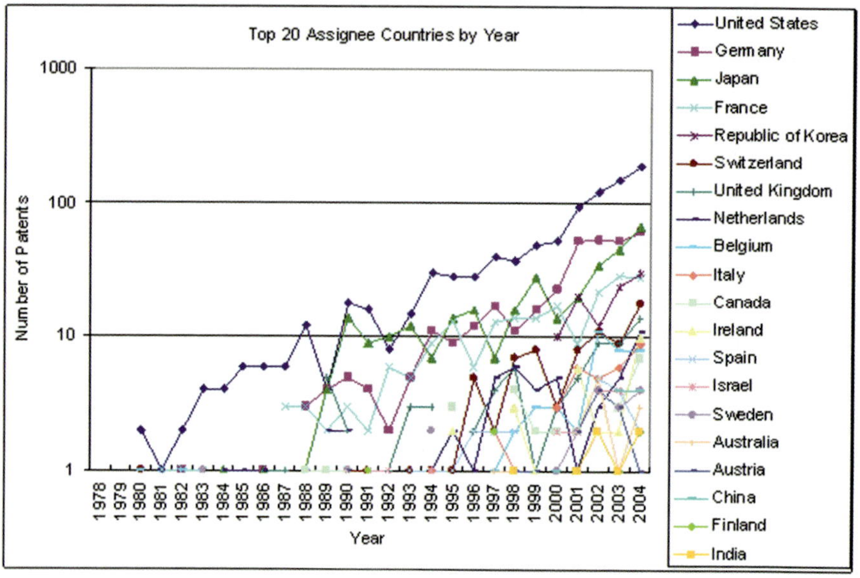

Figure 9-4. Top 20 nanotechnology assignee countries in EPO and their patents by year ("title-abstract" search, 1978-2004) (log scale).

4.1.2 Country Group Analysis

We organize and compare the patent publication status of four country groups: the United States, European group (25 countries in the European Union, plus Switzerland), Japan, and what we define as the Others group, which consists of all the other countries represented in these patent databases.

Table 9-7 and Figure 9-5 (in log scale) show the number of patents filed by the four country groups in the USPTO database from 1976 to 2004. In this database, the United States filed more patents than the other three groups. The European group, Japan, and the Others have similar numbers of nanotechnology patents each year. Table 9-8 and Figure 9-6 (in log scale) show the number of patents filed by the four country groups in the EPO database from 1978 to 2004. From the graph, we observe that the numbers of patents filed by the United States and European group countries were at the same level in EPO. The numbers of patents filed by Japan and Other countries were at the same level after 1998. In general, these latter two groups filed fewer patents than the United States and the European group.

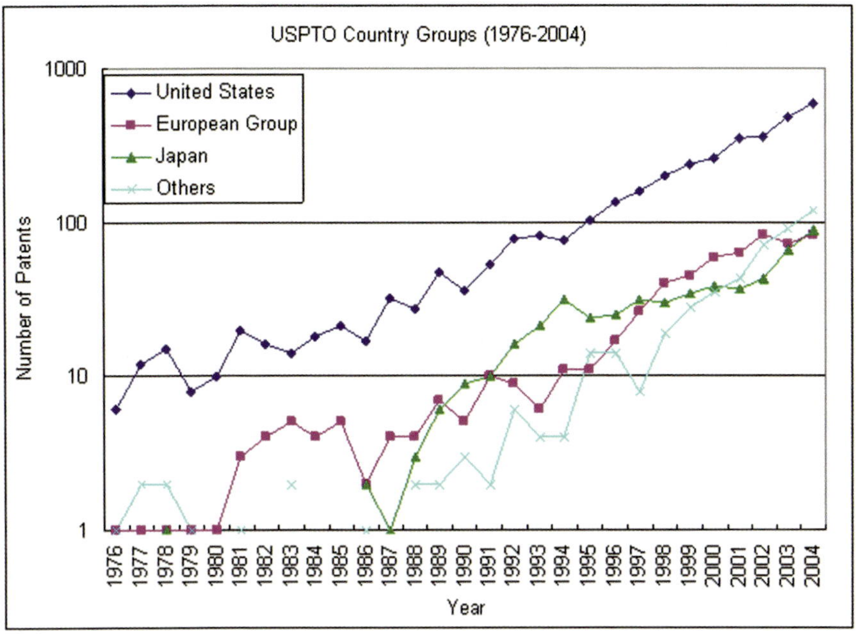

Figure 9-5. Patent publication trend by assignee country group in USPTO database ("title-abstract" search, 1976–2004) (log scale).

In both USPTO and EPO, the number of nanotechnology patents published by the four country groups increased. Comparing Figures 9-5 and 9-6 and Tables 9-7 and 9-8, we see that the United States filed many more nanotechnology patents in USPTO than in EPO. On the other hand, European group countries filed a few more patents in EPO than in USPTO. Japan and the Others countries filed similar numbers of patents in both databases. These phenomena show that inventors tend to file patents in their own countries' patent offices more often than in foreign countries' patent offices.

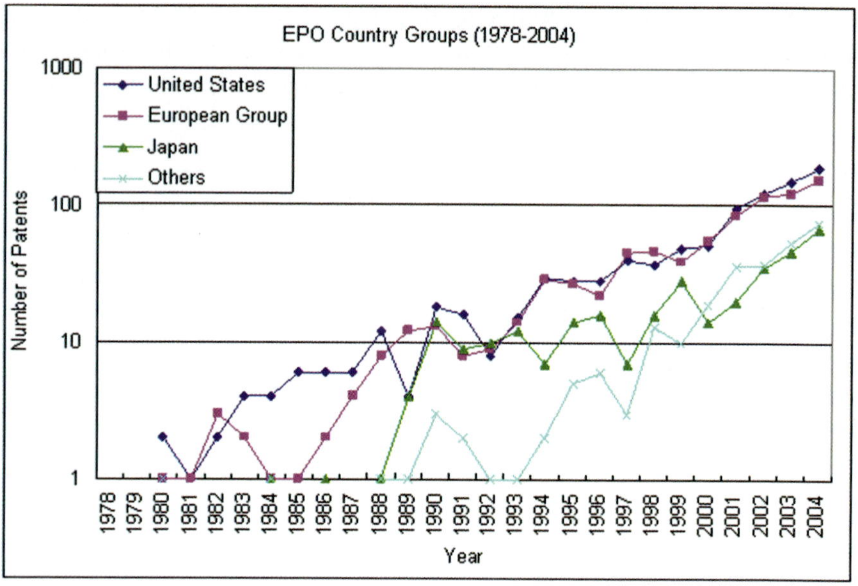

Figure 9-6. Patent publication trend by assignee country group in EPO database ("title-abstract" search, 1978-2004) (log scale).

Table 9-7. Patent publication trend by assignee country group in USPTO ("title-abstract" search, 1976–2004).

Year	United States	European Group	Japan	Others
1976	6	1	0	1
1977	12	1	0	2
1978	15	1	1	2
1979	8	1	0	1
1980	10	1	0	0
1981	20	3	0	1
1982	16	4	0	0
1983	14	5	0	2
1984	18	4	0	0
1985	21	5	0	0
1986	17	2	2	1
1987	32	4	1	0
1988	27	4	3	2
1989	47	7	6	2
1990	36	5	9	3
1991	53	10	10	2
1992	77	9	16	6
1993	82	6	21	4
1994	76	11	31	4
1995	102	11	24	14
1996	135	17	25	14
1997	157	26	31	8

Year	United States	European Group	Japan	Others
1998	196	40	30	19
1999	237	45	34	28
2000	262	59	38	35
2001	348	63	37	43
2002	358	83	43	71
2003	478	73	66	92
2004	590	83	89	120

Table 9-8. Patent publication trend by assignee country group in EPO database ("title-abstract" search, 1978-2004).

Year	United States	European Group	Japan	Others
1978	0	0	0	0
1979	0	0	0	0
1980	2	1	0	1
1981	1	1	0	0
1982	2	3	0	0
1983	4	2	0	0
1984	4	1	1	1
1985	6	1	0	0
1986	6	2	1	0
1987	6	4	0	0
1988	12	8	1	1
1989	4	12	4	1
1990	18	13	14	3
1991	16	8	9	2
1992	8	9	10	1
1993	15	14	12	1
1994	30	29	7	2
1995	28	27	14	5
1996	28	22	16	6
1997	40	45	7	3
1998	37	46	16	13
1999	49	39	28	10
2000	52	56	14	19
2001	95	85	20	36
2002	122	118	35	37
2003	150	123	46	54
2004	190	154	68	75

4.1.3 Analysis of Country Impact

We use the average number of cites measure to identify the high-impact nanotechnology assignee countries in USPTO and EPO. Tables 9-9 and 9-10 show the top 10 assignee countries based on the average number of cites measure. To keep the analysis meaningful, we only study the countries with more than 10 patents. In general, the average number of cites measures in USPTO are much higher than those in EPO, which may be due to the two patent offices' policies: USPTO requires inventors to cite previous works in their patents, while EPO doesn't (Bacchiocchi and Montobbio, 2004). Although five out of the ten high-impact countries (United States, Japan, Switzerland, France, and Israel) are the same in the two data sets, their rankings are significantly different. Compared with the top 20 assignee countries in Tables 9-5 and 9-6, we find that although the high-productivity countries are very similar in the two repositories, the high-impact countries are different. Among the high-impact countries, the United States and Japan published many of the patents with high average number of cites in both repositories, indicating their important roles in international nanotechnology development.

Table 9-9. USPTO top 10 countries with more than 10 patents based on the average number of cites measure ("title-abstract" search, 1976-2004).

Country Name	Number of Patents	Average Number of Cites
Australia	34	2.09
Japan	517	1.90
Switzerland	41	1.90
United States	3,450	1.81
Ireland	20	1.55
Canada	104	1.30
France	156	1.10
China (Taiwan)	71	0.89
Germany	204	0.89
Israel	21	0.86

Table 9-10. EPO top 10 countries with more than 10 patents based on the average number of cites measure ("title-abstract" search, 1978-2004).

Country Name	Number of Patents	Average Number of Cites
Japan	326	0.22
Spain	22	0.18
Belgium	42	0.17
France	201	0.15
United States	929	0.12
Republic of Korea	98	0.11
United Kingdom	72	0.11
Switzerland	77	0.10
Israel	20	0.10
Netherlands	51	0.10

4.2 Assignee Institution Analysis

Tables 9-11 through 9-13 report the 20 assignee institutions having the most nanotechnology patent publications in all three databases (the country attribute of JPO in Table 9-13 was identified manually). We also report the average patent ages (the average number of years the patents had been published, up to 2004) for each institution (Huang et al., 2003). A shorter average patent age indicates that the institution has become more active in recent years.

In the USPTO, International Business Machines Corp. (IBM) produced the most nanotechnology patents, followed by The Regents of the University of California, The United States of America as represented by the Secretary of the Navy, Eastman Kodak Co., and Minnesota Mining and Manufacturing Co. (3M Co.). The top USPTO assignee institutions, in general, had an average patent age of about five years, a long history in nanotechnology research. Micron Technology, Inc., Advanced Micro Devices, Inc., the California Institute of Technology, and Hewlett-Packard Development Company, L.P. had relatively smaller average patent ages (two years), but with significant numbers of patents published in the USPTO database. These four assignees were relatively more active in recent years.

Table 9-11. Top 20 USPTO patent assignee institutions by number of patents published ("title-abstract" search, 1976-2004).

Rank	Assignee Institution	Country	Number of Patents	Average Patent Age
1	IBM	US	171	6.65
2	The Regents of the University of California	US	123	4.30
3	The USA as represented by the Secretary of the Navy	US	82	5.15
4	Eastman Kodak Company	US	72	7.19
5	Minnesota Mining and Manufacturing Co.	US	59	5.92
6	Massachusetts Institute of Technology	US	56	3.61
7	Xerox Corporation	US	55	4.71
8	Micron Technology, Inc.	US	53	2.09
9	Matsushita Electric Industrial Co. Ltd.	Japan	45	6.89
10	L'Oreal	France	44	4.45
11	Texas Instruments, Incorporated	US	41	5.44
12	Motorola, Inc.	US	38	4.11
13	Advanced Micro Devices, Inc.	US	35	2.17
14	General Electric Company	Japan	31	4.42
15	Kabushiki Kaisha Toshiba	US	30	4.03
16	Allied Signal Inc.	Japan	28	3.29
17	California Institute of Technology	US	27	2.52
18	Hewlett-Packard Development Co., L.P.	US	27	0.30
19	Hitachi, Ltd.	Japan	27	6.96
20	Hyperion Catalysis International, Inc.	Japan	25	3.44

In the EPO, French cosmetic company L'Oreal held the most patents, followed by IBM, Rohm & Haas (an American special materials company), Eastman Kodak Co., and Samsung Electronics Co. Ltd. In general, the average age of the top 20 EPO assignee institutions' patents is four years. However, Samsung Electronics Co. Ltd., Japan Science and Tech Agency, and Henkel Kgaa had much smaller average patent ages than the other institutions, indicating their active innovations in recent years. IBM is the second largest assignee institution in terms of nanotechnology patent publication. However, its average patent age is 11.91 years, which shows its long history in the field. It has fewer patents published in recent years.

Table 9-12. Top 20 EPO patent assignee institutions by number of patents published ("title-abstract" search, 1978-2004).

Rank	Assignee Institution	Country	Number of Patents	Average Patent Age
1	L'Oreal	France	50	5.16
2	IBM	US	44	11.91
3	Rohm & Haas	US	41	3.41
4	Eastman Kodak Co	US	39	4.44
5	Samsung Electronics Co Ltd	Rep. of Korea	28	0.75
6	Japan Science & Tech Corp	Japan	27	2.67
7	BASF AG	Germany	25	6.56
8	Matsushita Electric Ind Co Ltd	Japan	24	7.67
9	Allied Signal Inc	US	21	4.33
10	Lucent Technologies Inc	US	21	3.24
11	Japan Science & Tech Agency	Japan	20	0.00
12	Bayer AG	Germany	19	5.11
13	Univ California	US	19	4.89
14	Centre Nat Rech Scient	France	18	4.44
15	Commissariat Energie Atomique	France	18	3.06
16	Inst Neue Mat Gemein Gmbh	Germany	18	5.72
17	Henkel Kgaa	Germany	17	1.41
18	Hewlett-Packard Co	US	17	2.94
19	Iljin Nanotech Co Ltd	Rep. of Korea	17	3.47
20	Samsung Sdi Co Ltd	Rep. of Korea	17	2.24

In the JPO, Nippon Electric Co. is the largest assignee institution, followed by Japan Science and Tech Corp., Agency of Industrial Science and Technology, Matsushita Electric Industrial Co. Ltd., and Tokyo Shibaura Electric Co. Most of the top 20 assignees in JPO had an average patent age of more than seven years. However, Japan Science and Tech Corp., National Institute for Materials Science, and National Institute of Advanced Industrial Science and Technology had an average patent age of fewer than three years with significant patent publications. Among these assignee institutions, Japan Science and Technology Corp. is the second largest assignee institution with 48 patents and an average patent age of 2.29 years, which indicates its active role in recent nanotechnology development in the JPO.

Table 9-13. Top 20 JPO patent assignee institutions by number of patents published ("title-abstract" search, 1976-2004).

Rank	Assignee Institution	Country	Number of Patents	Average Patent Age
1	Nippon Electric Co	Japan	103	7.17
2	Japan Science & Tech Corp	Japan	48	2.29
3	Agency Ind Science Techn	US	43	6.21
4	Matsushita Electric Ind Co Ltd	Japan	43	9.07
5	Tokyo Shibaura Electric Co	Japan	31	5.61
6	Sony Corp	Japan	30	8.10
7	Canon Kk	Japan	28	8.75
8	Seiko Instr Inc	Japan	26	8.35
9	Nat Inst for Materials Science	US	25	1.00
10	Nat Inst of Adv Ind & Technol	Japan	24	1.25
11	Sharp Kk	Japan	24	4.79
12	Japan Res Dev Corp	Japan	22	9.00
13	Hitachi Ltd	Japan	21	8.19
14	IBM	US	21	8.05
15	Jeol Ltd	Japan	21	9.19
16	L'Oreal	France	20	6.00
17	Nippon Telegraph & Telephone	Japan	19	7.32
18	Hitachi Metals Ltd	Japan	13	7.54
19	Fujitsu Ltd	Japan	12	8.75
20	Daiken Kagaku Kogyo Kk	Japan	11	3.36

Comparing the three tables, we observe that the top 20 assignee institutions are quite different among the three databases. However, those institutions that are in multiple repositories are important nanotechnology research centers. For example, IBM, L'Oreal and Hitachi, Ltd. are in both the USPTO and JPO top 20 assignee lists. Eastman Kodak Co., Hewlett-Packard Development Company, L.P., and The Regents of the University of California are in both the USPTO and EPO top 20 assignee lists. These institutions have broader impact on nanotechnology research than other institutions that mainly publish in a single repository.

We found that most of the top assignee institutions in USPTO (Table 9-11) and JPO (Table 9-13) are from the United States and Japan, respectively. This phenomenon is an indication of "home advantage" at the assignee institution level. In the EPO (Table 9-12), European institutions do not dominate the top 20 list, but the list still includes seven European companies: L'Oreal, BASF AG, Bayer AG, Centre National de la Recherche Scientifique (CNRS), Commissariat Energie Atomique, Institut für Neue Materialien, and Henkel Kgaa. In general, the top 20 assignee institutions in the three repositories are mainly from the United States, Japan, France, Germany, and the Republic of Korea.

We inspected the patent publication trends since 1976 for the top 10 institutions (http://ai.arizona.edu/research/nanomapping/USPTOEPOJPO_Analysis.htm) to determine the relative productivity of each institution in the nanotechnology field over time.

- In the USPTO, several institutions had a steady increase in patent publication, such as The Regents of the University of California and the United States of America as represented by the Secretary of the Navy. Some institutions showed a decrease in recent years, e.g., IBM (decreased after 2001) and Micron Technology, Inc. (decreased after 2001). Most assignee institutions started publishing nanotechnology patents in the 1990s, while IBM, The United States of America as represented by the Secretary of the Navy, Eastman Kodak Co., and Minnesota Mining and Manufacturing Company (3M Co.) started in the 1970s.
- In the EPO, Samsung Electronics Co. Ltd had a sharp increase in the number of patents it published after 2001; for most other institutions, such as L'Oreal and IBM, the number remained steady. For some institutions, the number decreased, such as Eastman Kodak Co. and Japan Science and Technology Corp. Rohm and Haas published a great number of patents around 2003, which raised its position to third overall of the EPO assignee institutions.
- In JPO, many assignee institutions have experienced a decrease in recent years, such as Nippon Electric Co., Agency of Industrial Science and Technology, Tokyo Shibaura Electric Co., etc. However, Japan Science and Technology Corp. and National Institute for Materials Science continued to have active patent publications.

High-impact institutions

Tables 9-14 and 9-15 show the high-impact assignee institutions in the USPTO and EPO according to the average number of citations their patents received from other nanotechnology patents. Similar to what is observed in the high-impact countries, the average numbers of citations of the top assignees in the USPTO were larger than those of the top assignees in the EPO. The two tables show that the sets of high-impact assignee institutions are quite different in the two repositories, indicating the different roles an institution may play in different repositories.

Table 9-14. USPTO top 10 assignee institutions with more than 10 patents based on average number of citations measure ("title-abstract" search, 1976–2004).

Assignee Institution	Number of Patents	Average No. of Cites
Digital Instruments, Inc.	24	9.42
AMCOL International Corporation	22	9.05
Agere Systems Guardian Corp.	11	9.00
Olympus Optical Co., Ltd.	15	7.47
Hyperion Catalysis International, Inc.	25	6.84
The Penn State Research Foundation	18	5.39
Allied Signal Inc.	28	5.21
The Board of Trustees of the Leland Stanford Junior Univ.	21	5.14
President & Fellows of Harvard College	25	5.12
Nano Systems L.L.C.	10	4.40

Table 9-15. EPO top 10 assignee institutions with more than 10 patents based on average number of citations measure ("title-abstract" search, 1978–2004).

Assignee Institution	Number of Patents	Average No. of Cites
Canon Kk	12	0.92
Lucent Technologies Inc	21	0.52
IBM	44	0.50
Hitachi Europ Ltd	15	0.47
Hitachi Ltd	15	0.40
L'Oreal	50	0.38
Matsushita Electric Ind Co Ltd	24	0.38
Rohm & Haas	41	0.37
Lee Cheol Jin	11	0.36
Max Planck Gesellschaft	10	0.30

4.3 Technology Field Analysis

In this research we use the third-level (subclass) IPC label to represent patent technology fields. Table 9-16 shows the top 10 USPTO technology fields according to the number of patents published between 1976 and 2004. Technology field "H01L: Semiconductor devices; electric solid state devices not otherwise provided for" had the most nanotechnology patents published, almost double the amount of the second largest technology field "A61K: reparations for medical, dental, or toilet purposes." The top technology fields in USPTO are mainly in biomedical research (e.g., technology fields A61K and C08K), material research (e.g., technology fields G01N, G01B, B32B, and B01D), and semiconductor research (e.g., technology fields H01L and H01J).

Table 9-16. Top 10 USPTO technology fields according to the number of patents published ("title-abstract" search, 1976–2004).

IPC Class	Class Name	Number of Patents
H01L	Semiconductor devices; electric solid state devices not otherwise provided for	748
A61K	Preparations for medical, dental, or toilet purposes	431
H01J	Electric discharge tubes or discharge lamps	364
B32B	Layered products, i.e. products built-up of strata of flat or non-flat, e.g. cellular or honeycomb	352
G01N	Investigating or analyzing materials by determining their chemical or physical properties	344
B05D	Performing operations transporting	211
C08K	Use of inorganic or non-macromolecular organic substances as compounding ingredients	182
G01B	Measuring length, thickness, or similar linear dimensions; measuring angles; measuring areas; measuring irregularities of surfaces or contours..	169
G02B	Optical elements, systems, or apparatus	147
B01D	Separation	139

Table 9-17 shows the top 10 technology fields according to the number of patents published between 1978 and 2004 in the EPO. Here, no one field dominates, as does H01L in the USPTO. The top 10 technology fields in the EPO are mainly related to biomedical research (e.g., technology fields A61K, C01B, and C08K), chemistry research (e.g., technology fields B01J and C08L), material research (e.g., technology fields G01N, C01B, and B01D), and semiconductor research (e.g., technology fields H01L and H01J).

Table 9-17. Top 10 EPO technology fields according to the number of patents published ("title-abstract" search, 1978–2004).

IPC Class	Class Name	Number of Patents
A61K	Preparations for medical, dental, or toilet purposes	347
H01L	Semiconductor devices; electric solid state devices not otherwise provided for	271
G01N	Investigating or analyzing materials by determining their chemical or physical properties	226
C01B	Non-metallic elements; compounds thereof	222
C08K	Use of inorganic or non-macromolecular organic substances as compounding ingredients	152
B01J	Chemical or physical processes, e.g. catalysis, colloid chemistry; their relevant apparatus	141
H01J	Electric discharge tubes or discharge lamps	119

IPC Class	Class Name	Number of Patents
C08L	Compositions of macromolecular compounds	90
G01B	Measuring length, thickness, or similar linear dimensions; measuring angles; measuring areas; measuring irregularities of surfaces or contours	87
B01D	Separation	81

The top 10 technology fields in JPO patents according to the number of patents published between 1976 and 2004 are presented in Table 9-18. The top 10 technology fields in JPO are mainly related to biomedical research (e.g., technology fields A61K, C01B, and C23C), material research (e.g., technology fields G01B, G01N, and B82B), and semiconductor research (e.g., technology fields H01L, H01J).

Table 9-18. Top 10 JPO technology fields according to the number of patents published ("title-abstract" search, 1976-2004).

IPC Class	Class Name	Number of Patents
H01L	Semiconductor devices; electric solid state devices not otherwise provided for	210
C01B	Non-metallic elements compounds thereof	153
H01J	Electric discharge tubes or discharge lamps	135
G01B	Measuring length, thickness, or similar linear dimensions; measuring angles; measuring areas; measuring irregularities of surfaces or contours	121
G01N	Investigating or analyzing materials by determining their chemical or physical properties	120
B01J	Chemical or physical processes, e.g. catalysis, colloid chemistry; their relevant apparatus	79
H01S	Devices using stimulated emission	56
B82B	Nano-structures manufacture or treatment thereof	51
A61K	Preparations for medical, dental, or toilet purposes	45
C23C	Coating metallic material coating material with metallic material surface treatment of metallic material by diffusion into the surface, by chemical conversion or substitution coating by vacuum evaporation, by sputtering, by ion implantation or by chemical vapour deposition, in general	45

Table 9-19 compares the top technology fields in the three databases. The three repositories have several top technology fields in common. For example, five technology fields appeared in all three repositories: "A61K: Preparations for medical, dental, or toilet purposes," "H01L: Semiconductor devices; electric solid state devices," "H01J: Electric discharge tubes or discharge lamps," "G01B: Measuring length, thickness, or similar linear dimensions; measuring angles; measuring areas; measuring irregularities of surfaces or contours," and "G01N: Investigating or analyzing materials by determining their chemical or physical properties." The USPTO and EPO have another two top technology fields in common: "B01D: Separation" and "C08K: Use of inorganic or non-macromolecular organic substances as compounding ingredients." The EPO and JPO have another two top technology fields in common: "B01J: Chemical or physical processes, e.g. catalysis, colloid chemistry; their relevant apparatus" and "C01B: Non-metallic elements; compounds thereof." Although the top 10 technology fields are very similar in the three repositories, their rankings (and number of patents published) have significant differences. This phenomenon indicates that although worldwide nanotechnology research has some common focus, different regions have their specific strengths in some technology fields.

Table 9-19. Top 10 technology fields in USPTO (1976-2004), EPO (1978-2004) and JPO (1976–2004) ("title-abstract" search).

Rank	USPTO		EPO		JPO	
	IPC Class	Number of Patents	IPC Class	Number of Patents	IPC Class	Number of Patents
1	H01L	748	A61K	347	H01L	210
2	A61K	431	H01L	271	C01B	153
3	H01J	364	G01N	226	H01J	135
4	B32B	352	C01B	222	G01B	121
5	G01N	344	C08K	152	G01N	120
6	B05D	211	B01J	141	B01J	79
7	C08K	182	H01J	119	H01S	56
8	G01B	169	C08L	90	B82B	51
9	G02B	147	G01B	87	A61K	45
10	B01D	139	B01D	81	C23C	45

We inspect the top 10 technology fields' patent publication trends since 1976 in the three repositories (http://ai.arizona.edu/research/nano mapping/USPTOEPOJPO_Analysis.htm). Most of the top 10 technology fields in the USPTO showed an increasing number of nanotechnology patents published since 1976. Among these technology fields, "H01L: Semiconductor devices; electric solid state devices" experienced the most rapid growth. The yearly patent publications in "A61K: Preparations for medical, dental, or toilet purposes" stabilized after 1996, which is different from the other technology fields.

In EPO, the top 10 technology fields also show increasing numbers. After

2000, "A61K: Preparations for medical, dental, or toilet purposes," "H01L: Semiconductor devices; electric solid state devices," and "C01B: Non-metallic elements; compounds thereof" showed faster growth than the other technology fields. Unlike for the USPTO trends, technology field H01L did not perform significantly better than A61K and G01N. The number of patents published in technology field "G01B: Measuring length, thickness, or similar linear dimensions; measuring angles; measuring areas; measuring irregularities of surfaces or contours" was consistent in recent years.

In the JPO, many of the technology fields experienced a decrease in recent years, which is significantly different from the technology field patent publication trends in USPTO and EPO. However, technology field "C01B: Non-metallic elements; compounds thereof" had a steady growth in patent publication.

The analysis also reveals changes in the three technology fields that are common among the repositories. For example, in recent years, the research on H01L was getting stronger in the USPTO, while research on A61K was becoming stable. In the EPO, research on A61K and H01L kept increasing. In the JPO, patent filings in almost all fields were significantly reduced in recent years.

High-impact technology fields

Tables 9-20 and 9-21 show the high-impact technology fields in the USPTO and EPO respectively, according to the average number of citations generated by the patents. The fields having the highest impact are quite different between the two repositories. However, technology fields "H01J: Electric discharge tubes or discharge lamps" and "G01B: Measuring length, thickness, or similar linear dimensions; measuring angles; measuring areas; measuring irregularities of surfaces or contours" appear in both top 10 lists with large numbers of patents and high average numbers of citations. These two technology fields have broad impact in worldwide nanotechnology research. In general, the USPTO high-impact technology fields covered topics on material research (e.g., technology fields B29B and G01B), electrical engineering research (e.g., technology fields H01J, C25D, and G01T), chemistry research (e.g., technology fields D01F and C08J), and biomedical research (e.g., technology fields C01B and C08K). The EPO high-impact technology fields covered topics on electrical engineering research (e.g., technology fields H01J and H01F), chemistry research (e.g., technology fields B01F, C09D, and C09B), physics research (e.g., technology fields G03C and G01R), and biomedical research (e.g., technology field A61K).

Table 9-20. USPTO top 10 technology fields with more than 10 patents based on the average number of citations measure ("title-abstract" search, 1976–2004).

IPC Class	Class Name	Number of Patents	Average Number of Cites
D01F	Chemical features in the manufacture of artificial filaments, threads, fibres, bristles, or ribbons; apparatus specially adapted for the manufacture of carbon filaments	80	4.31
B29B	Preparation or pretreatment of the material to be shaped; making granules or preforms; recovery of plastics or other constituents of waste material containing plastics	12	3.58
H01J	Electric discharge tubes or discharge lamps	364	3.40
C01B	Non-metallic elements; compounds thereof	136	3.39
G01B	Measuring length, thickness, or similar linear dimensions; measuring angles; measuring areas; measuring irregularities of surfaces or contours	169	3.12
C25D	Processes for the electrolytic or electrophoretic production of coatings; electroforming	30	3.03
G01T	Measurement of nuclear or X-ray radiation	12	3.00
C08J	Working-up general processes of compounding after-treatment not covered by subclasses	71	2.79
C08K	Use of inorganic or non-macromolecular organic substances as compounding ingredients	182	2.76
B44C	Producing decorative effects; mosaics; tarsia work; paperhanging	35	2.57

Table 9-21. EPO top 10 technology fields with more than 10 patents based on the average number of citations measure ("title-abstract" search, 1978–2004).

IPC Class	Class Name	Number of Patents	Average Number of Cites
G01B	Measuring length, thickness, or similar linear dimensions; measuring angles; measuring areas; measuring irregularities of surfaces or contours	87	0.39
H01J	Electric discharge tubes or discharge lamps	119	0.37
G03C	Photography cinematography; analogous techniques using waves other than optical waves; electrography holography	19	0.32
G01R	Measuring electric variables measuring magnetic variables	17	0.29

IPC Class	Class Name	Number of Patents	Average Number of Cites
B01F	Mixing, e.g. dissolving, emulsifying, dispersing	17	0.24
C09D	Coating compositions, e.g. paints, varnishes, lacquers, filling pastes, chemical paint or ink removers, inks, correcting fluids, woodstains, pastes or solids for colouring or printing; use of materials therefore	73	0.23
C09B	Organic dyes or closely related compounds for producing dyes; mordants; lakes	13	0.23
H01F	Magnets; inductances; transformers; selection of materials for their magnetic properties	51	0.20
G11B	Information storage based on relative movement between record carrier and transducer	57	0.18
A61K	Preparations for medical, dental, or toilet purposes	348	0.17

5. CONTENT MAP ANALYSIS

We generate a series of content maps of the three repositories for three time intervals: 1976(1978)-1989, 1990-1999, and 2000–2004. Each time interval's content map is compared with the previous time interval's content map to visualize the changes of topic areas. For each topic area, a growth rate is computed as the ratio between the number of documents in the current time period and that of the previous time period. A baseline growth rate is computed as the ratio between the total number of documents in the current time period and that of the previous time period. A topic region with similar growth rate to the base growth rate is assigned a green color. The topic region with higher (lower) growth rate is assigned a warmer (colder) color. If the topic is brand new, a red color is assigned to the region.

5.1 Content Map Analysis for 1976 (1978)-1989

Figure 9-7 presents the content map of USPTO nanotechnology patents published from 1976 to 1989. The major research topics in this time period included: "carbon atoms," "optical fibers," and "thin films." EPO and JPO had only 97 and 31 nanotechnology patents published from 1978 to 1989 and from 1976 to 1989, respectively. There are not enough patents in these two repositories to generate meaningful content maps for this time period.

Figure 9-7. USPTO content map ("title-abstract" search, 1976-1989).

5.2 Content Map Analysis for 1990-1999

Figures 9-8 to 9-10 present the content maps of nanotechnology patents published in USPTO, EPO, and JPO from 1990 to 1999, respectively. The USPTO content map also shows the comparison between the topics in 1976-1989 and 1990-1999, on which the topic region with higher (lower) growth rate was assigned a warmer (colder) color. The growth rate represented by each color is shown under the content map.

In USPTO several new research topics appeared, including: "aqueous solutions," "composite materials," "laser beams," "nucleic acids," "optical waveguide," "organic solvents," "reverse osmosis," "self-assembled monolayer," "semiconductor substrate," "silicon carbide," and "substrate surfaces."

In EPO the major topics of nanotechnology patents were "aqueous solutions," "atomic force," "carbon nanotubes," "magnetic core," "metal oxides," and "thin films." In JPO the major topics of nanotechnology patents were "atomic force microscope," "laser beams," "silicon substrate," and "thin films." By comparing the three content maps, we observe that USPTO had a broader coverage of nanotechnology topics than the other two databases in this time period.

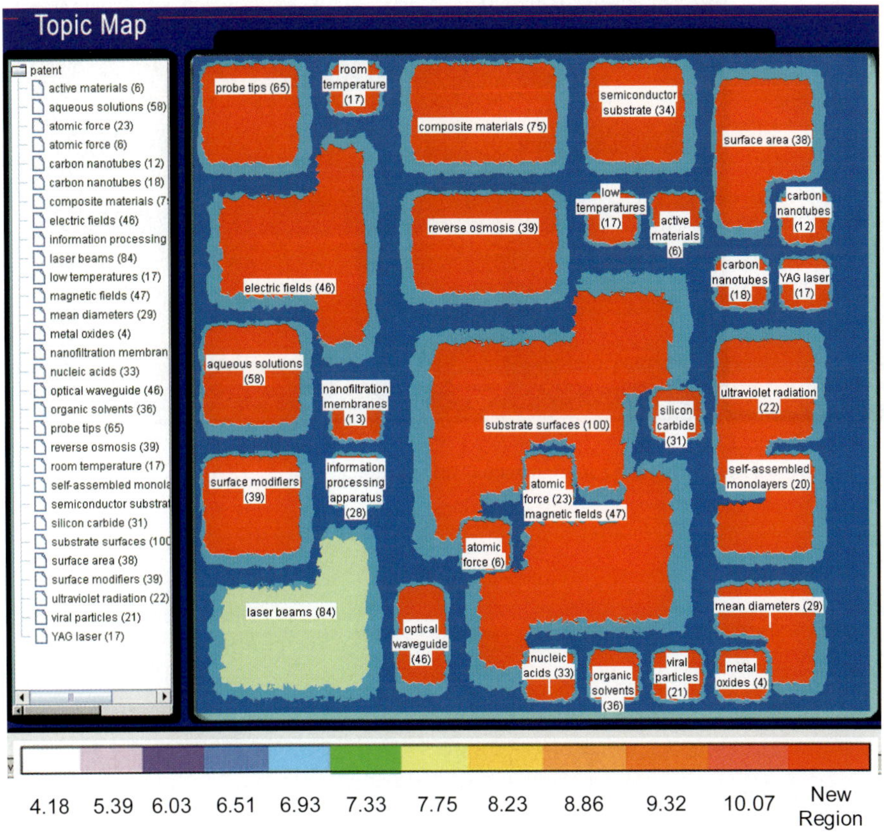

Figure 9-8. USPTO content map ("title-abstract" search, 1990-1999).

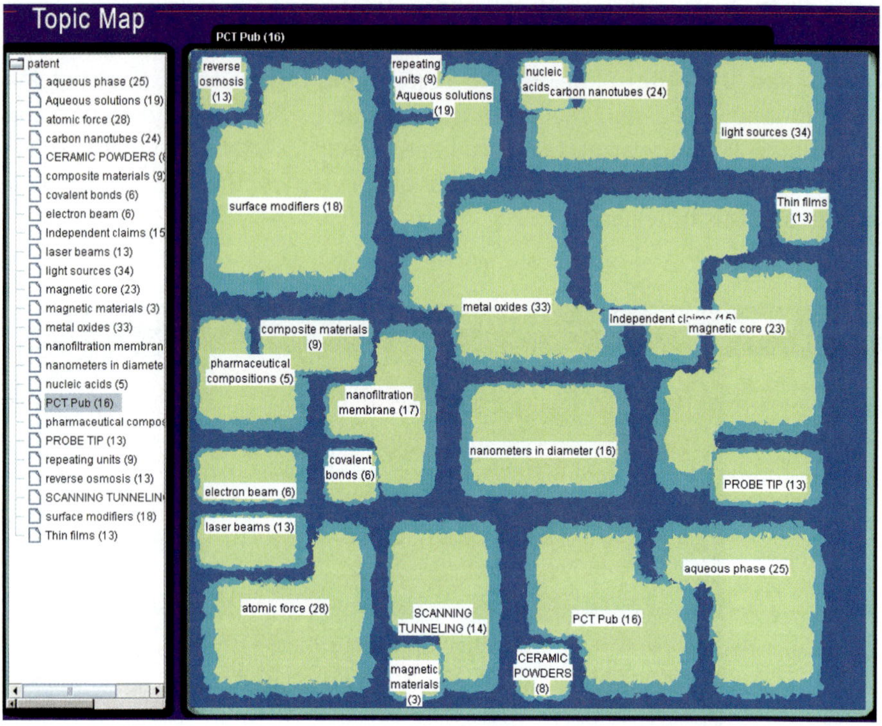

Figure 9-9. EPO content map ("title-abstract" search, 1990-1999).

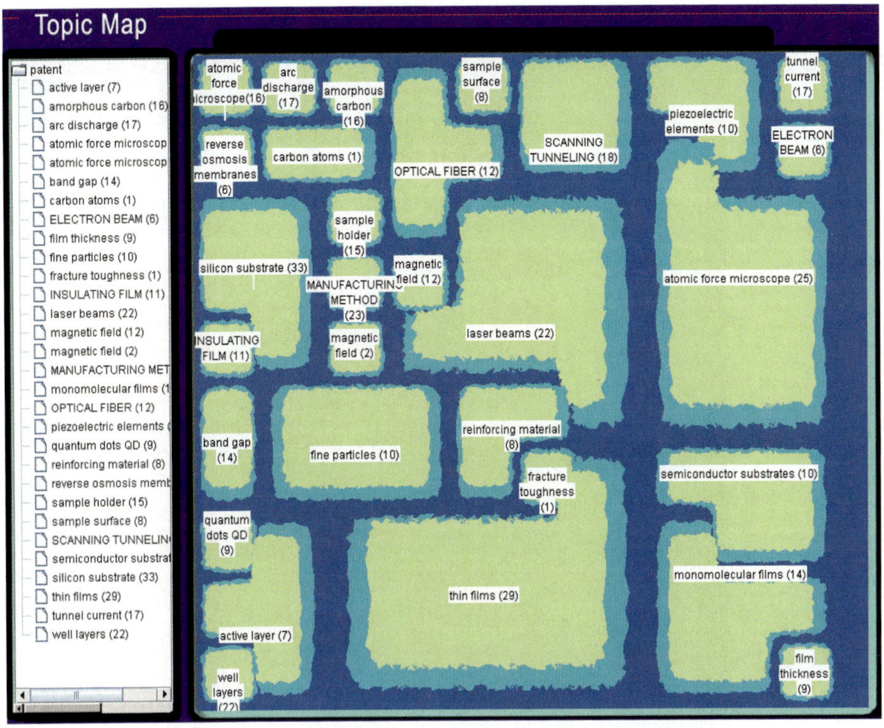

Figure 9-10. JPO content map ("title-abstract" search, 1990-1999).

5.3 Content Map Analysis for 2000-2004

Figures 9-11 to 9-13 present the content maps of nanotechnology patents published in USPTO, EPO, and JPO from 2000 to 2004.

In USPTO several topics had a significant increase in the numbers of patents published, such as "aqueous solutions," "composite materials," "carbon nanotubes," "nucleic acids," "self-assembled monolayer," and "thin films." Some new topics in this time period included "atomic force microscope," "clay materials," "dielectric layers," "nanocomposite materials," "naphtha stream," "polymeric materials," and "semiconductor devices."

In EPO there was a significant increase in the topics "aqueous solutions," "metal oxides," and "thin films." New major topics included "gate electrode," "low dielectric," "nanocomposite materials," "nanoparticulate compositions," and "polymer compositions." In JPO "atomic force microscope" and "thin films" were still major research topics. "Carbon nanofibers," "gate electrodes," "heat treatment," and "quantum dots" were the new topics in this time period.

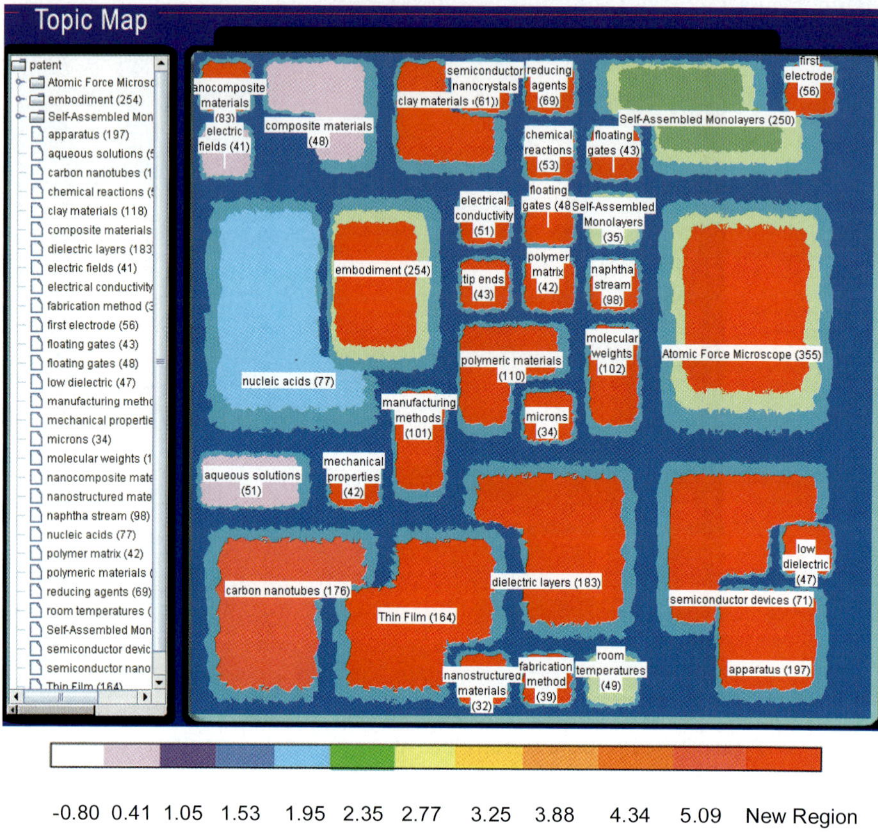

Figure 9-11. USPTO content map ("title-abstract" search, 2000-2004).

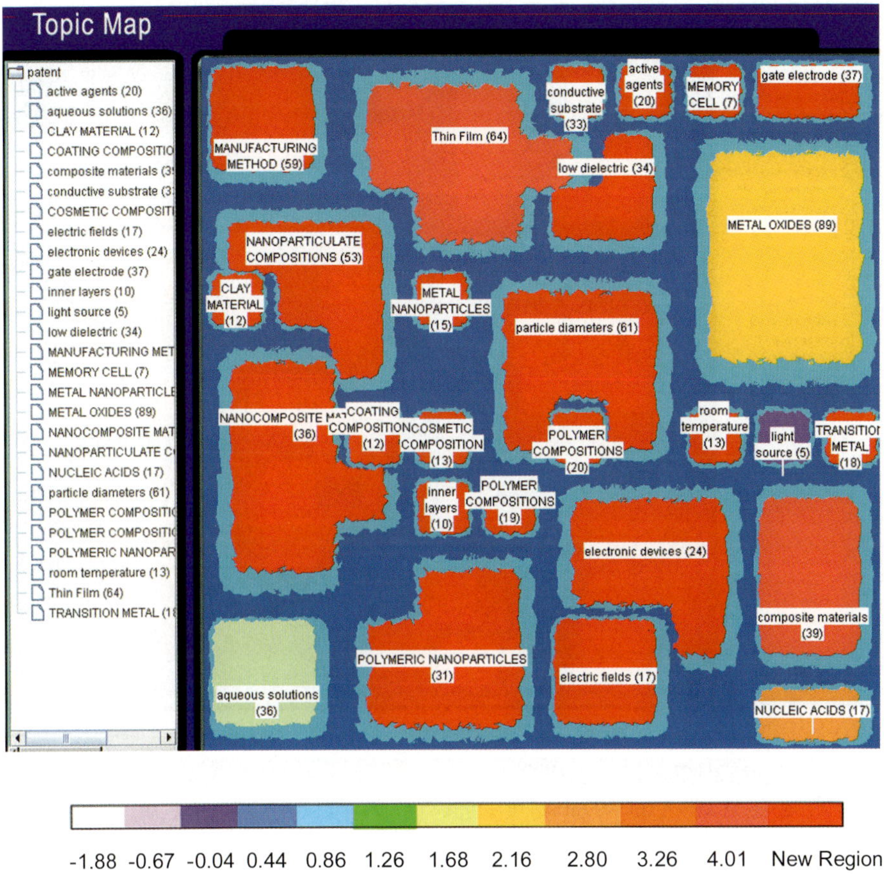

Figure 9-12. EPO content map ("title-abstract" search, 2000-2004).

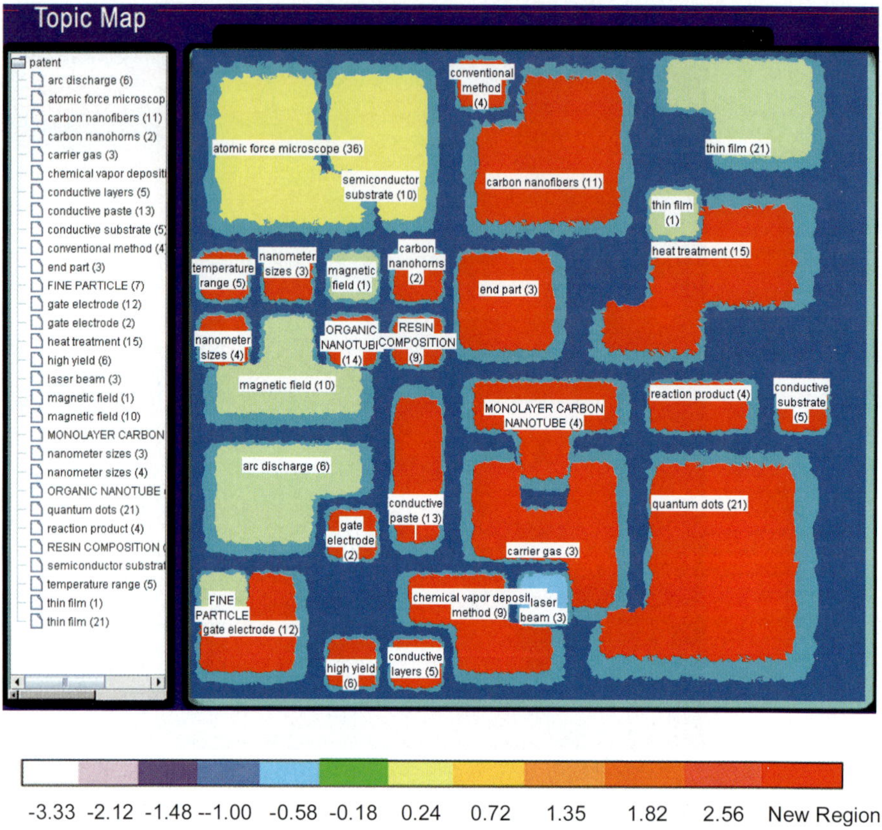

Figure 9-13. JPO content map ("title-abstract" search, 2000-2004).

By comparing the three repositories' nanotechnology patent content maps in different time spans, we found that USPTO patents had more topics than EPO and JPO. Many of the EPO and JPO topics were related to research tools/methods (e.g., "atomic force microscope," "thin films," and "scanning tunneling microscope") and physics research (e.g., "carbon nanotubes," "carbon nanofibers," "low dielectric," "magnetic core," and "metal oxides"). Many USPTO topics were related to physics research (e.g., "carbon nanotubes," "laser beams," "optical waveguide," and "self-assembled monolayer"), biomedical research (e.g., "nucleic acids," "organic solvents," "pharmaceutical compositions," and "reverse osmosis"), and electronic research (e.g., "dielectric layers," "semiconductor devices," and "semiconductor substrate").

6. CITATION NETWORK ANALYSIS

We adopted the patent citation network to explore knowledge diffusion in the nanotechnology domain at the three analytical unit levels: countries, assignee institutions, and technology fields. We use the top 100 links of each network (according to the number of citations between the nodes) to create core networks for analysis. These citation networks are visualized using an open source graph visualization software, Graphviz, provided by AT&T Labs (Gansner, 2000) (available at: http://www.research.att.com/sw/tools/graphviz/).

In the citation networks, the direction of the links represents the direction of the citations. For example, a link from "Country A" to "Country B" means that Country A's patents cited Country B's patents and the number beside the link is the total number of citations. It allows us to identify the salient citation patterns among the analytical units.

Due to the large numbers of assignee institutions and technology fields and the difficulty in displaying them clearly, we only illustrate and discuss country citation networks in this chapter. Figure 9-14 displays the USPTO country citation network between 1976 and 2004. The United States is the most significant citation center on the network, which cited and was cited by many other countries. Japan, Republic of Korea, the United Kingdom, China (Taiwan), and Germany are the secondary citation centers. Figure 9-15 represents the EPO country citation network between 1978 and 2004. The United States, France, Japan, Germany, and the United Kingdom are large citation centers on the network.

Comparing the USPTO and EPO citation networks, we observe that in both repositories the countries have close citation relationships, indicating effective knowledge diffusion in the nanotechnology domain at the country level. In EPO most assignee countries have more than one citing/cited country, indicating the complicated knowledge transfers between them. In USPTO several countries have a citation relationship only with the United States, while others had complicated citation relationships. Many of the countries that only cited United States patents were relatively new in the nanotechnology domain. The existence of these countries in the USPTO citation network shows the crucial role the United States plays in the nanotechnology patents filed in the USPTO. It also shows the broader impact of USPTO patents on nanotechnology research in other countries.

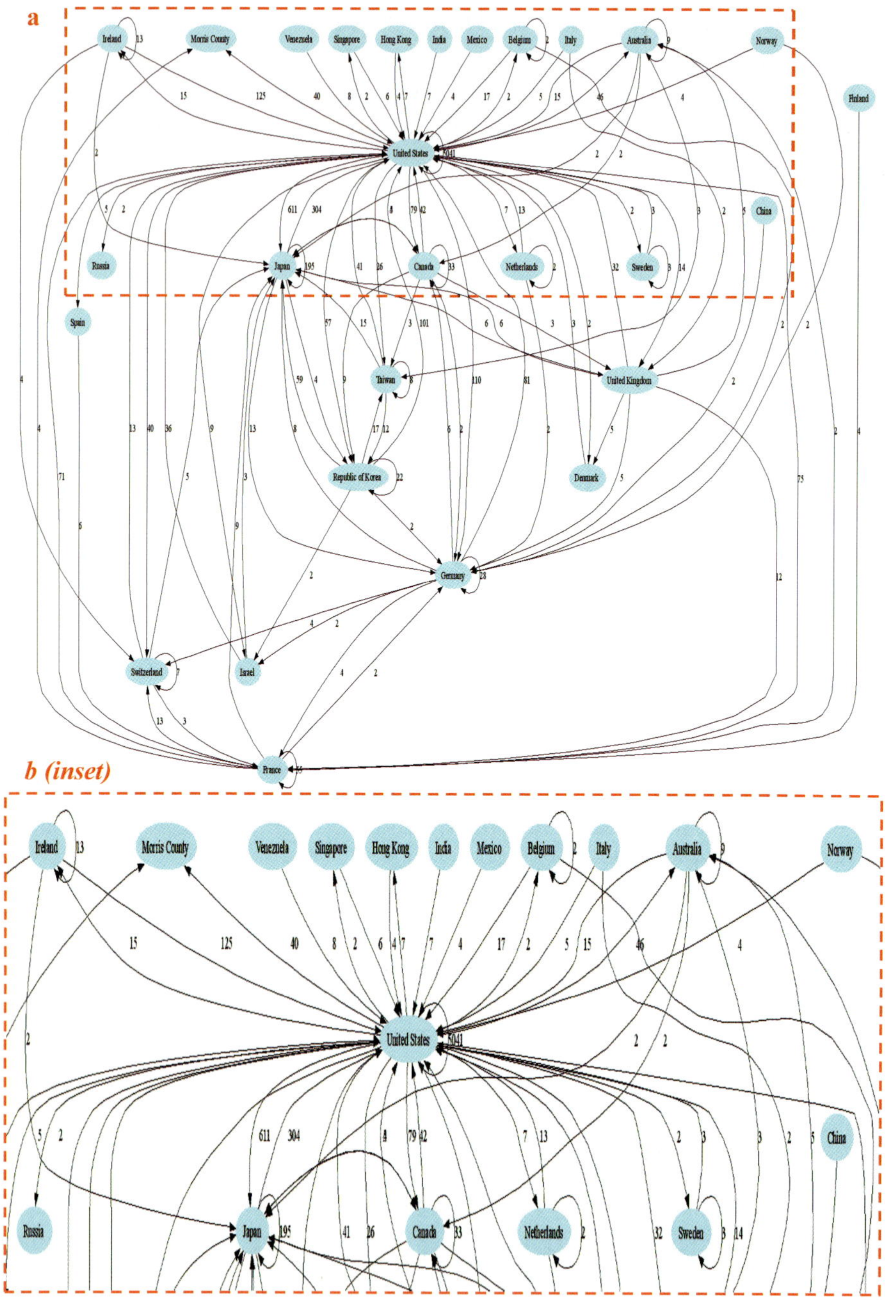

Figure 9-14: USPTO country citation network. a) Entire map ("title-abstract" search, 1976-2004). b) Inset showing detail of United States.

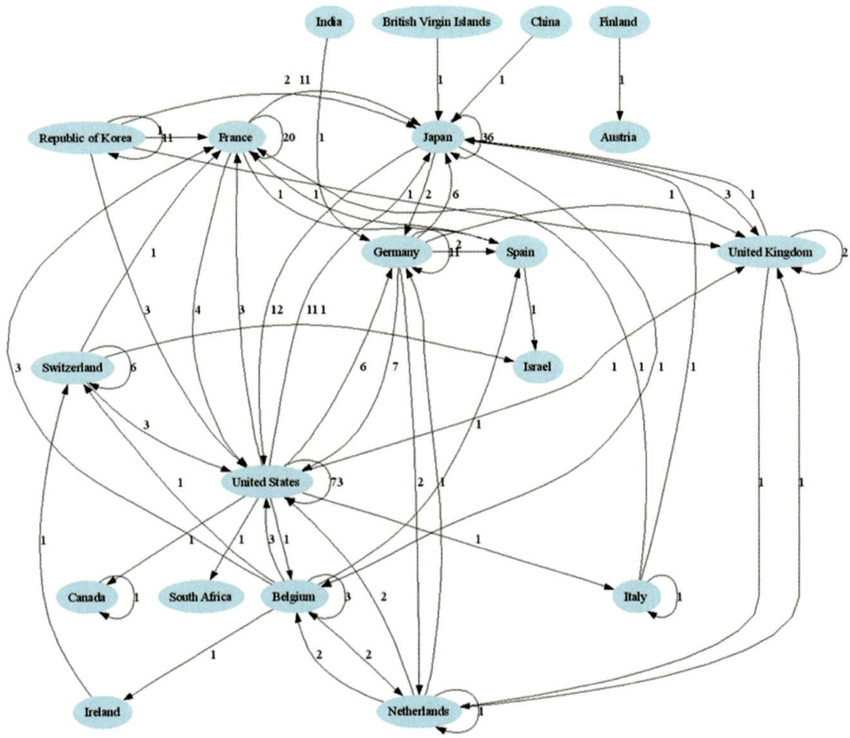

Figure 9-15. EPO country citation network ("title-abstract" search, 1978-2004).

7. CONCLUSIONS

The nanotechnology patents published in three repositories, the USPTO (1976–2004), EPO (1978–2004), and JPO (1976–2004), have been collected, parsed, and analyzed to assess nanotechnology's worldwide research and development status. We conducted three types of analyses (bibliographic, content map, and citation network) for countries, institutions, and technology fields. Key findings of our analyses are discussed in the following paragraphs.

The nanotechnology patents issued by both the USPTO and EPO experienced exponential growth in the past 30 years. The nanotechnology patents issued by the JPO followed the same trend until stabilizing after 1993.

In the USPTO and EPO, the high-productivity assignee countries and their rankings are very similar to each other. In both repositories, the United States filed the most nanotechnology patents.

The number of nanotechnology patents published by the four country groups (United States, European Group, Japan, and Others) increased in both the USPTO and EPO. The United States filed significantly more nanotechnology patents than the other three groups in the USPTO, while it filed a similar number of patents to European countries in EPO.

The top assignee institutions according to the number of patents published are significantly different among the USPTO, EPO, and JPO databases. However, IBM and L'Oreal are high-productivity assignee institutions that appear in all three repositories' top lists. Most of the top assignees in the USPTO and JPO are U.S. and Japanese institutions.

The content map analysis showed that USPTO patents cover more technology topic areas than EPO and JPO patents. Many of the USPTO research topics were related to physics research, biomedical research, and electronics research. Many of the EPO and JPO topics were related to research tools and methods and physics.

The assignee countries had close citation relationships in both USPTO and EPO country citation networks. In both networks, the United States was an important citation center. The USPTO patents, particularly the U.S. patents in the USPTO, had broader impact on worldwide nanotechnology development and attracted more citations from other countries than did EPO patents.

Future research will include the collaboration pattern of inventors in the three repositories. We also plan to extend our research framework to include patents from other local or regional patent offices, such as those in Germany, the People's Republic of China, South Korea, and France.

8. QUESTIONS FOR DISCUSSION

1. What are the major differences in the patent granting procedure among USPTO, EPO, and JPO?
2. What are other patent offices that could be considered for patent knowledge mapping research?
3. How can we handle the multilingual issues for non-English patents?
4. What are the most significant changes in the USTPO, EPO, and JPO repositories over the past 5 years with the increased government nanotechnology funding in different parts of the world?

Chapter 10

MAPPING NANOTECHNOLOGY KNOWLEDGE VIA LITERATURE DATABASE: A LONGITUDINAL STUDY, 1976-2004

CHAPTER OVERVIEW

This chapter analyzes nanotechnology papers published in the Thomson Science Citation Index (SCI) Expanded literature database to assess worldwide nanotechnology research status from 1976 to 2004. We identified 213,847 nanotechnology papers that were published in 4,175 journals from 1976 to 2004. These papers contain 120,687 unique first authors from 24,468 institutions in 156 countries/regions. Between 1976 and 2004, the United States authors generated the largest number of papers (61,068), followed by Japan (24,985), Germany (21,334), and China (20,389). Three countries that were not ranked in the top 20 USPTO patenting list showed significant productivity in academic paper publications: China (#4), Russian (#7), and India (#13). Most countries showed a similar publication growth pattern, except for China and South Korea. Both countries outpaced their competitors in recent years. All top 20 productive institutions were universities and national research centers rather than private companies. "Chinese Academy of Sciences" and "Russian Academy of Sciences" were the most productive institutions. They showed rapid growth since about 1998, outpacing other competing institutions. Both the country citation network and the institution citation network showed significant patterns in forming citation clusters.

1. INTRODUCTION

Scientific literature is often used to assess nanotechnology's research and development status, and generally reflects fundamental and applied scientific knowledge generated (mainly) from academia. Several previous studies have used bibliographic analysis of literature to assess the nanotechnology domain. Braun et al. identified an exponential growth pattern in nanotechnology literature publication (Braun et al., 1997). Porter and Cunningham found a divergence of research topics in nanotechnology literature (Porter and Cunningham, 1995). Kostoff and his colleagues reported key nanotechnology publication statistics for 2003, e.g., authors, journals, institutions, and countries (Kostoff et al., 2006a; Kostoff et al., 2006b). Meyer and Persson analyzed nanotechnology literature to assess the collaboration and knowledge diffusion patterns in nanotechnology research (Meyer and Persson, 1998). They found that the nanotechnology field has more interdisciplinary interactions than other areas of science.

Table 10-1 summarizes previous nanotechnology literature analysis research. Most previous research relied on the Thomson SCI database as their data source, with the only exception being the Porter and Cunningham study that used both INSPEC (more details to follow) and SCI. Most studies used keyword search on title, abstract, and the articles' keyword field. Several earlier studies were longitudinal in nature (e.g., 1986-1995 by Braun et al.); however, most recent studies only reported analysis results of selected years. The largest study that we were able to find used about 21,000 papers for their analyses (Kostoff et al., 2006a; Kostoff et al., 2006b). Considering the rapid advancement of nanotechnology in recent years, there is a critical and urgent need for a large-scale, longitudinal study of nanotechnology literature. In addition, most previous literature analyses replied on bibliographic analysis techniques (i.e., using basic bibliographic indicators). We also see an opportunity for adopting more advanced content analysis and citation analysis techniques to study the knowledge diffusion patterns in nanotechnology.

Table 10-1. A summary of previous nanotechnology literature analysis research.

Study	Data Source	Data Collection	Year	Number of Papers	Analysis Type
(Porter and Cunningham, 1995)	INSPEC & SCI	Keyword search on title, abstract, and keywords	1993, 1994 & 1986-1995	3,956 & 1,176	Bibliographic analysis
(Braun et al., 1997)	SCI	Keyword search on title	1986-1995	4,152	Bibliographic analysis
(Meyer and Persson, 1998)	SCI	Keyword search on title	1991-1996	5,430	Bibliographic analysis

Study	Data Source	Data Collection	Year	Number of Papers	Analysis Type
(Meyer, 2001)	SCI	Keyword search on title	1991-1996	5,430	Bibliographic analysis
(Schummer, 2004)	9 nano-related journals	Random sampling on the papers	2002	609	Co-authorship analysis
(Kostoff et al., 2006a)	SCI	Keyword search on title, abstract, and keywords	2003	21,474	Bibliographic analysis
(Kostoff et al., 2006b)	SCI	Keyword search on title, abstract, and keywords	2003	21,474	Bibliographic analysis, clustering analysis
This study	SCI	Keyword search on title, abstract, and keywords	1976-2004	213,847	Bibliographic analysis, content map analysis, and citation network analysis

In this chapter, we report a longitudinal study of the nanotechnology literature in the Thomson Science Citation Index (SCI) Expanded database. Our nanotechnology literature database, extracted from SCI, consists of 213,847 papers that were published in 4,175 journals from 1976 to 2004. We report results based on bibliographic analysis, content map analysis, and citation network analysis for major countries, institutions, and journals.

2. DATA DESCRIPTION

2.1 Data Source Comparison: SCI, Compendex, and INSPEC

Nanotechnology is a multidisciplinary research field, where papers are published in a variety of journals. Several large-scale, commercial science literature repositories cover nanotechnology, including SCI, Compendex, and INSPEC.

- The Thomson SCI database provides a broad coverage of journals in more than 150 disciplines, including Agriculture, Biology, Chemistry, Computer Science, Engineering, Materials Science, Medicine, Physics, Pharmacology, etc. The SCI database contains more than 5,900 journals. Each paper includes bibliographic information, abstracts, and citation information.
- The Compendex database includes 2,925 journals, 2,717 conferences, 63 monographs, and 57 book series. It mainly covers engineering and

applied science fields, such as Agricultural Engineering, Chemical Engineering, Computers, Materials, Applied Physics, etc. Compendex records contain bibliographic information and abstracts. Citation information is not included.

- The INSPEC database contains 4,030 journals, which focus on Computers, Electrical Engineering, Information Technology, Mechanical Engineering, and Physics. Similar to Compendex, INSPEC provides only bibliographic information and abstracts.

Comparing the journal coverage of the three databases, we found that SCI and Compendex have 1,476 journals in common (according to ISBN match); SCI and INSPEC have 1,224 journals in common (according to journal name match); and Compendex and INSPEC have 1,320 journals in common (according to journal name match). Overall, the SCI database has a much broader disciplinary coverage than the other two databases, which focus mainly on engineering disciplines. SCI covers a large number of major journals that were in Compendex and INSPEC -- about 50% of Compendex journals and 30% of INSPEC journals. Considering the diversity of nanotechnology research, the SCI database provides a more comprehensive coverage than Compendex and INSPEC. The most important difference, however, is that the SCI database provides citation information (citing and cited references) for each paper, which allows us to systematically study the citation impact of papers, authors, institutions, and countries. Because of this, we elected to extract nanotechnology papers from the SCI database for our study.

Data acquired from the SCI database are stored in the XML format. Some of the data fields, including journal names and institution names, have been converted to standardized abbreviations and are easy to recognize. However, the author field only contains complete last name and first initial, which causes a significant identification problem, especially concerning Asian researchers (e.g., "L. Zhang" is a common name in the Chinese Academy of Sciences).

2.2 Collection Statistics

Nanotechnology papers in the SCI database were identified by conducting keyword search in paper titles, keywords, and abstracts using the same list of nanotechnology keywords discussed in the earlier chapters (Huang et al., 2003; Huang et al., 2004). The citations to nanotechnology papers from other papers in the SCI database were also retrieved. In total, we identified 213,847 nanotechnology papers that were published in 4,175 journals from 1976 to 2004. These papers have 120,687 unique first authors from 24,468 institutions in 156 countries/regions.

Table 10-2. Number of nanotechnology papers in SCI database (1976-2004).

Year	Number of Papers
1976	90
1977	75
1978	77
1979	112
1980	113
1981	109
1982	104
1983	141
1984	138
1985	167
1986	275
1987	393
1988	619
1989	580
1990	1,060
1991	2,900
1992	3,818
1993	4,802
1994	6,422
1995	7,742
1996	9,743
1997	11,651
1998	13,601
1999	15,975
2000	18,085
2001	21,352
2002	25,697
2003	31,003

Table 10-2 presents the numbers of nanotechnology papers published in the SCI database each year from 1976 to 2004. Based on the log scale graph in Figure 10-1a, we observe that nanotechnology papers showed a significant growth pattern after about 1985 (possibly after the scanning tunneling microscope and atomic force microscope were developed). After 1990, nanotechnology papers continued to show an exponential growth pattern, which is indicated as a straight line on the log scale graph.

Figure 10-1b shows the annual nanotechnology paper publications by four major country groups: USA, Japan, European Countries, and Others. We observe that all country groups showed a significant growth pattern in paper publications. As the two major countries in nanotechnology, the U.S. and Japan produced a large number of papers. However, we notice that the group of European Countries has surpassed the U.S. in paper publication since 1996. In Figure 10-1b, the group of Others, which includes several new productive countries such as China and Korea, experienced an exceptional growth pattern after 1997. In 1999, the group of Others surpassed the U.S. in paper publication. In 2003 it outperformed the group of European Countries and took the first place.

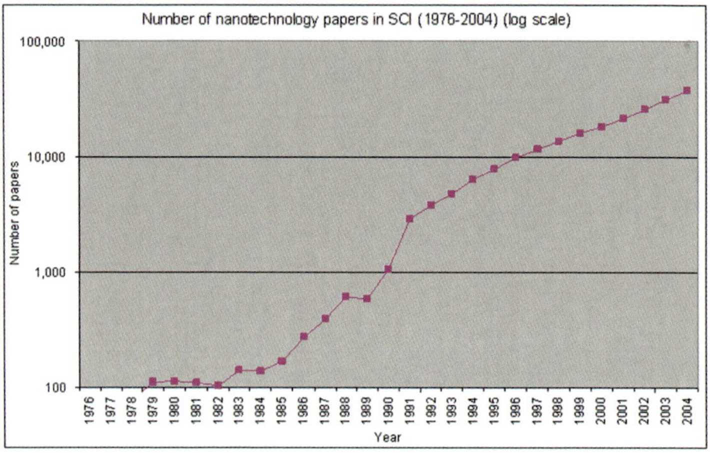

Figure 10-1a. Number of nanotechnology papers in SCI database (1976-2004) (log scale).

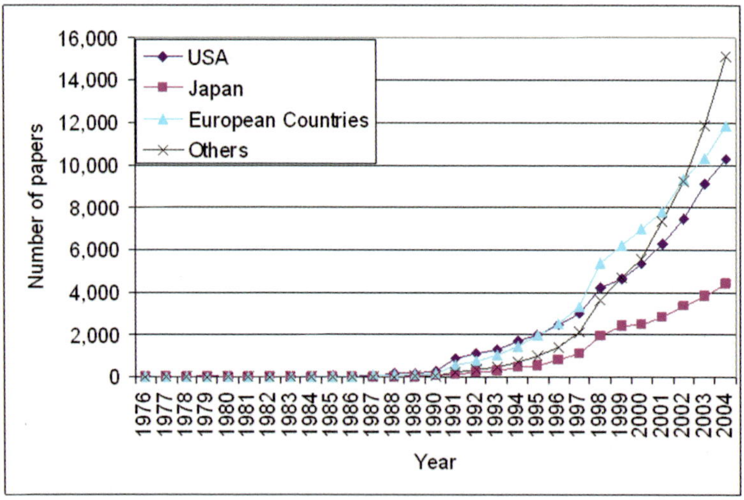

Figure 10-1b. Number of nanotechnology papers by country groups in SCI database (1976-2004) (normal scale).

Table 10-3 shows the top 20 countries or regions with the most nanotechnology papers published from 1976 to 2004. In the case of multiple authors from different countries (or institutions), the first author's country (or institution) affiliation was used to represent the paper since the first author is often the major contributor of the reported research. The United States authors generated the largest number of papers (61,068), followed by Japan (24,985), Germany (21,334), and China (P.R.) (20,389). In comparison with the top 20 countries with the most nanotechnology patents filed in the USPTO as reported in Chapter 4 (see Table 4-2, 1976-2002), three countries that were not ranked in the list showed significant productivity in paper publications: China (#4), Russia (#7), and India (#13). There are several possible reasons for this finding. All of these countries have had significant economic advances in the past decade. It is not surprising that their researchers are increasingly contributing to the fast-growing nanotechnology research field. It is expected that those academic contributions (as shown in SCI papers) will begin to be reflected in an increased number of patents and commercial products. In Chapter 7, India and China (P.R.) were ranked #17 and #19, respectively, for the number of USPTO patents filed from 2001 to 2004 (see Table 7-2).

Table 10-3. Top 20 countries and regions in nanotechnology paper publication (1976-2004).

Rank	Country/Region	Number of Papers
1	USA	61,068
2	Japan	24,985
3	Germany	21,334
4	Peoples R China	20,389
5	France	13,777
6	England	10,394
7	Russia	7,466
8	Italy	6,879
9	South Korea	6,679
10	Canada	5,017
11	Spain	4,941
12	Switzerland	4,280
13	India	3,869
14	Netherlands	3,635
15	Sweden	3,062
16	China(Taiwan)	2,886
17	Australia	2,788
18	Poland	2,703
19	Israel	2,509
20	Belgium	2,409

Table 10-4 shows the top 20 institutions in nanotechnology paper publication. All top institutions were universities and national research centers rather than private companies. This is significantly different from the top 20 assignees in the USPTO database, as reported in Chapter 4 (Table 4-5, 1976-2002). Only the Regents of the University of California and MIT were on the top 20 assignee list in the USPTO database. Among the top 20 institutions in nanotechnology paper publication, six are in the United States, five are in Japan, three are in China (P.R.), and the remainder are in Europe. "Chinese Academy of Sciences" and "Russian Academy of Sciences" were the most productive by a significant margin. The concentration of research capabilities in large academies is a particularity of the respective two countries.

Table 10-4. Top 20 institutions in nanotechnology paper publication (1976-2004).

Rank	Institution	Country/Region	Number of Papers
1	Chinese Acad Sci	Peoples R China	5,858
2	Russian Acad Sci	Russia	3,720
3	Univ Tokyo	Japan	2,465
4	CNRS[1]	France	2,395
5	Univ Paris	France	2,374
6	Osaka Univ	Japan	2,134
7	Tohoku Univ	Japan	2,123
8	Univ Illinois	USA	1,960
9	Univ Calf Berkeley	USA	1,809
10	MIT	USA	1,595
11	Tokyo Inst Technol	Japan	1,477
12	Univ Cambridge	England	1,451
13	Univ Sci & Technol China	Peoples R China	1,445
14	CSIC[2]	Spain	1,439
15	Univ Texas	USA	1,438
16	Tsing Hua Univ	Peoples R China	1,387
17	Univ Calif Santa Barbara	USA	1,322
18	Kyoto Univ	Japan	1,309
19	Harvard Univ	USA	1,292
20	CNR[3]	Italy	1,266

Notes:
[1]CNRS (France) = Centre national de la recherche scientifique (National Scientific Research Center).
[2]CSIC (Spain) = Consejo Superior de Investigaciones Científicas (Spanish National Research Council).
[3]CNR (Italy) =Consiglio Nazionale delle Ricerche (National Research Council).

Table 10-5 shows the journals in which the largest number of nanotechnology papers was published. *Physical Review B*, *Abstracts of Papers of the American Chemical Society*, and *Applied Physics Letters* were the top three journals that published the most nanotechnology papers. Among the top 20 journals, most of them focused on physics, chemistry, and material science in the 29 year interval. However, other journals with focus on electronics, photonics, engineering and other fields emerged after 2000.

Table 10-5. Top 20 journals in nanotechnology paper publication (1976-2004).

Rank	Journal	Number of Papers
1	Physical Review B	8,207
2	Abstracts of Papers of the American Chemical Society	7,636
3	Applied Physics Letters	7,056
4	Langmuir	5,167
5	Journal of Applied Physics	4,732
6	Surface Science	4,362
7	Journal of Physical Chemistry B	3,754
8	Physical Review Letters	3,353
9	Journal of The American Chemical Society	2,809
10	Thin Solid Films	2,710
11	Journal of Vacuum Science & Technology B	2,583
12	Japanese Journal of Applied Physics Part 1-Regular Papers Short Notes & Review Papers	2,317
13	Applied Surface Science	2,189
14	Journal of Crystal Growth	1,969
15	Chemistry of Materials	1,930
16	Journal of Magnetism And Magnetic Materials	1,807
17	Chemical Physics Letters	1,786
18	Journal of Chemical Physics	1,723
19	Macromolecules	1,578
20	Advanced Materials	1,495

3. BIBLIOGRAPHIC ANALYSIS

In order to assess the publication trend of nanotechnology papers, we used bibliographic analysis to analyze papers at different analytical unit levels, including countries/regions, institutions, and journals. This section presents our analysis of publication activity by country, institution, and journal.

3.1 Country Publication Trend

Figure 10-2 shows the number of publications per year (in log scale) of the top 10 most productive countries/regions in nanotechnology in the interval 1976-2004. Overall, the United States published the most nanotechnology papers. In addition, we observe a fast and consistent growth pattern for most countries after 1990. Using a normal scale, Figure 10-3 shows the publication trend of the top ten countries (without the United States). Most countries showed a similar growth pattern, except for China and South Korea. Both countries outpaced their competitors in recent years.

China surpassed Japan in nanotechnology publication in 2003 and has been the second most prolific country since then. South Korea also showed rapid development after 2000. It surpassed Italy, Russia, and England and became the sixth most prolific country in nanotechnology publication in 2004.

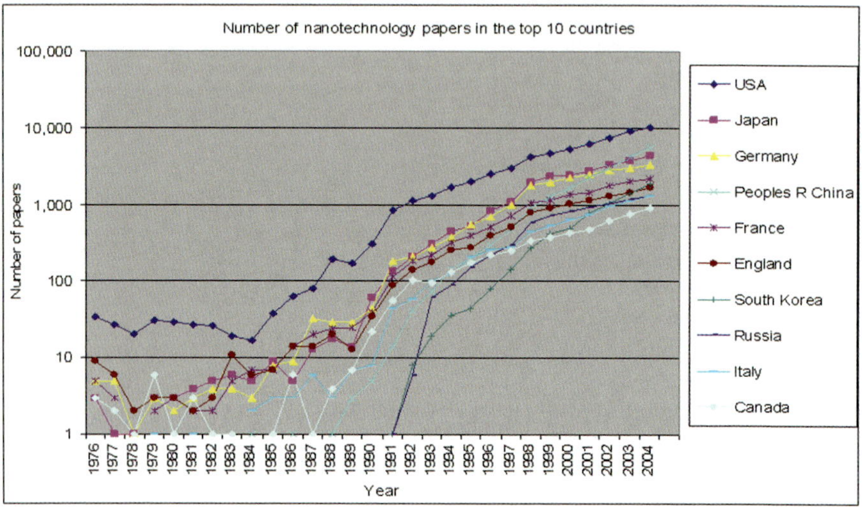

Figure 10-2. Top 10 countries/regions in nanotechnology paper publication per year (1976-2004) (log scale).

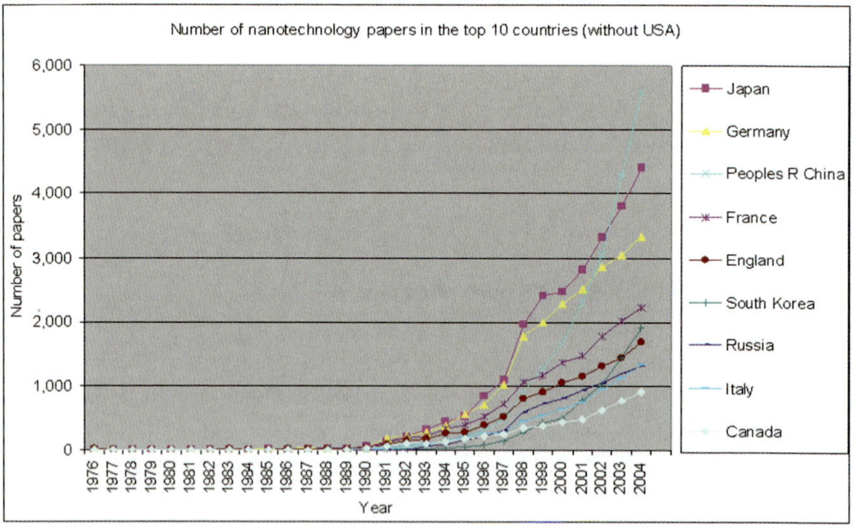

Figure 10-3. Top 10 countries/regions (without USA) in nanotechnology paper publication per year (1976-2004).

3.2 Institution Publication Trend

Figure 10-4 shows the publication trend of the top 10 most productive institutions. Most of the top 10 institutions had a steady growth pattern. "Chinese Academy of Sciences" and "Russian Academy of Sciences" showed the fastest growth since about 1998, outpacing other competing institutions.

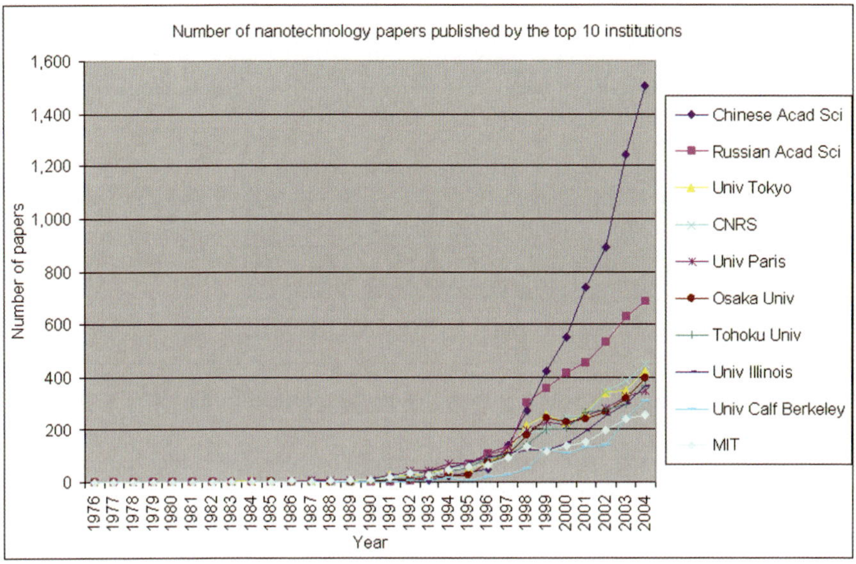

Figure 10-4. Top 10 institutions in nanotechnology paper publication (1976-2004).

3.3 Journal Publication Trend and High-impact Journals

Figure 10-5 shows the publication trend in the top 10 journals that publish nanotechnology papers. Most of them exhibited a growth pattern. Similar to the statistics shown in Table 10-5, *Physical Review B*, *Abstracts of Papers of the American Chemical Society*, and *Applied Physics Letters* were the three journals that published the most nanotechnology papers.

We list the top 20 journals (with at least 20 nanotechnology-related papers each) based on the average number of cites from our nanotechnology papers in Table 10-6. Unlike the ranking in Figure 10-5, *Nucleic Acids Research*, *Journal of Molecular Liquids*, and *Organic & Biomolecular Chemistry* were the top three journals based on the average number of cites from other nanotechnology papers. The total number of cites from papers covering all topical areas is much larger in the case of highly cited papers (up to one or two order of magnitude larger for nanotechnology papers published in journals such as *Science* and *Nature*). This is an indication of the interdisciplinary nature of nanoscale science and engineering.

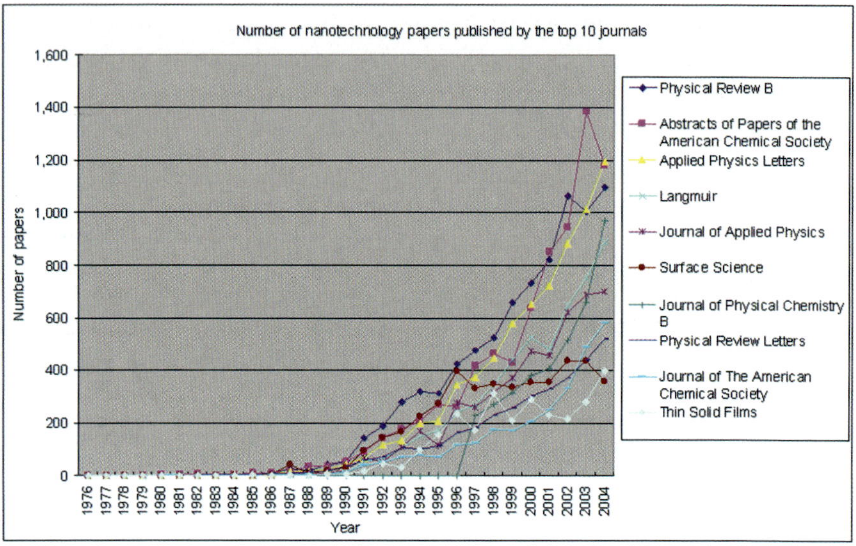

Figure 10-5. Top 10 journals in nanotechnology paper publication (1976-2004).

Table 10-6. Top 20 journals (with more than 20 papers) based on the average number of cites from other nanotechnology papers in the SCI database.

Rank	Journal	No. of Papers	Average No. of Cites
1	Nucleic Acids Research	190	0.49
2	J. of Molecular Liquids	40	0.48
3	Organic & Biomolecular Chemistry	54	0.41
4	J. of Analytical Atomic Spectrometry	47	0.34
5	Pharmaceutical Research	60	0.28
6	Connective Tissue Research	25	0.28
7	Chirality	46	0.20
8	Micron	109	0.17
9	Technology Review	44	0.16
10	Polymer Degradation & Stability	94	0.15
11	Chemical Record	31	0.13
12	Physical Review D	130	0.12
13	Integrated Ferroelectrics	123	0.11
14	Microelectronic Engineering	837	0.11
15	Advances in Chemistry Series	28	0.11
16	Analytical Sciences	116	0.10
17	Materials & Manufacturing Processes	42	0.10
18	JSME Intl. J. Series A-Solid Mechanics & Material Eng.	23	0.09
19	J. of the Optical Society of America B-Optical Physics	97	0.08
20	J. of Chemical Education	75	0.08

4. CONTENT MAP ANALYSIS

We generated content maps using the titles and abstracts of the identified nanotechnology papers. We divided our collection into three time periods based on the history of nanotechnology development: 1976-1989, 1990-1999, and 2000-2004. Since the records in the SCI database from 1976 to 1989 do not contain abstracts, we were unable to generate a content map for that time period. Thus, we report and compare the content maps for 1990-1999 and 2000–2004. To show the change of topics, we compute a growth rate for each topic area as the ratio between the number of documents in the interval 2000-2004 and that of the interval 1990-1999 for the same topic. The growth rates are visualized in the content map using different colors. Defining a baseline growth rate as the ratio between the total number of documents in 2000-2004 and that of 1990-1999, a topic region with similar growth rate to the baseline growth rate is assigned a green color. A topic region with a higher (lower) growth rate is assigned a warmer (colder) color. If the topic is brand new, a red color is assigned to the region.

4.1 Content Map Analysis for 1990-1999

Figure 10-6 shows the content map for the papers from 1990 to 1999. Important topics include: research tools (e.g., "Scanning Tunneling Microscopies," "Transmission Electron Microscopy," "Atomic Force Microscope"), physical phenomena (e.g., "Quantum Dots," "Single Crystals," "Self-Assembled Monolayers," "Nanomolar Concentrations," "Porous Silicon," "Surface Morphologies"), and experiment environments (e.g., "Electric Fields," "X-Ray Diffraction," "Temperature Dependences," "Activation Energies"). These topics were mentioned in thousands of papers and reflected the experiments conducted at that time and the phenomena the researchers observed.

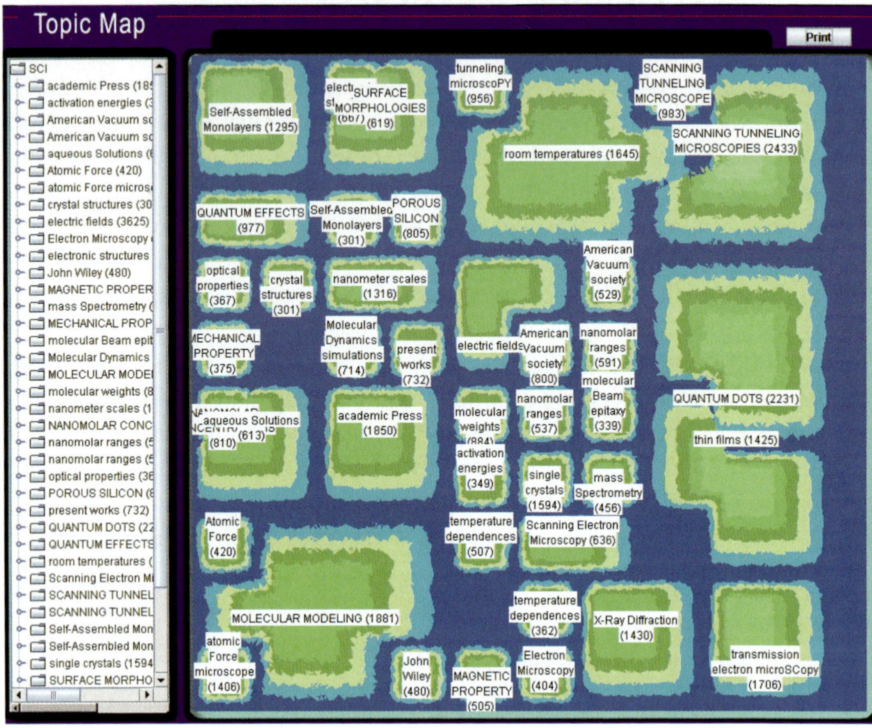

Figure 10-6. SCI nanotechnology papers' content map (1990-1999).

4.2 Content Map Analysis for 2000-2004

Figure 10-7 shows the content map of nanotechnology papers published between 2000 and 2004. It also visualizes the topic growth rates as compared to the previous interval (1990-1999) in different colors.

Table 10-7 compares the topics of the two time periods shown on the content maps. It arranges the topics in three sections: (1) the emerging topics, which are sorted in descending order by the number of papers in each region; (2) the developing topics, which are sorted in descending order by their growth rate; and (3) the disappearing topics, which are sorted in ascending order by the number of papers in the region. We observe that research in the two time periods shared several topics, which were related to research tools (e.g., "Scanning Tunneling Microscopies," "Transmission Electron Microscopy," "Atomic Force Microscope"), physical phenomena (e.g., "Quantum Dots," "Self-Assembled Monolayers"), and experiment environments (e.g., "X-Ray Diffraction"). In general, the number of papers in these topics showed a pattern of growth. However, some research topics related to specific results and experiment environments slowed or disappeared, such as: "Electric Fields," "Single Crystals," "Molecular Dynamics Simulations," etc. Some new topics introduced to the field during 2000-2004 included: "Carbon Nanotubes," "Quantum Effects," "Heat Treatments," "X-Ray Photoelectron Spectroscopy," and "Magnetic Fields."

Figure 10-7. SCI papers' content map (2000-2004).

Table 10-7. Topic changes of nanotechnology papers from (1990-1999) to (2000-2004).
* N/A indicates that there are too few papers to report on the content map or the growth rate cannot be computed

	Region Label	# of papers in the region (1990-1999)	# of papers in the region (2000-2004)	Growth Rate
Emerging topics	Carbon Nanotubes	N/A*	4,185	N/A
	Quantum Effects	N/A	2,706	N/A
	Heat Treatments	N/A	2,215	N/A
	Average Diameters	N/A	2,111	N/A
	X-Ray Photoelectron Spectroscopy	N/A	1,697	N/A
	Magnetic Fields	N/A	1,060	N/A
	Magnetic Property	505	6,426	11.72
	Molecular Beam Epitaxy	339	1,754	4.17
	Optical Properties	367	1,323	2.60

	Region Label	# of papers in the region (1990-1999)	# of papers in the region (2000-2004)	Growth Rate
Developing Topics	Self-Assembled Monolayers	1,596	5,627	2.53
	Nanomolar Concentrations	810	2,541	2.14
	Transmission Electron Microscopy	1,706	5,321	2.12
	Electron Microscopy	404	1,236	2.06
	Atomic Force microscope	1,406	3,663	1.61
	Thin Films	1,425	3,513	1.47
	Scanning Electron Microscopy	636	1,482	1.33
	X-Ray Diffraction	1,430	2,832	0.98
	Molecular Modeling	1,881	3,642	0.94
	Molecular Weights	884	1,190	0.35
	Scanning Tunneling Microscopies	3,416	4,182	0.22
	Crystal Structures	301	340	0.13
	Quantum Dots	2,231	2,003	-0.10
	Nanometer Scales	1,316	1,027	-0.22
	Electric Fields	3,625	968	-0.73
	Single Crystals	1,594	379	-0.76
Vanishing topics	Activation Energies	349	N/A	N/A
	Mechanical Property	375	N/A	N/A
	Atomic Force	420	N/A	N/A
	Mass Spectrometry	456	N/A	N/A
	Aqueous Solutions	613	N/A	N/A
	Surface Morphologies	619	N/A	N/A
	Electronic Structures	667	N/A	N/A
	Molecular Dynamics Simulations	714	N/A	N/A
	Present Works	732	N/A	N/A
	Porous Silicon	805	N/A	N/A
	Temperature Dependences	869	N/A	N/A
	Tunneling Microscopy	956	N/A	N/A
	Quantum Effects	977	N/A	N/A
	Nanomolar Ranges	1,128	N/A	N/A
	American Vacuum Society	1,329	N/A	N/A
	Room Temperatures	1,645	N/A	N/A
	Academic Press	1,850	N/A	N/A
	Baseline Growth Rate			0.766

5. CITATION NETWORK ANALYSIS

We analyzed the nanotechnology paper citation networks for major countries and institutions. We conducted network topological analysis for all literature published between 1976 and 2004.

5.1 Country Citation Network

Table 10-8 reports the topological measures of the country citation network and institution citation network. The country citation network consists of 66 countries, 348 inter-country citation relations, and 9 self-citation relations. The network contains only one component. In other words, the researchers in every country directly or indirectly affect each other through citations. The country citation network's clustering coefficient (0.693) is much larger than that of a random network (0.162) of the same size, which indicates that countries tend to form citation clusters in nanotechnology research.

Table 10-8. The topological attributes of the country citation network and institution citation network (1976-2004).

Measures	Country Citation Network	Institution Citation Network
l: average path length	2.019	4.050
l_{rand}*: average path length of a random network*	1.779	4.440
C: clustering coefficient	0.693	0.069
C_{rand}*: clustering coefficient of a random network*	0.162	0.004
D: network diameter	4	11
N_C*: number of components*	1	20
$Node_c$*: number of nodes in the largest component*	66 (100%)	1,185 (95.8%)
$Link_c$*: number of links in the largest component*	348(100%)	3,041 (98.9%)

Figure 10-8 shows the country citation network from 1976 to 2004, which contains the top 100 most frequent citation relations. In this graph, the United States was the largest citation center. Germany, Japan, France, China, and Russia were the secondary citation centers. These major citation centers had close citation relationships among them. In comparison to the USPTO country citation network shown in Figure 4-15 (for 1976-2002), China and Russia were new citation centers not reported in previous patent networks.

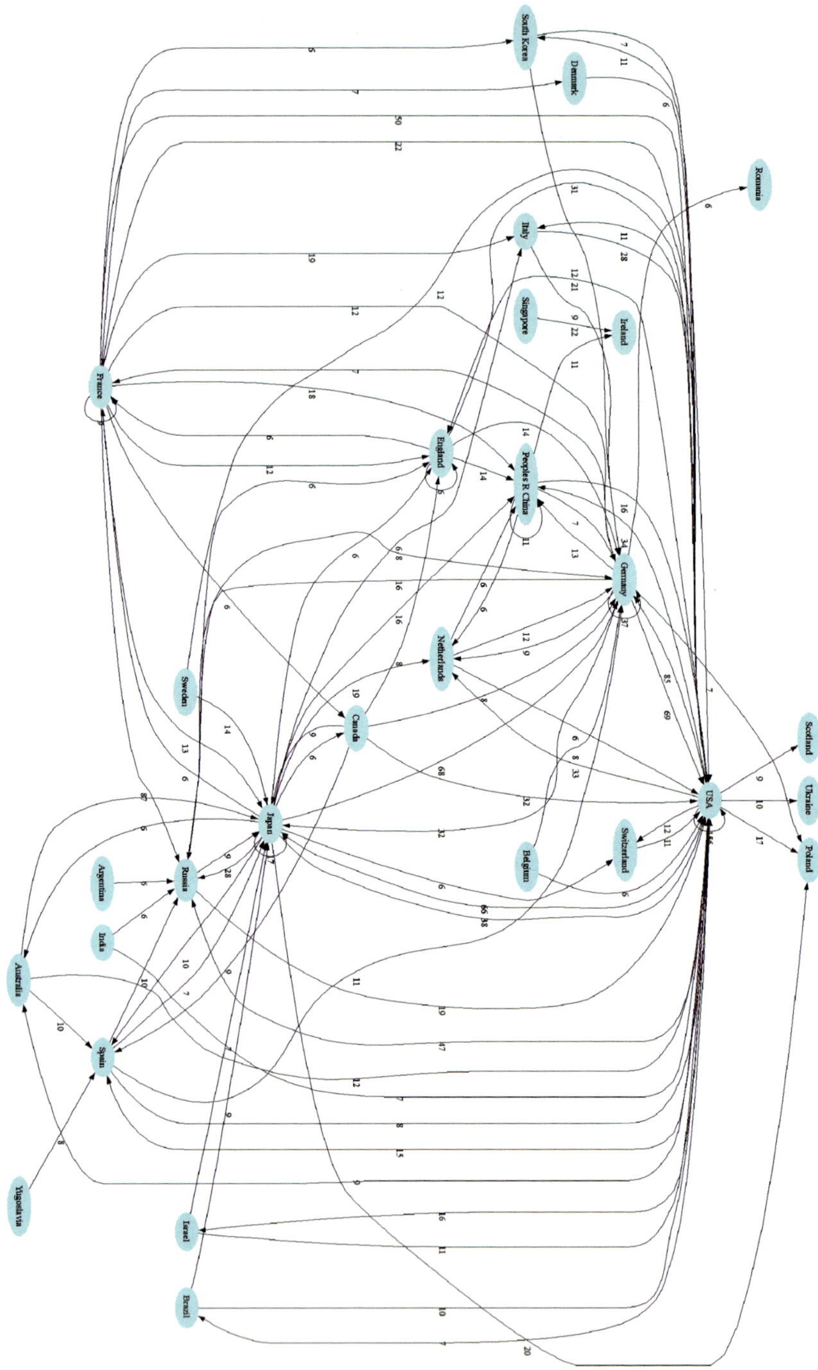

Figure 10-8. SCI nanotechnology papers' country citation network (1976-2004).

5.2 Institution Citation Network

The institution citation network of papers published between 1976 and 2004 consists of 1,237 institutions, 3,075 inter-institution citation relations, and 7 self-citation relations. As shown in Table 10-8, the institution citation network consists of 20 disconnected components. The largest one contains 1,185 (95.8%) institutions and 3,041 (98.9%) relations. Thus, most institutions working in the nanotechnology field interact with others directly or indirectly through citation relations. Similar to the country citation network, the institution citation network also has a much larger clustering coefficient (0.069) than the random network (0.004) of the same size, which indicates that research institutions in the nanotechnology field have a very high tendency to form citation clusters. Different from the country citation network, the institution citation network has a smaller average path length (4.050) than a random network (4.440) of the same size. Such a small-world characteristic suggests that knowledge can transfer between institutions more easily in this network than in a random network.

Figure 10-9 shows the institution citation network for 1976-2004. In this network, "University of Houston," "Baylor College of Medicine," and "Triplex Pharmaceutical Corporation" were the largest citation centers and created the largest citation cluster. "Moscow Mv Lomonosov State University" and "Eindhoven University of Technology" were also large citation centers and created another citation cluster. The "Chinese Academy of Sciences" and the "Russian Academy of Sciences" (the two most productive institutions in 2003-2004) were not among the major citation centers. This may be because their publication surges are quite recent. We expect they will become major citation centers over time as new papers are more frequently cited.

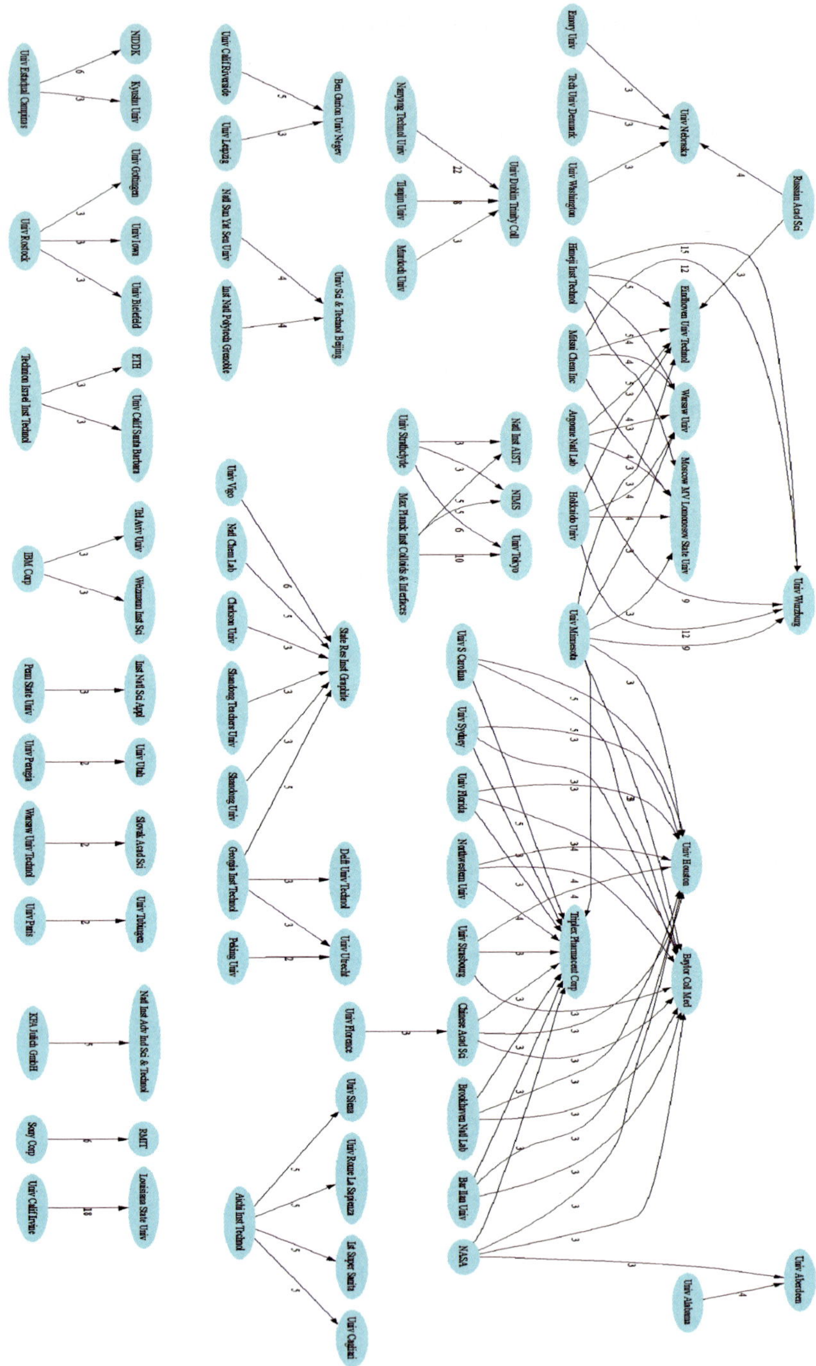

Figure 10-9. SCI nanotechnology papers' core institution citation network (1976-2004).

6. CONCLUSIONS

Using nanotechnology papers in the SCI database from 1976 to 2004, we conducted a large-scale, longitudinal study using bibliographic, content map, and citation network analyses. The investigation was made on 213,847 nanotechnology papers that were published in 4,175 journals in this interval. Key findings are:

- Nanotechnology papers have a significant growth pattern after 1985. They continued to show an exponential growth pattern after 1990. Between 1976 and 2004, the United States authors generated the largest number of papers (61,068), followed by Japan (24,985), Germany (21,334), and China (20,389).
- Three countries not ranked in the top 20 countries with the most nanotechnology patents filed in the USPTO show significant productivity in paper publications: China (#4 in ranking after papers), Russia (#7), and India (#13).
- Most countries present a similar publication growth pattern, except for China and South Korea. These two countries outpaced other countries in recent years. China surpassed Japan in nanotechnology publication in 2003 and has been the second most prolific country since then. South Korea also showed rapid development after 2000. It surpassed Italy, Russia, and England and became the sixth most prolific country in nanotechnology publication in 2004.
- All top 20 productive institutions were universities and national research centers rather than private companies. The "Chinese Academy of Sciences" and "Russian Academy of Sciences" were the most productive and exhibit the fastest growth. The concentration of research efforts in institutes of the Academy is a particularity of those countries.
- *Physical Review B*, *Abstracts of Papers of the American Chemical Society*, and *Applied Physics Letters* were the top three journals that published the most nanotechnology papers.
- Using content map analysis on two time periods (1990-1999 and 2000-2004), we observe that several research topics are shared, particularly concerning research tools (e.g., "Scanning Tunneling Microscopies," "Transmission Electron Microscopy," "Atomic Force Microscope"), physical phenomena (e.g., "Quantum Dots," "Self-Assembled Monolayers"), and experiment environments (e.g., "X-Ray Diffraction"). In addition, some new topics were introduced to the field during 2000-2004 such as: "Carbon Nanotubes," "Quantum Effects," "Heat Treatments," "X-Ray Photoelectron Spectroscopy," and "Magnetic Fields."

- Using citation network analysis, the United States was found to be the largest citation center. Germany, Japan, France, China, and Russia were the secondary citation centers. These major citation centers had close citation relationships among them. China and Russia are citation centers only for papers that are not in the patent citation networks based on the USPTO database.
- In the institution citation networks, "University of Houston," "Baylor College of Medicine," and "Triplex Pharmaceutical Corporation" were the largest citation centers and created the largest citation cluster.

7. QUESTIONS FOR DISCUSSION

1. What are the most important countries in academic nanotechnology research in recent years? What areas of nanotechnology research are they focusing on?
2. What institutions are most prominent in academic nanotechnology research in recent years? What areas of research are they focusing on?
3. What are the obvious linkages and spillovers between academic papers and patents for major nanotechnology researchers, institutions, and countries?
4. What are the most likely recent academic nanotechnology innovations that could have the highest patenting and commercial success?

Chapter 11

THE NANO MAPPER SYSTEM: ACCESSING AND VISUALIZING NANOTECHNOLOGY PATENTS AND GRANTS

CHAPTER OVERVIEW

Nanotechnology research has experienced rapid growth in recent years. Advances in information technology enabled us to analyze the publication, content, and relationships of nanotechnology-related documents to understand the domain's development status. This chapter presents our effort to build a knowledge mapping system, Nano Mapper (http://nanomapper.eller.arizona.edu), which integrates the analysis of nanotechnology patents and grants into a Web-based platform. The Nano Mapper system contains nanotechnology-related patents from the United States Patent and Trademark Office (USPTO), European Patent Office (EPO), and Japan Patent Office (JPO), and grant documents from the National Science Foundation (NSF). It provides simple search functionalities and makes available a set of analysis and visualization tools that can be applied on different levels of analytical units at different time periods for the patent and grant data. Using the statistics, trend graphs, citation networks, and content maps generated by Nano Mapper, we report nanotechnology innovations in USPTO patents in 2005-2006. Nano Mapper simplifies the patent/grant analysis processes and demonstrates the feasibility of building large-scale Web-based cross-database knowledge mapping systems to access and analyze innovations in various science and engineering disciplines.

1. INTRODUCTION

Nanotechnology has been recognized as an important indicator of a country's technological competence. More than 65 countries have adopted national projects or programs to promote nanotechnology research.

Analyzing scientific documents has become one of the major methods for technology assessment. Patents have been used to analyze the innovations with potential commercial value to in nanotechnology (Meyer, 2001; Huang et al., 2003; Huang et al., 2004), grant documents to study the effect of public funding on nanotechnology (Huang et al., 2005; Roco, 2005), and academic literature to represent the research efforts in academia (Schummer, 2004; Kostoff et al., 2006). In the past 30 years of nanotechnology development, a large number of scientific documents have been generated and are available in various databases around the world. However, previous studies mainly focused on applying certain analytical techniques on specific data sets (in specific time periods and for specific countries/regions) to answer specific research questions. Few of them aimed to make the analytical tools and data sets available to the public for various uses.

Web-based knowledge mapping systems have the potential to support different technology assessment applications. A number of technical challenges need to be addressed for such systems:

- *Distributed collection of data and documents:* Patents are published by patent offices in different countries and have distinct format requirements. Academic literature are published in various journals and collected in different databases. Collecting nanotechnology-related documents from multiple databases under a single group of indicators is a non-trivial task.
- *Non-structured data formats:* Although digitalized documents have been widely used in the storage of patents, grants, and other types of documents, such documents usually contain different data fields. Significant efforts are needed for data parsing and preprocessing in order to make the non-structured data ready for analysis.
- *Implementation of analysis tools:* Depending on document characteristics and data fields, analysis tools may need to be tailored accordingly. The underlying algorithms may need to be optimized in order to analyze large volume data sets in a real-time fashion.

Due to these challenges, few knowledge mapping systems for scientific document analysis are available. In our research, we proposed a framework to build such knowledge mapping systems to analyze nanotechnology's development status. We develop a prototype system, Nano Mapper, which provides integrated Web access to a variety of visualization and analytical tools for nanotechnology patents from the United States Patent and Trademark Office (USPTO), European Patent Office (EPO), and Japan Patent Office (JPO), and grants from the National Science Foundation (NSF). In the current system, we do not include academic literature for copyright reasons.

In this chapter, we briefly review previous patent and grant analysis studies and summarize existing nanotechnology Web portal research. Then, we present the Nano Mapper system architecture and its major functionalities. Lastly, we present a case study of the nanotechnology patents published in the USPTO in 2005-2006 using Nano Mapper.

2. RESEARCH BACKGROUND

2.1 Patent Analysis

Patents contain rich contents regarding technology innovations. A large number of patents published in the patent offices worldwide are publicly available. As an important indicator of technological advance, patents have been widely used to assess the research and development (R&D) status of different domains (Narin, 1994; Karki, 1997; Oppenheim, 2000), including nanotechnology (Huang et al., 2003) and gastroenterology (Lewison, 1998). In the nanotechnology domain, Meyer studied the interrelationships between academia and industry using patents from the USPTO and scientific literature from the Thomson Science Citation Index database (Meyer, 2001). Hullmann et al. used bibliometric measures on both patents and literature to assess nanotechnology's status in the 1980s and 1990s (Hullmann and Meyer, 2003). Huang et al. extended previous studies and developed a patent analysis framework that included bibliometric analysis, content analysis, and citation analysis to assess nanotechnology development at the country, institution, and technology field levels (Huang et al., 2003; 2004; 2005; 2006).

Patents are managed differently by patent offices around the world. Using data from a single office may lead to biased analysis results. Previous research found that domestic applicants tend to file more patents with their home country patent office than foreign applicants do ("home advantage" effect) (European Commission, 1997). Such "home advantage" effect affects the composition of patents in different patent databases (Ganguli, 1998; Criscuolo, 2006). In addition, different patent offices have different examination procedures and policies, which may also affect patent application results and contents. Patents from various major patent offices worldwide should be considered for a comprehensive global assessment (Li et al., 2007b).

2.2 Grant Analysis

A significant amount of public funding has been devoted to nanotechnology R&D worldwide after 2000 as shown in Chapter 1. More than 5% of the National Science Foundation (NSF) budget was dedicated to support nanotechnology research in 2005 (Roco, 2005), while the overall U.S. government funding for nanotechnology was about 1% of the total federal R&D. The major portion of nanotechnology funding in Europe is from the European Commission and individual country governments (Hullmann, 2006). Japan and Korea have the largest national R&D investments in nanotechnology.

Previous research has studied the impact of public funding on research and innovation in different domains by analyzing grant documents. Many of these studies used scientific publications as indicators of research output (Adams and Griliches, 1998; Arora and Gambardella, 1998; Narin, 1998; Payne and Siow, 2003) and found that the impact of public funding is dependent on the particular technology field. In the nanotechnology domain, Huang et al. (2005, 2006) studied the relationship between NSF funding and patent publications (see Chapters 5 and 7). It was found that the patents published by NSF-funded researchers had significantly higher impact in the nanotechnology domain compared with other reference groups.

2.3 Web Portals for Nanotechnology

Many Web portals have been built to collect nanotechnology-related information (Table 11-1). Some of these focus on providing nanotechnology-related news articles, interviews, and research reports, such as: "Nanotechnology Now," "Nano Tsunami," and "Nano Science & Technology Institute." Some aim to build a hub of URLs to nanotechnology online resources, including: Websites, forums, books, journals, and databases, e.g., "ENS Nanotechnology Portal" and "Nano Scout." Some sites provide the history of nanotechnology development and an introduction to the domain, e.g., Wikipedia Nanotechnology Portal. Due to the importance of nanotechnology in various fields, we anticipate other such portals to emerge in the near future.

Table 11-1. Selected nanotechnology Web portals.

Web Portal	Focus
ENS Nanotechnology Portal http://www.ensbio.com/nanotechnologyPortal.html	URLs to online resources
Nanotechnology Now http://www.nanotech-now.com/	News and research reports
Nano Science & Technology Institute http://www.nsti.org/	News and academic conference information
Nano Scout http://www.nanoscout.de/	URLs to online resources
Nano Tsunami http://www.nano-tsunami.com/	News
Wikipedia Nanotechnology Portal http://en.wikipedia.org/wiki/Portal:nanotechnology	Nanotechnology history and introduction

These Web portals can help researchers find general nanotechnology-related information and news; however, they do not provide access to high-quality nanotechnology-related scientific documents (e.g., scientific literature and patents) or provide support for various technology assessment functions. To the best of our knowledge, advanced knowledge mapping functions have not been implemented in existing nanotechnology Web portals. Building online portals with patent and grant search and analysis functionalities may better assist nanotechnology researchers and policy makers in making more timely decisions.

3. NANO MAPPER SYSTEM DESIGN

In this research, we propose a framework to build knowledge mapping systems for patent analysis and grant analysis in the nanotechnology domain. As shown in Figure 11-1, the framework contains three components: data acquisition, parsing, and system building. Our prototype system is called the Nano Mapper (http://nanomapper.eller.arizona.edu).

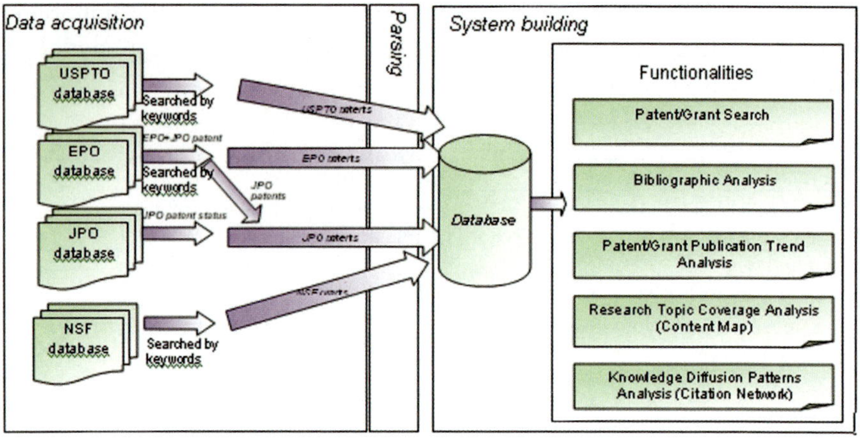

Figure 11-1. A framework for building nanotechnology knowledge mapping systems.

3.1 Data Acquisition

In the data acquisition step, we use keyword search to collect nanotechnology-related patents and grants. As shown in Table 11-2, a list of nanotechnology keywords provided by domain experts is used to search and retrieve documents from the online interface of the databases (data through 2006). In Nano Mapper, patents are collected from USPTO, EPO, and JPO due to their important role in nanotechnology research. Grants are retrieved from the NSF.

- USPTO provides online full-text access to patents issued since 1976, which can be searched using almost any of a patent's data fields.
- EPO's database, esp@cenet, provides access to European patents issued since 1978, which can be searched based on title, abstract, and some bibliographic information. Esp@cenet also stores patent applications from more than 80 countries.
- The JPO patent database, Patent Abstracts of Japan (PAJ), contains the patents issued since 1976. This system is difficult to search due to system limitations. We chose to retrieve JPO patent applications from esp@cenet and check their publication status (application or registered patent) through PAJ. Only registered patents are kept in our database.
- NSF provides online access to grant abstracts, which can be searched using almost any data field.

Table 11-2. Nanotechnology keywords and the number of patents collected from USPTO, EPO, JPO and grants collected from NSF.

Keywords	USPTO patents (1976-2006)			EPO patents (1978 - 2006)	JPO patents (1976 - 2006)	NSF grants (1991 -2006)
	Title-abstract search	Title-claims search	Full-text search	Title-abstract search	Title-abstract search	Title-abstract search
atomic force microscope	277	465	3,020	71	67	241
atomic force microscopic	2	6	91	2	1	16
atomic force microscopy	91	143	2,347	23	8	430
atomic-force-microscope	0	0	6	0	0	40
atomic-force-microscopy	0	0	5	0	0	67
atomistic simulation	0	0	10	0	0	107
biomotor	0	1	8	1	0	0
molecular device	9	22	230	5	3	371
molecular electronics	5	5	422	4	3	384
molecular modeling	34	51	2,365	3	1	1255
molecular motor	2	3	99	4	0	135
molecular sensor	0	9	48	2	1	185
molecular simulation	2	2	73	1	1	449
nano*	6,352	15,973	90,093	3,248	847	8,121
quantum computing	28	41	144	4	1	471
quantum dot*	160	267	988	64	90	524
quantum effect*	40	65	699	18	67	435
scanning tunneling microscope	148	218	1,284	47	80	190
scanning tunneling microscopic	0	1	25	0	1	8
scanning tunneling microscopy	28	52	996	8	0	326

Keywords	USPTO patents (1976-2006)			EPO patents (1978 - 2006)	JPO patents (1976 - 2006)	NSF grants (1991 -2006)
	Title-abstract search	Title-claims search	Full-text search	Title-abstract search	Title-abstract search	Title-abstract search
scanning-tunneling-microscope	0	0	24	0	0	11
scanning-tunneling-microscopy	0	1	1	0	0	24
selfassembl*	3	4	31	1	0	13
self assembly	161	268	2,692	46	7	316
self assembled	251	460	2,672	38	1	241
self assembling	131	208	1,237	57	5	187
self-assembled	233	426	2,506	0	0	570
self-assembling	120	189	1,127	0	0	286
self-assembly	142	239	2,478	0	5	772
Total	8,219	19,119	115,721	3,647	1189	16,175
Unique Total	**7,406**	**17,544**	**97,509**	**3,596**	**1150**	**10,114**

Different databases provide different search interfaces and functions. All four databases support keyword search on document titles and abstracts ("title-abstract" search). USPTO supports more complex search functions. We also search USPTO patents by matching the keywords in patent title, abstract, and claims ("title-claims" search) and on the entire patent document ("full-text" search) (Huang et al., 2003). In general, "title-abstract" search provides more accurate results while the other two search methods provide better coverage of nanotechnology-related patents. Table 11-2 shows the number of documents collected with each nanotechnology keyword from the four databases by different search methods.

3.2 Parsing

The documents retrieved from online databases are usually in the html format. These documents need to be parsed and stored in a relational database for Web access. In general, each data source needs a separate parser. However, since the search interfaces seldom change, the parsers can be reused to update data collections for the system periodically. In the Nano Mapper system, the patent parsers extract patent identification information (patent ID, patent application number, patent priority number), bibliographic information (publication date, inventor name, applicant name), classification information (international classification, United States classification,

European classification), citation information, and content information (title, abstract, claims, and description) from patents. The grant parsers extract grant ID, bibliographic information (start and expiration date, grant amount, principal investigator), funding agent information (NSF organization, program, and directorate), and content information (title, abstract) from grants.

3.3 System Building

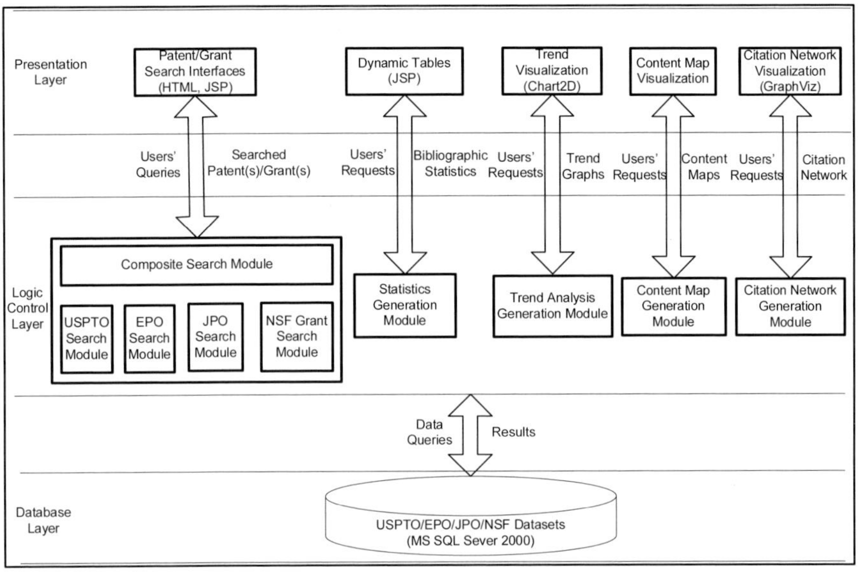

Figure 11-2. Nano Mapper system architecture.

After parsing the collected documents into a database, a knowledge mapping system can be built based on the architecture shown in Figure 11-2. The system is based on a conventional Web-based, three-layer architecture which contains a presentation layer, a logic control layer, and a database layer.

The presentation layer implements the user interface and provides Web access to five types of functions: search function, basic statistics, trend analysis, citation network analysis, and content map analysis. The search and statistics functions are implemented with JSP dynamic pages. The visualizations are implemented using several Java Applet packages. We customized an open source Java library - Chart 2D (see URL: http://chart2d.sourceforge.net) - to visualize patent and grant publication trends in charts and an open source graph drawing software - Graphviz, provided by AT&T Labs (http://www.research.att.com/sw/tools/graphviz) (Gansner, 2000) - to visualize the citation networks. We used the Topic Map package developed by the Artificial Intelligence Lab at the University of Arizona (Chen et al., 1996) to visualize the content maps of nanotechnology-related patents and grants.

At the logic control layer, (SQL) is used to perform search and analytical functions. To handle large data sets and provide online analysis of statistics, trends, and citation networks, some pre-computing is conducted and the publication statistics and citation statistics are summarized on a yearly basis. For content analysis, we identify major technology topics from the nanotechnology documents and generate content maps using the self-organization map (SOM) algorithm (Chen et al., 1996; Ong et al., 2005). This is a time-consuming process, so only content maps for selected time periods are made available. At the database layer, we use Microsoft SQL Server 2000 to store parsed patent and grant data.

3.4 Nano Mapper System Functionalities

3.4.1 Search Functions

The Nano Mapper system provides three functions to search patents and grants:

- searching using patent/grant identifiers;
- searching keywords in title, abstract, or (patent) claims; and
- searching by a combination of criteria on different patent/grant data fields (implemented as Advanced Search).

Nano Mapper also provides a combined search function, which is conducted on all four data sets. The results from the four databases are shown together in one interface and can be browsed and compared.

Figure 11-3 illustrates the advanced search function using the USPTO data set. This interface enables users to input criteria on most data fields (Figure 11-3a). For USPTO patents, these include patent title, examiner, inventor, assignee, assignee country, classification code, abstract, claims, etc. For some categorical data fields, e.g., assignee country, the interface provides lookup functions to help find appropriate search criteria. The result set for a query is sorted by publication date in a reverse order (Figure 11-3b). The user can browse the results using the navigation bar at the bottom. The user can also access the details of any documents, including all data fields in our system and the URLs to their original Web sites (Figure 11-3c).

Figure 11-3a. Nano Mapper Advanced Search interface

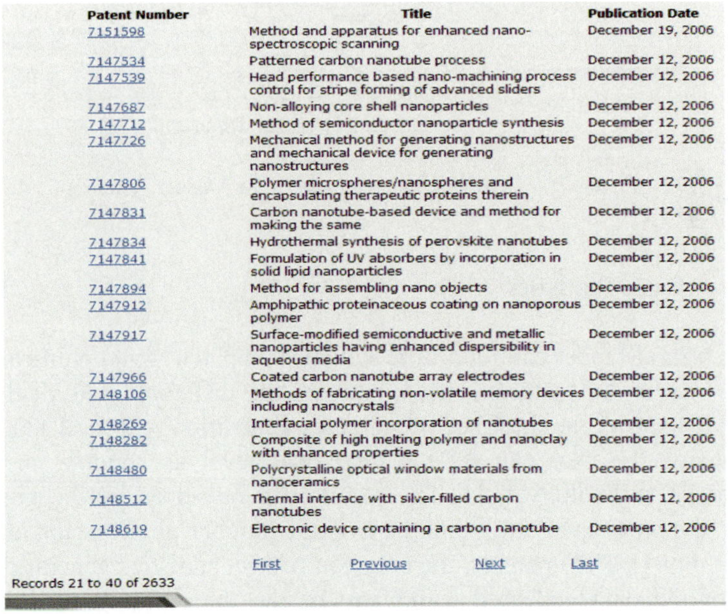

Patent Number	Title	Publication Date
7151598	Method and apparatus for enhanced nano-spectroscopic scanning	December 19, 2006
7147534	Patterned carbon nanotube process	December 12, 2006
7147539	Head performance based nano-machining process control for stripe forming of advanced sliders	December 12, 2006
7147687	Non-alloying core shell nanoparticles	December 12, 2006
7147712	Method of semiconductor nanoparticle synthesis	December 12, 2006
7147726	Mechanical method for generating nanostructures and mechanical device for generating nanostructures	December 12, 2006
7147806	Polymer microspheres/nanospheres and encapsulating therapeutic proteins therein	December 12, 2006
7147831	Carbon nanotube-based device and method for making the same	December 12, 2006
7147834	Hydrothermal synthesis of perovskite nanotubes	December 12, 2006
7147841	Formulation of UV absorbers by incorporation in solid lipid nanoparticles	December 12, 2006
7147894	Method for assembling nano objects	December 12, 2006
7147912	Amphipathic proteinaceous coating on nanoporous polymer	December 12, 2006
7147917	Surface-modified semiconductive and metallic nanoparticles having enhanced dispersibility in aqueous media	December 12, 2006
7147966	Coated carbon nanotube array electrodes	December 12, 2006
7148106	Methods of fabricating non-volatile memory devices including nanocrystals	December 12, 2006
7148269	Interfacial polymer incorporation of nanotubes	December 12, 2006
7148282	Composite of high melting polymer and nanoclay with enhanced properties	December 12, 2006
7148480	Polycrystalline optical window materials from nanoceramics	December 12, 2006
7148512	Thermal interface with silver-filled carbon nanotubes	December 12, 2006
7148619	Electronic device containing a carbon nanotube	December 12, 2006

First Previous Next Last

Records 21 to 40 of 2633

Figure 11-3b. Summary list of patent search results.

METHOD AND APPARATUS FOR ENHANCED NANO-SPECTROSCOPIC SCANNING

Patent Number	7151598 [Link to USPTO]
Publication Date	December 19, 2006
Inventor	Vladimir - Poponin (United States)
Assignee	Poponin; Vladimir (United States)
Application Number	10/959,238
Filed	October 5, 2004
Primary Examiner	Evans; F. L.
Assistant Examiner	-
Attorney, Agent or Firm	Perkins Coie LLP
Current U.S. Class	356/301 , 435/287.2 , 435/288.7
Internation Class	G01J 3/44 , G01N 21/65

Reference

References Cited	[Referenced By]	
U.S. Patent Documents	**Patent Number**	**Publication date**
	6002471	December 1, 1999
	6441359	August 1, 2002
	5479024	December 1, 1995
	6850323	February 1, 2005
Other References		
	Copy of International Search Report from PCT Application No. PCT/US04/10544. cited by other.	

Abstract

Apparatus and method for examining the identity of chemical groups in a sample are disclosed. The apparatus has a substrate having a plasmon resonant surface on which the sample is supported, a source of a beam of light, and a lens assembly having a tip region and a nanolens composed of one or more plasmon resonance particles (PRPs) on the tip region. The PRPs are arranged to produce near-field electromagnetic gap modes in a space between the nanolens and a confronting detection region on the substrate surface when the gap between the nanolens and substrate is 30 nm or less. A focusing mechanism in the apparatus operates to move the lens assembly toward and away from the substrate surface, with a gap of less than 30 nm, to produce electromagnetic gap modes that enhance the Raman spectroscopy signals produced by the sample in the detection region.

Claim

The invention claimed is:

Figure 11-3c. Details of a patent document.

Figure 11-3. Advanced search in the Nano Mapper system.

3.4.2 Basic Statistics

The Nano Mapper can report the statistics on patent/grant publication and citation status for selected time periods on different analytical levels. Figure11-4 shows the interface of statistics generation with USPTO patents. For patents, the user can set the analytical level as country, institution, inventor, or technology field. The results can be sorted by the number of patents, the number of cites, and the average number of cites each analytical unit got. For USPTO patents, the user can restrict statistics generation in the range of the data collected using any of the three search methods. The statistics can be downloaded in CSV format for further offline analysis.

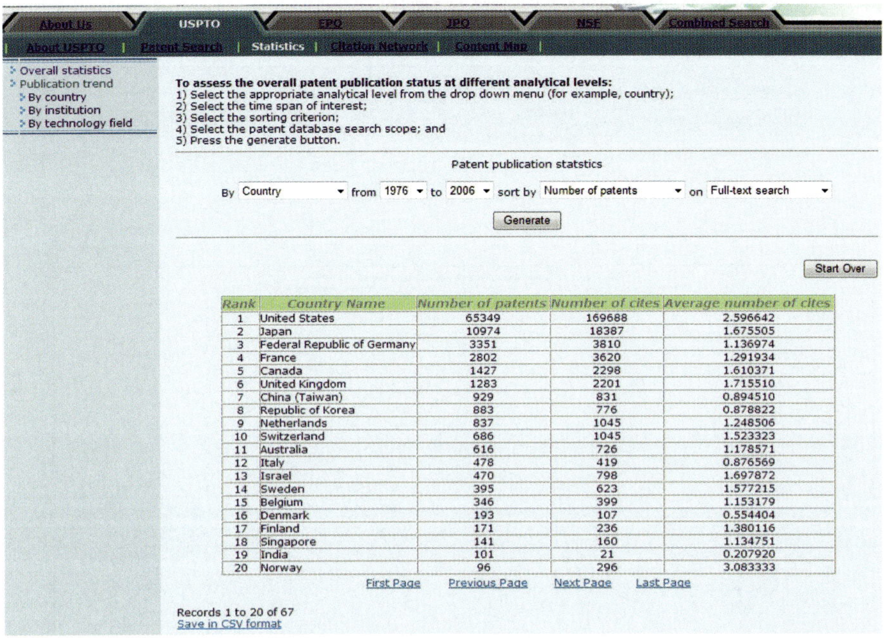

Rank	Country Name	Number of patents	Number of cites	Average number of cites
1	United States	65349	169688	2.596642
2	Japan	10974	18387	1.675505
3	Federal Republic of Germany	3351	3810	1.136974
4	France	2802	3620	1.291934
5	Canada	1427	2298	1.610371
6	United Kingdom	1283	2201	1.715510
7	China (Taiwan)	929	831	0.894510
8	Republic of Korea	883	776	0.878822
9	Netherlands	837	1045	1.248506
10	Switzerland	686	1045	1.523323
11	Australia	616	726	1.178571
12	Italy	478	419	0.876569
13	Israel	470	798	1.697872
14	Sweden	395	623	1.577215
15	Belgium	346	399	1.153179
16	Denmark	193	107	0.554404
17	Finland	171	236	1.380116
18	Singapore	141	160	1.134751
19	India	101	21	0.207920
20	Norway	96	296	3.083333

Figure 11-4. Country level statistics for nanotechnology-related patents in USPTO.

3.4.3 Publication Trend Analysis

Nano Mapper can visualize and compare the annual publication trends of patents and grants at different analytical levels. Figure 11-5 shows an example of the country level analysis on USPTO patents for the United States, Japan, Germany, France, and Canada. To add analytical units of interest, the user can search for names in a pop-up window. The interface also provides shortcuts to add the top 10 or the top 11 to 20 most productive analytical units into the comparison. The analysis results include a line chart and a table of statistics showing different units' number of publications in each year. The statistics can be downloaded in CSV file format for further offline analysis.

To compare various countries' nanotechnology patent publication trends:
1) Select countries of interest using the buttons on the left and right of the list;
2) Select the patent database search scope by using the drop down menu (the difference between the three search methods); and
3) Press the "Compare" button.

Select countr(ies) to add ->
Add top 10 countries ->
Add next top 10 countries ->
Add groups ->

Please add the countries you want to compare:
United States
Japan
Federal Republic of Germany
France

-> Select and remove

Compare

on Full-text search

Start Over

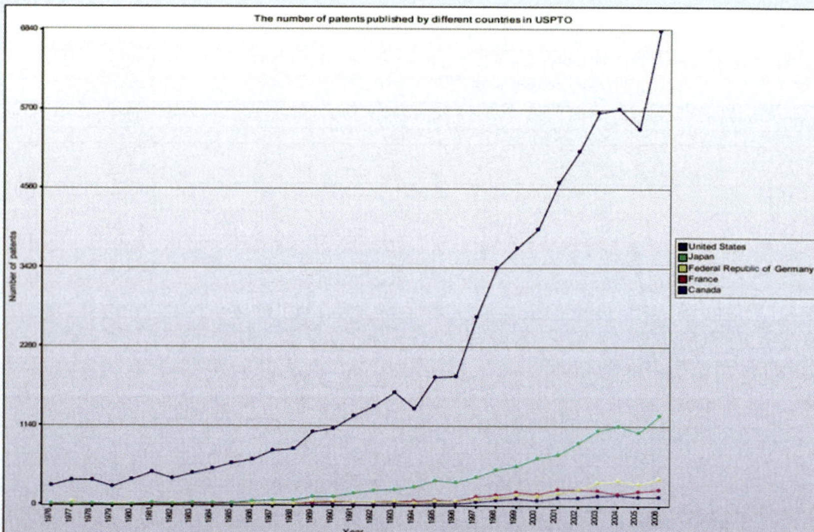

Year	United States	Japan	Federal Republic of Germany	France	Canada
1976	291	38	20	13	6
1977	363	25	52	11	3
1978	372	34	38	17	1
1979	272	22	30	12	2
1980	386	25	32	20	5
1981	486	42	44	16	6
1982	396	28	43	19	4
1983	466	39	45	30	5
1984	535	49	43	14	10
1985	607	44	47	41	8
1986	663	70	30	30	12
1987	794	86	31	34	14
1988	822	87	33	37	12
1989	1059	138	67	47	13
1990	1113	133	67	58	18
1991	1295	185	59	57	23
1992	1445	225	55	62	24
1993	1632	243	66	65	29
1994	1402	273	46	67	22
1995	1853	362	72	88	37
1996	1866	339	67	73	37
1997	2714	421	95	140	52
1998	3430	525	128	155	84
1999	3710	577	157	194	108
2000	3985	696	151	171	81
2001	4656	746	188	240	120
2002	5104	912	242	236	119
2003	5654	1091	341	228	149
2004	5706	1150	362	179	157
2005	5433	1063	311	210	130
2006	6839	1306	389	238	136

Save in CSV format

Figure 11-5. Country level publication trend analysis of nanotechnology-related patents in USPTO.

3.4.4 Citation Network Analysis

As shown in Figure 11-6, Nano Mapper enables users to visualize patent citation networks at different analytical levels for different time periods. To emphasize the more important citation relationships, the top 100 relationships between analytical units with the largest number of citations are visualized. In the citation networks, the direction of a link represents the direction of the citations between two nodes. For example, a link from "United States" to "Germany" means that the United States' patents cited German patents. Each link is labeled with the total number of citations.

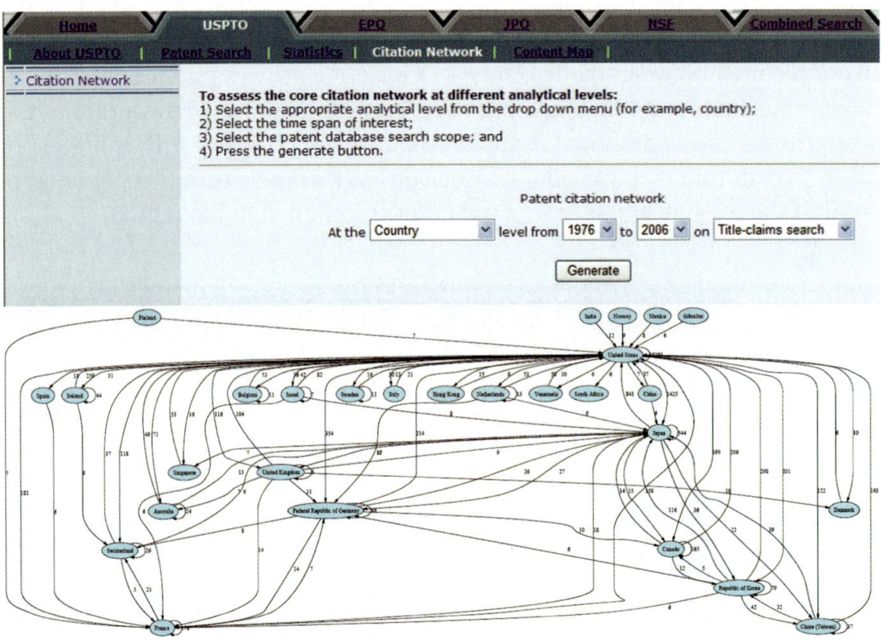

Figure 11-6. Country citation network of USPTO nanotechnology-related patents ("title-claims" search, 1976-2006).

3.4.5 Content Map Analysis

Nano Mapper adopts the content map technology to identify and visualize major nanotechnology topics at different time periods in the document titles and abstracts. The research topics are represented by noun phrases and extracted from patent/grant documents using a Natural Language Processing tool, the Arizona Noun Phraser. The topics are organized by the multi-level self-organization map algorithm (Chen et al., 1996; Ong et al., 2005) and visualized by the topic map interface. As Figure

11-7 shows, the topic map interface contains two components: a folder tree and a hierarchical content map. The folder tree displays the topics identified from nanotechnology-related patents/grants. The hierarchical content map displays corresponding topic regions in the map. Each topic region is labeled with the topic keyword and the number of documents. The size of a topic region is proportional to the number of documents related to that topic. If the user clicks a topic region, the sub-topics will be expanded on the interface. If there are no sub-topics, the documents related to the selected topic will be shown.

Since generating a content map is time-consuming, we pre-generate a set of content maps for a sequence of time periods for each data set. For the content maps of two continuous time periods, we compute the growth rate of each topic area between the two maps. A baseline growth rate is computed at the entire content map level. A topic region with a similar growth rate to the base growth rate is assigned a green color. A topic region with a higher or lower growth rate is assigned a warmer or colder color, respectively (Figure 11-7). If the topic is brand new, a red color is assigned to the region.

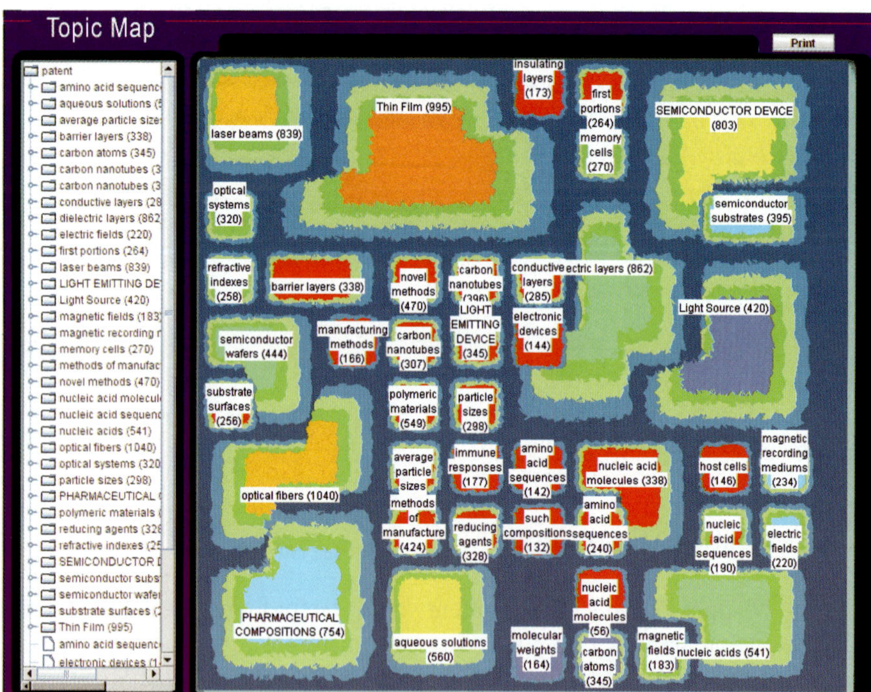

Figure 11-7. The content map for topics in USPTO nanotechnology-related patents from 2000 to 2004.

4. CASE STUDY: NANOTECHNOLOGY DEVELOPMENT IN USPTO (2005-2006)

The Nano Mapper enables us to perform dynamic, timely, and highly customized patent and grant analysis. Using the system, we present a case study of the nanotechnology development status in USPTO patents between 2005 and 2006. This is an extension of our previous USPTO longitudinal studies (Huang et al., 2003; 2004; 2005; 2006).

In the Nano Mapper system we collected nanotechnology patents issued by the USPTO from 1976 to 2006 using three search methods. (The output files have been re-formatted and enlarged in tables and figures for clarity.) Figure 11-8 shows the annual publication of nanotechnology-related patents in the USPTO from 1976 to 2006. Although the three search methods have different coverage, they show a similar growth pattern of nanotechnology development. In 2005-2006, the rapid growth of nanotechnology patent publication continues with a slight fluctuation (in 2005).

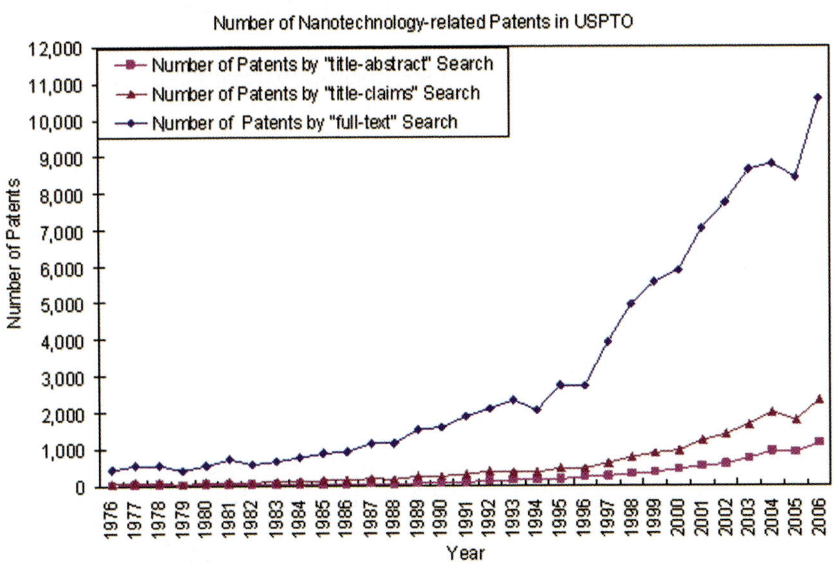

Figure 11-8. Number of nanotechnology patents in USPTO using three types of search methods (1976-2006).

4.1 Country Analysis: (1976-2004) vs. (2005-2006)

Table 11-3 presents the top 10 most productive nanotechnology assignee countries in the USPTO in 1976-2004 and 2005-2006 using "title-abstract" search. The United States and Japan continued to be the top 2 countries in 2005-2006. China (Taiwan), Republic of Korea, and Netherlands had rapid growth. Their ranks improved significantly among all the countries. China (P.R.) and Israel entered the top 10 assignee countries lists in 2005-2006.

Table 11-3. Most productive assignee countries by "title-abstract" search (USPTO nanotechnology-related patents, 1976-2004 and 2005-2006).

	1976-2004		2005-2006	
Rank	Assignee country	No. of patents	Assignee country	No. of patents
1	United States	3,450	United States	1,322
2	Japan	517	Japan	205
3	Federal Rep. of Germany	204	China (Taiwan)	95
4	France	156	Republic of Korea	89
5	Republic of Korea	131	Federal Rep. of Germany	66
6	Canada	104	Canada	34
7	China (Taiwan)	71	France	30
8	United Kingdom	60	Netherlands	22
9	Netherlands	54	China	21
10	Switzerland	41	Israel	14

4.2 Institution Analysis: (1976-2004) vs. (2005-2006)

Table 11-4 shows the top 10 assignee institutions that have published the largest number of nanotechnology patents in the USPTO.

Table 11-4. Most productive assignees by "title-abstract" search (USPTO nanotechnology-related patents, 1976-2004 and 2005-2006).

	1976-2004		2005-2006	
Rank	Assignee institution	No. of patents	Assignee institution	No. of patents
1	IBM Corp.	171	The Regents of the Univ. of CA	61
2	The Regents of the Univ. of CA	123	Hewlett-Packard Dev. Co., L.P.	40
3	The USA as represented by the Secretary of the Navy	82	IBM Corp.	38
4	Eastman Kodak Company	72	William Marsh Rice University	37
5	Minnesota Mining and Manufacturing Company	59	Intel Corporation	36
6	Mass. Institute of Technology	56	Samsung Electronics Co., Ltd.	33
7	Xerox Corporation	55	Industrial Tech. Research Inst.	27
8	Micron Technology, Inc.	53	Micron Technology, Inc.	22
9	Matsushita Electric Industrial Co. Ltd.	45	Nanosys, Inc.	20
10	L'Oreal	44	Mass. Institute of Technology	20

International Business Machines Corporation, The Regents of the University of California, and Micron Technology, Inc. continued to be the most productive institutions in 2005-2006, as in 1976-2004. Several institutions, including Hewlett-Packard, Rice University, Samsung Electronics, and Intel Corporation, had a significant increase in nanotechnology patent publication in 2005-2006.

5. CONCLUSIONS

This chapter presents our effort in developing a knowledge mapping system to assess nanotechnology's development status based on patent and grant analysis. We developed a prototype system, Nano Mapper, which contains nanotechnology-related patents from the USPTO, EPO, and JPO and grants from NSF. The Nano Mapper system provides search functions, statistics, trend analysis, citation network analysis, and content map analysis for nanotechnology patents and grants.

Using Nano Mapper, we analyzed nanotechnology patents published in 2005-2006 in the USPTO. We found that nanotechnology patent publication continues the previous growth trend. In 2005-2006, China (Taiwan), Republic of Korea, and Netherlands had rapid growth in patent publication in USPTO. International Business Machines Corporation, The Regents of the University of California, and Micron Technology, Inc. continued to be productive institutions. Hewlett-Packard, Rice University, Samsung Electronics, and Intel Corporation had a significant increase in nanotechnology publication.

The Nano Mapper system provides a Web-based portal infrastructure for researchers and policy makers. In our future research, we will update the data sets in the system annually to provide a platform for researchers to track the latest developments in nanotechnology. We also plan to incorporate other types of scientific documents into our framework and introduce additional analytical and visualization methods. The general framework proposed here for nanotechnology can be applied to other technology fields.

6. QUESTIONS FOR DISCUSSION

1. What are some other major nanotechnology-related conferences and online resources for researchers, policy makers, and industry practitioners?
2. What other data sources could be considered for inclusion in Nano Mapper? What other functionalities would be useful to incorporate into a knowledge mapping system such as Nano Mapper?
3. What other types of analysis could be useful for researchers, policy makers, and industry practitioners?

APPENDIX

In order to provide up-to-date statistics for the recent trend in patent publications, we collected and analyzed the patent publications in the USPTO, EPO, and JPO from 2005-2007. Recent patent publications in the Thomson SCI database were not available to us at the time of writing.

For USPTO, the data was collected from the USPTO Web site on 01/11/2008. At that time the USPTO Web site contained data up to 01/08/2008. We extracted the patents published from 01/01/2005 to 12/31/2007. The latest patent we got was published on 12/25/2007.

For EPO and JPO, the data was collected from their respective Web sites on 01/21/2008. We extracted the patents published from 01/01/2005 to 12/31/2007. The latest EPO patent we got was published on 12/26/2007. However, the latest JPO patent was published on 04/25/2007, as JPO has a slower process of making patent data available on their Web site.

A. KEY COUNTRIES, INSTITUTIONS, AND INVENTORS IN USPTO DATABASE, 2005-2007

A1. Full-Text Search

Table A-1. Nanotechnology patent assignee countries in USPTO ("full-text" search, 2005-2007).

Rank	Assignee Country	Number of Patents	Rank	Assignee Country	Number of Patents
1	United States	13,506	27	Liechtenstein	13
2	Japan	2,653	28	Barbados	13
3	Fed. Rep. of Germany	836	29	British Virgin Islands	7
4	France	534	30	New Zealand	7
5	China (Taiwan)	428	31	South Africa	6
6	Rep. of Korea	406	32	Gibraltar	5
7	Canada	333	33	Cayman Islands	4
8	Netherlands	325	34	Netherlands Antilles	3
9	Australia	276	35	Malaysia	3
10	United Kingdom	258	36	Hungary	3
11	Switzerland	193	37	Mexico	2
12	Israel	163	38	Poland	2
13	Sweden	108	39	Portugal	2
14	Belgium	106	40	Saudi Arabia	1
15	Italy	82	41	St. Christopher-Nevis-Anguilla	1
16	Singapore	70	42	Thailand	1
17	China	66	43	The Bahamas	1
18	Denmark	56	44	Western Samoa	1
19	Finland	51	45	Niger	1

Rank	Assignee Country	Number of Patents	Rank	Assignee Country	Number of Patents
20	India	39	46	Panama	1
21	Hong Kong	33	47	Iceland	1
22	Bermuda	28	48	Macao	1
23	Ireland	26	49	Argentina	1
24	Austria	24	50	Brazil	1
25	Norway	23	51	Greece	1
26	Spain	15	52	Cyprus	1

Table A-2. Top 50 USPTO patent assignee institutions by number of patents published ("full-text" search, 2005-2007).

Rank	Assignee Institution	Country	Number of Patents	Average Patent Age
1	Micron Technology, Inc.	United States	446	1.45
2	International Business Machines Corp.	United States	329	1.42
3	Intel Corporation	United States	300	1.46
4	Hewlett-Packard Development Co., L.P.	United States	267	1.50
5	The Regents of the Univ. of Calif.	United States	236	1.43
6	Silverbrook Research Pty Ltd	Australia	206	1.08
7	Advanced Micro Devices, Inc.	United States	178	1.52
8	3M Innovative Properties Co.	United States	170	1.43
9	Kabushiki Kaisha Toshiba	Japan	165	1.40
10	Canon Kabushiki Kaisha	Japan	164	1.45
11	Genentech, Inc.	United States	161	1.19
12	Xerox Corporation	United States	153	1.37
13	Eastman Kodak Company	United States	151	1.38
14	General Electric Company	United States	126	1.44
15	Samsung Electronics Co., Ltd.	Rep. of Korea	126	1.33
16	Applied Materials, Inc.	United States	124	1.54
17	Matsushita Electric Industrial Co., Ltd.	Japan	123	1.41
18	Hitachi, Ltd.	Japan	119	1.57
19	Infineon Technologies AG	Fed. Rep. of Germany	118	1.41
20	Sony Corporation	Japan	117	1.37
21	Massachusetts Institute of Technology	United States	96	1.46
22	Fuji Photo Film Co., Ltd.	Japan	92	1.44
23	E. I. du Pont de Nemours and Company	United States	91	1.46
24	Seiko Epson Corporation	Japan	87	1.52
25	Freescale Semiconductor, Inc.	United States	86	1.48
26	Agilent Technologies, Inc.	United States	76	1.37
27	Fujitsu Limited	Japan	74	1.48
28	California Institute of Technology	United States	73	1.47
29	Sharp Kabushiki Kaisha	Japan	70	1.23
30	The USA as represented by the Secretary of the Navy	United States	66	1.55
31	Seagate Technology LLC	United States	65	1.47
32	Hitachi Global Storage Technologies Netherlands B.V.	Netherlands	63	1.43

Rank	Assignee Institution	Country	Number of Patents	Average Patent Age
33	ASML Netherlands B.V.	Netherlands	59	1.32
34	Honeywell International Inc.	United States	58	1.38
35	Bristol-Myers Squibb Company	United States	56	1.36
36	Industrial Technology Research Inst.	China (Taiwan)	56	1.41
37	Finisar Corporation	United States	54	1.41
38	Lucent Technologies Inc.	United States	54	1.39
39	Taiwan Semiconductor Manufacturing Co., Ltd.	China (Taiwan)	54	1.48
40	Board of Regents, The Univ. of Texas System	United States	53	1.47
41	Texas Instruments Incorporated	United States	52	1.33
42	Kimberly-Clark Worldwide, Inc.	United States	51	1.37
43	Metrologic Instruments, Inc.	United States	51	1.51
44	Samsung SDI Co., Ltd.	Rep. of Korea	51	1.20
45	Acushnet Company	United States	49	1.27
46	L'Oreal	France	49	1.55
47	Corning Incorporated	United States	47	1.43
48	Sharp Laboratories of America, Inc.	United States	47	1.40
49	Pfizer Inc.	United States	46	1.41
50	Semiconductor Energy Laboratory Co., Ltd.	Japan	46	1.45

Table A-3. Top 50 USPTO patent inventors by number of patents published ("full-text" search, 2005-2007).

Rank	Last Name	First Name	Middle Initial	No. of Patents	Rank	Last Name	First Name	Middle Name	No. of Patents
1	Silverbrook	Kia	-	143	26	Zhu	Xiaoxun	-	27
2	Forbes	Leonard	-	132	27	Huibers	Andrew	G.	26
3	Wood	William	I.	119	28	Kodas	Toivo	T.	26
4	Goddard	Audrey	-	118	29	Hsu	Sheng Teng	-	26
5	Gurney	Austin	L.	115	30	Wang	Haihong	-	26
6	Godowski	Paul	J.	109	31	Hampden-Smith	Mark	J.	25
7	Ahn	Kie	Y.	74	32	Hu	Chenming	-	25
8	Desnoyers	Luc	-	59	33	Nawaki	Masaru	-	25
9	Knowles	Harry	C.	43	34	Tsikos	Constantine	J.	25
10	Yu	Bin	-	41	35	Colbert	Daniel	T.	24
11	Gilton	Terry	L.	39	36	Hillan	Kenneth	J.	24
12	Watanabe	Colin	K.	38	37	King	Tsu-Jae	-	24
13	Moore	John	T.	35	38	Williams	Stanley	R.	24
14	Iwata	Hiroshi	-	32	39	Botstein	David	-	24
15	Smalley	Richard	E.	32	40	Morikawa	Yoshinao	-	23
16	Roy	Margaret Ann	-	31	41	Islam	Mohammed	N.	22
17	Shibata	Akihide	-	30	42	Yeo	Yee-Chia	-	22

Rank	Last Name	First Name	Middle Initial	No. of Patents	Rank	Last Name	First Name	Middle Name	No. of Patents
18	Walmsley	Simon Robert	-	30	43	Ouderkirk	Andrew	J.	22
19	Bhattacharyya	Arup	-	29	44	Ahmed	Shibly	S.	21
20	Sreenivasan	Sidlgata	V.	29	45	Pan	James	-	21
21	Lapstun	Paul	-	29	46	Rueckes	Thomas	-	21
22	Sasaki	Yoshitaka	-	29	47	Den	Tohru	-	21
23	Yamazaki	Shunpei	-	29	48	Gunn, III	Lawrence	C.	20
24	Campbell	Kristy	A.	28	49	Dai	Hongjie	-	20
25	Ferrara	Napoleone	-	27	50	Sharma	Manish	-	20

A2. Title-Abstract Search

Table A-4. Nanotechnology patent assignee countries in USPTO ("title-abstract" search, 2005-2007).

Rank	Assignee Country	Number of Patents	Rank	Assignee Country	Number of Patents
1	United States	1,457	17	Ireland	9
2	Japan	233	18	India	6
3	China (Taiwan)	106	19	Singapore	5
4	Republic of Korea	104	20	Hong Kong	4
5	Fed. Rep. of Germany	80	21	Finland	4
6	Canada	39	22	Gibraltar	2
7	France	35	23	Denmark	2
8	China	27	24	Austria	1
9	Netherlands	25	25	Barbados	1
10	Israel	19	26	Brazil	1
11	United Kingdom	16	27	Cyprus	1
12	Italy	14	28	New Zealand	1
13	Belgium	12	29	Norway	1
14	Australia	11	30	Liechtenstein	1
15	Sweden	10	31	Macao	1
16	Switzerland	9	32	Mexico	1

Table A-5. Top 50 USPTO patent assignee institutions by number of patents published ("title-abstract" search, 2005-2007).

Rank	Assignee Institution	Country	No. of Patents	Average Patent Age
1	The Regents of the Univ. of California	United States	61	1.50
2	Hewlett-Packard Development Co., L.P.	United States	51	1.49
3	International Business Machines Corp.	United States	40	1.39
4	William Marsh Rice Univ.	United States	39	1.33
5	Intel Corporation	United States	36	1.50
6	Samsung Electronics Co., Ltd.	Rep. of Korea	36	1.31

Rank	Assignee Institution	Country	No. of Patents	Average Patent Age
7	Industrial Technology Research Inst.	China (Taiwan)	27	1.37
8	Micron Technology, Inc.	United States	25	1.33
9	General Electric Company	United States	21	1.20
10	Massachusetts Institute of Technology	United States	20	1.30
11	Nanosys, Inc.	United States	20	1.20
12	Nantero, Inc.	United States	19	1.32
13	Eastman Kodak Company	United States	18	1.39
14	The USA as represented by the Secretary of the Navy	United States	17	1.59
15	Sharp Laboratories of America, Inc.	United States	15	1.47
16	Agilent Technologies, Inc.	United States	14	1.29
17	Canon Kabushiki Kaisha	Japan	14	1.50
18	Hon Hai Precision Ind. Co., Ltd.	China (Taiwan)	14	1.29
19	Nanosphere, Inc.	United States	14	1.50
20	The Board of Trustees of the Leland Stanford Junior Univ.	United States	14	1.74
21	3M Innovative Properties Company	United States	13	1.15
22	Rohm and Haas Company	United States	13	1.62
23	Freescale Semiconductor, Inc.	United States	12	1.50
24	Hitachi, Ltd.	Japan	12	1.33
25	Infineon Technologies AG	Fed. Rep. of Germany	12	1.33
26	Kabushiki Kaisha Toshiba	Japan	12	1.38
27	Tsinghua Univ.	China	11	1.36
28	Fuji Xerox Co., Ltd.	Japan	10	1.30
29	Hyperion Catalysis International, Inc.	United States	10	1.50
30	Samsung SDI Co., Ltd.	Rep. of Korea	10	1.60
31	The Board of Trustees of the Univ. of Illinois	United States	10	1.20
32	General Motors Corporation	United States	9	1.44
33	Honeywell International Inc.	United States	9	1.44
34	Matsushita Electric Industrial Co., Ltd.	Japan	9	1.56
35	Motorola, Inc.	United States	9	1.56
36	NanoProducts Corporation	United States	9	1.67
37	UT-Battelle, LLC	United States	9	1.44
38	Advanced Micro Devices, Inc.	United States	8	1.75
39	Battelle Memorial Institute	United States	8	1.38
40	California Institute of Technology	United States	8	1.50
41	Georgia Tech Research Corporation	United States	8	1.75
42	Northwestern Univ.	United States	8	1.38

Rank	Assignee Institution	Country	No. of Patents	Average Patent Age
43	Sandia Corporation	United States	8	1.75
44	Seagate Technology LLC	United States	8	1.25
45	Xerox Corporation	United States	8	1.47
46	Daiken Chemical Co., Ltd.	Japan	7	1.14
47	Fuji Photo Film Co., Ltd.	Japan	7	1.32
48	National Taiwan Univ.	China (Taiwan)	7	1.71
49	Philip Morris USA Inc.	United States	7	1.29
50	ASML Holding N.V.	Netherlands	6	1.00

Table A-6. Top 50 USPTO patent inventors by number of patents published ("title-abstract" search, 2005-2007).

Rank	Last Name	First Name	Middle Name	No. of Patents	Rank	Last Name	Middle Name	First Name	No. of Patents
1	Smalley	Richard	E.	32	26	Romano	T.	Linda	9
2	Colbert	Daniel	T.	24	27	Storhoff	J.	James	9
3	Rueckes	Thomas	-	21	28	Watanabe	-	Hiroyuki	9
4	Dai	Hongjie	-	20	29	Yadav	-	Tapesh	9
5	Hafner	Jason	H.	19	30	Zhou	Z.	Otto	9
6	Rinzler	Andrew	G.	19	31	Choi	-	Won-bong	8
7	Segal	Brent	M.	18	32	Fan	-	Shoushan	8
8	Guo	Ting	-	17	33	Hauge	H.	Robert	8
9	Liu	Jie	-	17	34	Jaiprakash	C.	Venkatachalam	8
10	Nikolaev	Pavel	-	17	35	Moy	-	David	8
11	Thess	Andreas	-	17	36	Rohrbaugh	-	Robert Henry	8
12	Mirkin	Chad	A.	16	37	Shimizu	-	Masaaki	8
13	Smith	Kenneth	A.	16	38	Wang	-	Xingwu	8
14	Bertin	Claude	L.	14	39	Chen	-	Jian	7
15	Niu	Chunming	-	14	40	Duan	-	Xiangfeng	7
16	Williams	Stanley	R.	13	41	Fink	-	Richard Lee	7
17	Letsinger	Robert	L.	11	42	Harada	-	Akio	7
18	Brock	Darren	K.	10	43	Liu	-	Liang	7
19	Den	Tohru	-	10	44	McDonald	-	Michael Ray	7
20	Elghanian	Robert	-	10	45	Mukherjee	K.	Amiya	7
21	Forbes	Leonard	-	10	46	Nakayama	-	Yoshikazu	7
22	Chen	Yong	-	9	47	Taton	A.	Thomas	7
23	Empedocles	Stephen	-	9	48	Zhang	-	Yuegang	7
24	Hsu	Sheng Teng	-	9	49	Anazawa	-	Kazunori	6
25	Mucic	Robert	C.	9	50	Bawendi	G.	Moungi	6

B. KEY COUNTRIES, INSTITUTIONS, AND INVENTORS IN THE EPO DATABASE, 2005-2007

Table B-1. Nanotechnology patent assignee countries in EPO ("title-abstract" search, 2005-2007).

Rank	Assignee Country	No. of Patents	Rank	Assignee Country	No. of Patents
1	United States	795	22	Barbados	8
2	Japan	318	23	Denmark	8
3	Fed. Republic of Germany	265	24	Finland	8
4	France	133	25	Turkey	5
5	Republic of Korea	133	26	Iran	4
6	Netherlands	68	27	Brazil	3
7	Italy	66	28	Mexico	3
8	Switzerland	62	29	New Zealand	3
9	United Kingdom	54	30	Norway	3
10	Spain	31	31	Poland	2
11	Ireland	31	32	British Virgin Islands	2
12	Israel	30	33	Greece	2
13	Belgium	29	34	Hungary	2
14	Sweden	19	35	Iceland	1
15	Canada	18	36	Portugal	1
16	China (Taiwan)	17	37	Jordan	1
17	Australia	14	38	Malaysia	1
18	Singapore	13	39	Saudi Arabia	1
19	China	12	40	Slovenia	1
20	India	12	41	Thailand	1
21	Austria	9			

Table B-2. Top 50 EPO patent assignee institutions by number of patents published ("title-abstract" search, 2005-2007).

Rank	Assignee Institution	Country	No. of Patents	Average Patent Age
1	Japan Science & Tech Agency	Japan	47	0.96
2	Commissariat Energie Atomique	France	38	0.82
3	Koninkl Philips Electronics NV	Netherlands	30	0.63
4	Samsung Electronics Co Ltd	Rep. of Korea	30	0.77
5	3M Innovative Properties Co	United States	29	0.52
6	Nanosys Inc	United States	29	0.86
7	Centre Nat Rech Scient	France	27	1.00
8	Elan Pharma Int Ltd	Ireland	27	0.81
9	Du Pont	United States	26	1.19
10	Gen Electric	United States	25	0.84
11	Samsung SDI Co Ltd	Rep. of Korea	20	0.85
12	Nantero Inc	United States	20	0.90

Rank	Assignee Institution	Country	No. of Patents	Average Patent Age
13	Univ California	United States	20	0.85
14	Univ Rice William M	United States	20	0.90
15	Agilent Technologies Inc	United States	19	1.11
16	Hewlett-Packard Development Co.	United States	16	1.44
17	Siemens AG	Fed. Rep. of Germany	13	0.46
18	Honda Motor Co Ltd	Japan	13	1.23
19	BASF AG	Fed. Rep. of Germany	12	0.58
20	Oreal	France	12	1.25
21	Fiat Ricerche	Italy	12	1.25
22	LG Electronics Inc	Rep. of Korea	12	0.75
23	Massachusetts Inst Technology	United States	12	0.83
24	Fuji Xerox Co Ltd	Japan	11	1.73
25	Nissin Kogyo KK	Japan	11	1.00
26	IBM	United States	11	0.55
27	Intel Corp	United States	11	1.45
28	Procter & Gamble	United States	11	0.64
29	Matsushita Electric Ind Co Ltd	Japan	10	1.10
30	Nat Inst of Advanced Ind Sci	Japan	10	1.10
31	Agency Science Tech & Res	Singapore	10	0.00
32	Fraunhofer Ges Forschung	Fed. Rep. of Germany	9	1.00
33	St Microelectronics Srl	Italy	9	0.67
34	DSM IP Assets BV	Netherlands	9	1.00
35	Consejo Superior Investigation	Spain	9	1.00
36	Boston Scient Ltd	Barbados	8	0.50
37	Centre Nat Rech Scient	France	8	0.00
38	Arkema France	France	8	0.00
39	Yissum Res Dev Co	Israel	8	0.88
40	Hitachi Ltd	Japan	8	1.00
41	Korea Electronics Telecomm	Rep. of Korea	8	1.00
42	Korea Inst Science Technology	Rep. of Korea	8	0.88
43	LG Chemical Ltd	Rep. of Korea	8	0.63
44	California Inst of Tech	United States	8	0.75
45	Eastman Kodak Co	United States	8	0.13
46	Lucent Technologies Inc	United States	8	1.88
47	Hitachi Software Eng	Japan	7	1.43
48	Toray Industries	Japan	7	0.57
49	Ciba SC Holding AG	Switzerland	7	1.14
50	Dow Global Technologies Inc	United States	7	1.14

Table B-3. Top 50 EPO patent inventors by number of patents published ("title-abstract" search, 2005-2007).

Rank	Last Name	First Name	Middle Name	No. of Patents	Rank	Last Name	First Name	Middle Name	No. of Patents
1	Segal	Brent	M.	18	26	Bertin	Claude	L.	7
2	Rueckes	Thomas	-	16	27	Kolb	Brant	U.	7
3	Niu	Chunming	-	14	28	Mascolo	Danilo	-	7
4	Jenkins	Scott	-	12	29	Okada	Shinsuke	-	7
5	Liversidge	Gary	-	12	30	Sahi	Vijendra	-	7
6	Magario	Akira	-	11	31	Samuelson	Lars	I.	7
7	Noguchi	Toru	-	11	32	Stumbo	David	-	7
8	Parce	Wallace	J.	10	33	Bawendi	Moungi	G.	6
9	Anazawa	Kazunori	-	9	34	Cerofolini	Gianfranco	-	6
10	Hirakata	Masaki	-	9	35	Chen	Jian	-	6
11	Romano	Linda	T.	9	36	Empedocles	Stephen	-	6
12	Ward	Jonathan	W.	9	37	Isozaki	Takashi	-	6
13	Watanabe	Hiroyuki	-	9	38	Kawakatsu	Hideki	-	6
14	Yasuda	Akio	-	9	39	Kishi	Kentaro	-	6
15	Barrera	Enrique	V.	8	40	Ma	Jun	-	6
16	Chhabra	Rajeev	-	8	41	Nakayama	Yoshikazu	-	6
17	Duan	Xiangfeng	-	8	42	Nie	Shuming	-	6
18	Innocenti	Gianfranco	-	8	43	Ooma	Shigeki	-	6
19	Isele	Olaf	E.A.	8	44	Pan	Yaoling	-	6
20	Manabe	Chikara	-	8	45	Pullini	Daniele	-	6
21	Repetto	Piermario	-	8	46	Ryde	Tuula	-	6
22	Shigematsu	Taishi	-	8	47	Sato	Keiichi	-	6
23	Watanabe	Miho	-	8	48	Schroder	Kurt	A.	6
24	Bakkers	Erik	P.A.M.	7	49	Star	Alexander	-	6
25	Baran, Jr	Jimmie	R.	7	50	Wyland	Chris	-	6

C. KEY INSTITUTIONS AND INVENTORS IN THE JPO DATABASE, 2005-2007

Table C-1. Top 50 JPO patent assignee institutions by number of patents published ("title-abstract" search, 2005-2007).

Rank	Assignee Institution	Country	No. of Patents	Average Patent Age
1	Nat Inst for Materials Science	United States	34	2.00
2	Nat Inst of Adv Ind & Technol	Japan	11	2.00
3	Japan Science & Tech Agency	Japan	10	1.90
4	Tokyo Shibaura Electric Co	Japan	7	1.57

Rank	Assignee Institution	Country	No. of Patents	Average Patent Age
5	Korea Inst Science Technology	Rep. of Korea	6	1.83
6	Hitachi Shipbuilding Eng Co	Japan	5	2.00
7	Nissei Plastics Ind Co	Japan	4	1.75
8	IBM	United States	4	1.75
9	Univ Meijo	Japan	4	1.25
10	Nakayama Yoshikazu	Japan	3	2.00
11	Samsung Sdi Co Ltd	Japan	3	1.33
12	Shinshu Univ	Japan	3	1.67
13	Tohoku Techno Arch Co Ltd	Japan	3	1.67
14	Daiken Kagaku Kogyo KK	Japan	2	2.00
15	Fujitsu Ltd	Japan	2	1.50
16	Horie Senko KK	Japan	2	2.00
17	Ind Tech Res Inst	China (Taiwan)	2	1.50
18	Inatome Boseki KK	Japan	2	2.00
19	Kawamura Inst Chem Res	Japan	2	2.00
20	Matsushita Electric Works Ltd	Japan	2	1.50
21	Nissin Kogyo KK	Japan	2	2.00
22	Nippon Telegraph & Telephone	Japan	2	2.00
23	Nippon Electric Co	Japan	2	2.00
24	Nikkiso Co Ltd	Japan	2	2.00
25	Nat Inst of Information & Comm	Japan	2	2.00
26	Noritake Co Ltd	Japan	2	2.00
27	Okayama Univ	Japan	2	2.00
28	Samsung Corning Co Ltd	Japan	2	2.00
29	Sharp KK	Japan	2	0.50
30	Takeyari KK	Japan	2	2.00
31	Univ California	United States	2	1.00
32	Univ Nagoya	Japan	2	2.00
33	Univ Shinshu	Japan	2	2.00
34	Univ Tokyo	Japan	2	1.50
35	Nissan Motor	Japan	1	2.00
36	Nat Inst of Natural Sciences	Japan	1	2.00
37	Nara Inst of Science & Technol	Japan	1	2.00
38	Konica Minolta Holdings Inc	Japan	1	1.00
39	Kazusa Dna Kenkyusho	Japan	1	0.00
40	Nagano Prefecture	Japan	1	1.00
41	Murano Shunji	Japan	1	2.00
42	Mitsui Chemicals Inc	Japan	1	2.00

Rank	Assignee Institution	Country	No. of Patents	Average Patent Age
43	Mitsubishi Rayon Co	Japan	1	2.00
44	Mitsubishi Heavy Ind Ltd	Japan	1	1.00
45	Matsushita Electric Ind Co Ltd	Japan	1	1.00
46	Lucent Technologies Inc	United States	1	2.00
47	Kyushu Inst of Technology	Japan	1	1.00
48	Kwangju Inst of Science & Tech	Rep. of Korea	1	1.00
49	Kawada Hiroaki	Japan	1	2.00
50	Kansai Tlo KK	Japan	1	1.00

Table C-2. Top 50 JPO patent inventors by number of patents published ("title-abstract" search, 2005-2007).

Rank	Last Name	First Name	Middle Name	No. of Patents	Rank	Last Name	Middle Name	First Name	No. of Patents
1	Bando	Yoshio	-	24	26	En	-	Kengun	2
2	Jintsui	Fuu	-	6	27	Furukawa	-	Toshiharu	2
3	Ando	Yoshinori	-	5	28	Futamata	-	Masayuki	2
4	Jung	Hee	T.	5	29	Haensler	-	Jean	2
5	Aida	Takuzo	-	4	30	Hakey	C.	Mark	2
6	Dmitri	Golberg	-	4	31	Han	-	Rogun	2
7	Daiku	Hiroyuki	-	4	32	Harada	-	Akio	2
8	Chengchun	Tang	-	4	33	Harada	-	Minoru	2
9	Cho	Shinraku	-	4	34	Hayashi	-	Hidenaru	2
10	Fukushima	Takanori	-	4	35	Horak	V.	David	2
11	Inoue	Sakae	-	4	36	Inatome	-	Takashi	2
12	Inoue	Tetsuya	-	4	37	Ishibe	-	Jirou	2
13	Sato	Tomoyuki	-	4	38	Ishikawa	-	Mitsuru	2
14	Endo	Morinobu	-	3	39	Jonathan	P.	Hill	2
15	Lee	Rim	S.	3	40	Jun	F.	Dee	2
16	Li	Yubao	-	3	41	Kamimura	-	Sashiro	2
17	Nakayama	Yoshikazu	-	3	42	Kanayama	-	Masaru	2
18	Ogawa	Atsuko	-	3	43	Kawanishi	-	Yuji	2
19	Otsu	Genichi	-	3	44	Kawazoe	-	Tadashi	2
20	Zhu	Ying-Chun	-	3	45	Kin	-	Jinka	2
21	Ajiri	Masafumi	-	2	46	Kin	-	Takematsu	2
22	Akita	Seiji	-	2	47	Kishimoto	-	Shigeru	2
23	Akasaka	Yoshihiro	-	2	48	Ko	K.	Yon	2
24	Arai	Susumu	-	2	49	Koburger,Iii	W	Charles	2
25	Chang	Hyuk	-	2	50	Kurachi	-	Hiroyuki	2

REFERENCES

Adams, J. and Griliches, Z. (1998). "Research Productivity in a System of Universities," *Annals of INSEE*, 49(50), 127-162.

Adderley, R. and Musgrove, P. B. (2001). "Data Mining Case Study: Modeling the Behavior of Offenders Who Commit Serious Sexual Assaults," in F. Provost and R. Srikant (Eds.), *Proceedings of the 7th ACM SIGKDD International Conference on Knowledge Discovery and Data Mining*, Association for Computing Machinery, New York, 215-220.

Albert, R. and Barabasi, A.-L. (2002). "Statistical Mechanics of Complex Networks," *Reviews of Modern Physics*, 74, 47-97.

Albert, R., Jeong, H. and Barabasi, A.-L. (1999). "Diameter of the World-Wide Web," *Nature*, 401, 130-131.

Amaral, L. A. N., Scala, A., Barthelemy, M. and Stanley, H. E. (2000). "Classes of Small-World Networks," *Proceedings of the National Academy of Sciences*, 97, 11149-11152.

Andrews, K. (1995). "Visualizing Cyberspace: Information Visualization in the Harmony Internet Browser," *Proceedings of InfoVis'95, IEEE Symposium on Information Visualization*, New York, 97-104.

Arora, A. and Gambardella, A. (1998). The Impact of NSF Support for Basic Research in Economics. Economics Working Paper Archive at WUSTL.

Aureka (2002). Aureka Online Services, Aureka Systems, available at: http://www.aurigin.com/aureka_online.html.

Bacchiocchi, E. and Montobbio, F. (2004). EPO vs. USPTO Citation Lags. Working Paper CESPRI, 161.

Balconi, M., Breschi, S., and Lissoni, F. (2004). "Networks of Inventors and the Role of Academia: An Exploration of Italian Patent Data," *Research Policy*, 33, 127-145.

Balconi, M. and Laboranti, A. (2004). The Multidimensionality of the Academic Performance in the Applied Sciences and Engineering: Evidence from a Case Study. University di Pavia.

Barabasi, A.-L. and Albert, R. (1999). "Emergence of Scaling in Random Networks," *Science*, 286(5439), 509-512.

Bilke, S. and Peterson, C. (2001). "Topological Properties of Citation and Metabolic Networks," *Physical Review E*, 64, 036106.

Börner, K., Chen, C. and Boyack, K. W. (2003). "Visualizing Knowledge Domains," *Annual Review of Information Science and Technology*, 37.

Braun, T., Schubert, A. and Zsindely, S. (1997). "Nanoscience and Nanotechnology on the Balance," *Scientometrics*, 38(2), 321-325.

Burt, R. S. (1976). "Positions in Networks," *Social Forces*, 55(1), 93-122.

Callon, M., Law, J., and Rip, A. (Eds.) (1986). *Mapping the Dynamics of Science and Technology: Sociology of Science in the Real World*, Macmillan, London.

Card, S. K., Moran, T. P., and Newell, A. (1983). *The Psychology of Human-Computer Interaction*, Lawrence Erlbaum Associates, Hillsdale, NJ.

Card, S. K., Robertson, G. G. and York, W. (1996). "The WebBook and the WebForager: An Information Workspace for the World Wide Web," *Proceedings of the ACM SIGCHI Conference on Human Factors in Computing Systems (CHI'96)*, ACM Press, New York, 111-117.

Card, S. K. and Mackinlay, J. D. (1997). "The Structure of the Information Visualization Design Space," *Proceedings of IEEE Symposium on Information Visualization 1997 (InfoVis'97)*, Phoenix, AZ, 92-99.

Card, S. K., Mackinlay, J. D. and Shneiderman, B. (1999). *Readings in Information Visualization: Using Vision to Think*, Morgan Kaufmann Publishers, Inc., San Francisco, CA.

Chakrabarti, A. K., Dror, I. and Eakabuse, N. (1993). "Interorganizational Transfer of Knowledge - an Analysis of Patent Citations of a Defense Firm," *IEEE Transactions on Engineering Management,* 40, 91-94.

Chen, C. (2003). *Mapping Scientific Frontiers,* Springer-Verlag, London, UK.

Chen, C. and Paul, R. J. (2001). "Visualizing a Knowledge Domain's Intellectual Structure," *IEEE Computer*, 34(3), 65-71.

Chen, H. (2001). *Knowledge Management Systems: A Text Mining Perspective*, University of Arizona, Tucson, AZ.

Chen, H. and Chau, M. (2004). "Web Mining: Machine Learning for Web Applications," *Annual Review of Information Science and Technology (ARIST),* 38, 289-329.

Chen, H., Houston, A. L., Sewell, R. R. and Schatz, B. R. (1998). "Internet Browsing and Searching: User Evaluation of Category Map and Concept Space Techniques," *Journal of the American Society for Information Science*, 49(7), 582-603.

Chen, H., Lally, A., Zhu, B. and Chau, M. (2003). "HelpfulMed: Intelligent Searching for Medical Information over the Internet," *Journal of the American Society for Information Science and Technology (JASIST)*, 54(7), 683-694.

Chen, H., Martinez, J., Ng, T. D. and Schatz, B. R. (1997). "A Concept Space Approach to Addressing the Vocabulary Problem in Scientific Information Retrieval: An Experiment on the Worm Community System," *Journal of the American Society for Information Science,* 48(1), 17-31.

Chen, H., Schatz, B. R., Ng, T. D., Martinez, J., Kirchhoff, A. J. and Lin, C. (1996). "A Parallel Computing Approach to Creating Engineering Concept Spaces for Semantic Retrieval: The Illinois Digital Library Initiative Project," *IEEE Transactions on Pattern Analysis and Machine Intelligence*, 18(8), 771-782.

Chen, H., Schuffels, C. and Orwig, R. (1996). "Internet Categorization and Search: A Self-Organizing Approach," *Journal of Visual Communication and Image Representation*, 7(1), 88-102.

Chernoff, H. (1973). "The Use of Faces to Represent Points in K-Dimensional Space Graphically," *Journal of the American Statistical Association*, 68(342), 361-368.

Chinchor, N. A. (1998). "Overview of MUC-7/MET-2," *Proceedings of the Seventh Message Understanding Conference (MUC-7).*

Cowan, R. and Jonard, N. (2004). "Network Structure and the Diffusion of Knowledge," *Journal of Economic Dynamics and Control,* 28, 1557-1575.

Cox, T. F. and Cox, M. A. (1994). *Multidimensional Scaling*, Chapman & Hall, London.

Crane, D. (1972). *Invisible Colleges: Diffusion of Knowledge in Scientific Communities*, University of Chicago Press, Chicago, IL.

Criscuolo, P. (2005). "The 'Home Advantage' Effect and Patent Families. A Comparison of OECD Triadic Patents, the USPTO and the EPO," *Scientometrics*, 66, 23-41.

Deerwester, S., Dumais, S., Furnas, G. W., Landauer, T. K. and Harshman, R. (1990). "Indexing by Latent Semantic Analysis," *Journal of the Society for Information Science,* 41(6), 391-407.

Donath, J. (2002). "Supporting Community and Building Social Capital: A Semantic Approach to Visualizing Online Conversations," *Communications of the ACM,* 45(4), 45-49.

Eades, P. (1984). "A Heuristic for Graph Drawing," *Congressus Numerantium*, 42, 149-209.

Eick, S. G., Steffen, J. L. and Sumner, E. E. (1992). "Seesoft – A Tool for Visualizing Line-Oriented Software," *IEEE Transactions on Software Engineering*, 18(11), 11-18.

European Commission (1997). Second European Report on S&T Indicators. Bruxelles.

Everitt, B. S. (1974). *Cluster Analysis*, Wiley, New York, NY.

Forsythe, G. E., Malcolm, M. A., and Moler, C. B. (1977). *Computer Methods for Mathematical Computations*, Prentice-Hall, Englewood Cliffs, NJ.

Furnas, G. W. (1986). "Generalized Fisheye Views, Human Factors in Computing Systems," *Proceedings of CHI'86, ACM Conference on Human Factors in Computing Systems*, New York, 16-23.

Ganguli, P. (1998). "Intellectual Property Rights in Transition," *World Patent Information*, 20, 171-180.

Gansner, E. and North, S. (2000). "An Open Graph Visualization System and Its Applications to Software Engineering," *Software - Practice and Experience*, 30(11), 1203-1233.

Garfield, E. (1964). "Science Citation Index-New Dimension in Indexing - Unique Approach Underlies Versatile Bibliographic Systems for Communicating and Evaluating Information," *Science*, 144(361), 649-654.

Garfield, E. (1972). "Citation Analysis as a Tool in Journal Evaluation - Journals Can Be Ranked by Frequency and Impact of Citations for Science Policy Studies," *Science*, 178(4060), 471-479.

Garfield, E., Revesz, G. S. and Batzig, J. H. (1973). "Synthetic Chemical Literature from 1960-to-1969," *Nature*, 242(5396), 307-309.

Garfield, E. (1978a). "The 100 Articles Most Cited By Social Sciences," *Essays of an Information Scientist*, 3, 563-572.

Garfield, E. (1978b). "The 300 Most-cited Authors, 1961-1976," *Essays of an Information Scientist*, 3, 538-550.

Garfield, E. (1979). *Citation Indexing: Its Theory and Application in Science, Technology and Humanities*, John Wiley, New York.

Garfield, E. (1995). "Citation Indexes for Science: A New Dimension in Documentation through Association of Ideas," *Science*, 122, 108-111.

Grotzinger, S. J. and Witzgall, C. (1984). "Projections onto Order Simplexes," *Applied Mathematics and Optimization*, 12(1), 247-270.

Hearst, M. (1995) "TileBars: Visualization of Term Distribution Information in Full Text Information Access," *Proceedings of the ACM SIGCHI Conference on Human Factors in Computing Systems (CHI)*, Denver, CO, 59-66.

Hearst, M. A. and Karadi, C. (1997). "Cat-a-Cone: An Interactive Interface for Specifying Searches and Viewing Retrieval Results Using a Large Category Hierarchy," *The Proceedings of SIGIR'97*, 246-255.

Hitchcock, S., Brody, D. B. T., Gutteridge, C., Carr, L., Hall, W., Lagoze, C. and Harnad, S. (2002). "Open Citation Linking: The Way Forward," *D-Lib Magazine.*

Hu, D., Chen, H., Huang, Z. and Roco, M. C. (2007). "Longitudinal Study on Patent Citations to Academic Research Articles in Nanotechnology," *Journal of Nanoparticle Research,* 9, 529-542.

Huang, I.-A., Hung, J. M. H., Kao, H. and Lin, S. (2004). "Extracting Citation Metadata from Online Publication Lists Using BLAST," *Pacific-Asia Conference on Knowledge Discovery and Data Mining (PAKDD).*

Huang, M. H., Chiang, L. Y. and Chen, D. Z. (2003a). "Constructing a Patent Citation Map using Bibliographic Coupling: A Study of Taiwan's High-tech Companies," *Scientometrics,* 58, 489-506.

Huang, Z., Chen, H., Yip, A., Ng, D., Guo, F., Chen, Z.-K. and Roco, M. C. (2003). "Longitudinal Patent Analysis for Nanoscale Science and Engineering: Country, Institution and Technology Field," *Journal of Nanoparticle Research,* 5, 333-363.

Huang, Z., Chen, H., Chen, Z. and Roco, M. C. (2004). "International Nanotechnology Development in 2003: Country, Institution and Technology Field Analysis Based on USPTO Database," *Journal of Nanoparticle Research,* 6(4), 325-354.

Huang, Z., Chen, H., Yan, L, and Roco, M. C. (2005). "Longitudinal Nanotechnology Development (1991-2002): National Science Foundation Funding and Its Impact on Patents," *Journal of Nanoparticle Research*, 7(4-5), 343-376.

Huang, Z., Chen, H., Li, X. and Roco, M. C. (2006). "Connecting NSF Funding to Patent Innovation in Nanotechnology (2001-2004)," *Journal of Nanoparticle Research,* 8(6), 859-879.

Hullmann, A. (2006). "Who is Winning the Global Nanorace?" *Nature Nanotechnology*, 1(2), 81-83.

Hullmann, A. and Meyer, M. (2003). "Publications and Patents in Nanotechnology - An Overview of Previous Studies and the State of the Art," *Scientometrics*, 58(3), 507-527.

Jaffe, A. and Trajtenberg, M. (2002). *Patents, Citations & Innovations*, The MIT Press, Cambridge, MA.

Jain, A. K. and Dubes, R. C. (1988). *Algorithms for Clustering Data*, Prentice-Hall, Englewood Cliffs, NJ.

Jeong, H., Neda, Z. and Barabasi, A.-L. (2003). "Measuring Preferential Attachment for Evolving Networks," *Europhysics Letters*, 61, 567-572.

Jeong, H., Tombor, B., Albert, R., Oltvai, Z. N. and Barabasi, A.-L. (2000). "The Large-Scale Organization of Metabolic Networks," *Nature*, 407, 651-654.

Jewell, M. (2004). *ParaTools v1.10.*

Johnson B. and Shneiderman, B. (1991). "Tree-Maps: A Space-Filling Approach to the Visualization of Hierarchical Information Structures," *Proceedings of IEEE Visualization'91 Conference*, San Diego, 284-291.

Jolliffe, I. T. (2002). *Principal Component Analysis*, Springer.

Kamada, T. and Kawai, S. (1989). "An Algorithm for Drawing General Undirected Graphs," *Information Processing Letters*, 31, 7-15.

Karki, M. M. (1997). "Patent Citation Analysis: A Policy Analysis Tool," *World Patent Information,* 19, 269-272.
Kleinberg, J. (1998). "Authoritative Sources in a Hyperlinked Environment," *Proceedings of ACM-SIAM Symposium on Discrete Algorithms*, San Francisco, CA, 668-677.
Klerks, P. (2001). "The Network Paradigm Applied to Criminal Organizations: Theoretical Nitpicking or a Relevant Doctrine for Investigators? Recent Developments in the Netherlands," *Connections*, 24(3), 53-65.
Kochen, M. (1989). *The Small World*, Ablex, Norwood, NJ.
Kohonen, T. (1989). *Self-Organization and Association Memory*, Springer, New York.
Kohonen, T. (1995). *Self-Organizing Maps*, Springer-Verlag, Berlin, Heidelberg.
Kohonen, T., Kaski, S., Lagus, K., Salojarvi, J., Honkela, J., Paatero, V. and Saarela, A. (2000). "Self Organization of a Massive Document Collection," *IEEE Transactions on Neural Networks*, 11(3), 574-585.
Kostoff, R. N., Stump, J. A., Johnson, D., Murday, J. S., Lau, C. G. and Tolles, W. M. (2006a). "The Structure and Infrastructure of the Global Nanotechnology Literature," *Journal of Nanoparticle Research*, 8(3-4), 301-321.
Kostoff, R. N., Murday, J. S., Lau, C. G. and Tolles, W. M. (2006b). "The Seminal Literature of Nanotechnology Research," *Journal of Nanoparticle Research,* 8(2), 193-213.
Kostoff, R. N., Koytcheff, R. G. and Lau, C. G. (2007). "Technical Structure of the Global Nanoscience and Nanotechnology Literature," *Journal of Nanoparticle Research,* 9(5), 701-724.
Kowalski, T. J., Maschio, A. and Megerditchian, S. H. (2003). "Dominating Global Intellectual Property: Overview of Patentability in the USA, Europe and Japan," *Journal of Commercial Biotechnology*, 9, 305-331.
Kruskal, J. B. (1964). "Nonmetric Multidimensional Scaling: A Numerical Method," *Psychometrika*, 29(2), 115-129.
Kumar, R., Raghavan, P., Rajagopalan, S., Sivakumar, D., Tomkins, A. and Upfal, E. (2000). "The Web as a Graph," *Proc. 19th ACM SIGACT-SIGMOD-AIGART Symposium of Principles of Database Systems*, 1-10.
Lamping, J., Rao, R. and Pirolli, P. (1995). "A Focus + Context Technique Based on Hyperbolic Geometry for Visualizing Large Hierarchies," *Proceedings of the ACM SIGCHI Conference on Human Factors in Computing Systems (CHI)*, Denver, CO, 401-408.
Lawrence, S., Giles, C. L. and Bollacker, K. (1999). "Digital Libraries and Autonomous Citation Indexing," *IEEE Computer*, 32(6), 67-71.
Lewison, G. (1998). "Gastroenterology Research in the United Kingdom: Funding Sources and Impact," *GUT*, 43, 288-293.
Li, X., Chen, H., Huang, Z. and Roco, M. C. (2007a). "Patent Citation Network in Nanotechnology (1976-2004)," *Journal of Nanoparticle Research,* 9(3), 337-352.
Li, X., Lin, Y., Chen, H. and Roco, M. C. (2007b). "Worldwide Nanotechnology Development: A Comparative Study of USPTO, EPO, and JPO Patents (1976-2004)," *Journal of Nanoparticle Research*, 9(6), 977-1002.
Lin, C., Chen, H. and Nunamaker, J. F. (2000). "Verifying the Proximity Hypothesis for Self-Organizing Maps," *Journal of Management Information Systems*, 16(3), 57-70.

Lorrain, F. P. and White, H. C. (1971). "Structural Equivalence of Individuals in Social Networks," *Journal of Mathematical Sociology,* 1, 49-80.

Lukach, R. and Plasmans, J. (2001). "A Study of Knowledge Spill-overs from the Compatible EPO and USPTO Patent Datasets for Belgian Companies," in Clarysse, M.C.a.B. (Ed),.*Belgian Report on Science, Technology and Innovation 2001 - Volume II: The Belgian Innovation System: Lessons and Challenges*, Federal Office for Scientific, Technical and Cultural Affairs, 241-267.

Lux Research, Inc. (2005). *The Nanotech Report 2004: Investment Overview and Market Research for Nanotechnology* (Third Edition), Lux Research, Inc., New York.

Lux Research, Inc. (2007). *Personal Communication.*

Mackinlay, J. D., Rao, R. and Card, S. K. (1995). "An Organic User Interface for Searching Citation Links," *Proceedings of CHI'95, ACM Conference on Human Factors in Computing Systems*, New York, 67-73.

Mane, K. K. and Börner K. (2003). "Mapping Topics and Topic Bursts in PNAS," *Proceedings of the National Academy of Sciences*, 101.

McAndrew, D. (1999). "The Structural Analysis of Criminal Networks," in D. Canter and L. Alison (Eds.), *The Social Psychology of Crime: Groups, Teams, and Networks, Offender Profiling Series,* III, Aldershot, Dartmouth.

McQuaid, M. J., Ong, T., Chen, H. and Nunamaker, J. F. (1999). "Multidimensional Scaling for Group Memory Visualization," *Decision Support Systems*, 27(1), 163-176.

Meyer, M. and Persson, O. (1998). "Nanotechnology - Interdisciplinarity, Patterns of Collaboration and Differences in Application," *Scientometrics*, 42(2), 195-205.

Meyer, M. (2000). "Patent Citations in a Novel Field of Technology - What Can They Tell about Interactions between Emerging Communities of Science and Technology?" *Scientometrics,* 48, 151-178.

Meyer, M. (2001). "Patent Citation Analysis in a Novel Field of Technology: An Exploration of Nano-science and Nano-technology," *Scientometrics,* 51, 163-183.

Morone, P. and Taylor, R. (2004). "Knowledge Diffusion Dynamics and Network Properties of Face-to-Face Interactions," *Journal of Evolutionary Economics,* 14, 327-351.

Munzner, T. (2000). Interactive Visualization of Large Graphs and Networks. Ph.D. Dissertation, Stanford University, June, 2000.

Murtagh, F. (1995). *Multidimensional Clustering Algorithms*, Physica-Verlag, Vienna, AU.

Narin, F. (1994). "Patent Bibliometrics," *Scientometrics*, 30(1), 147-155.

Narin, F. (1998). "Patents and Publicly Funded Research," in *Assessing the Value of Research in the Chemical Sciences*, National Academy Press, 59-72.

Narin, F. (2000). "Tech-Line Background Paper," CHI Research, Inc., available at: http://www.chiresearch.com/techline/tlbp.pdf.

National Science and Technology Council (2004). The National Nanotechnology Initiative Strategic Plan.

Newman, M. E. J. (2001a). "The Structure of Scientific Collaboration Networks," *Proceedings of the National Academy of Sciences,* 98, 404-409.

Newman, M. E. J. (2001b). "Clustering and Preferential Attachment in Growing Networks," *Physical Review E*, 64(2).

Newman, M. E. J. (2003). "Coauthorship Networks and Patterns of Scientific Collaboration," *Proceedings of the National Academy of Sciences,* 101, 5200-5205.

NSET (2004). NNI Strategic Planning, Washington, D.C.

NSF (1997). Partnership in Nanotechnology: Synthesis, Processing, and Utilization of Functional Nanostructures, NSF 98-20, Arlington, Virginia (Program solicitation).

Ong, T.-H., Chen, H., Sung, W.-K. and Zhu, B. (2005). "Newsmap: A Knowledge Map for Online News," *Decision Support Systems*, 39(4), 583-598.

Oppenheim, C. (2000). "Do Patent Citations Count?" in B. Cromin and H. B. Atkins (Eds.), *The Web of Knowledge*, Information Today, Inc., Medford.

Orwig, R., Chen, H. and Nunamaker, J. F. (1997). "A Graphical, Self-Organizing Approach to Classifying Electronic Meeting Output," *Journal of the American Society for Information Scienc*e, 4(2), 157-170.

Pakes, A. and Sokoloff, K. L. (1995). "Science, Technology, and Economic Growth," *Proceedings of the National Academy of Sciences*, 93, 12655-12657.

Payne, A. A. and Siow, A. (2003). "Does Federal Research Funding Increase University Research Output?" *Advances in Economic Analysis and Policy*, 3(1), Article 1.

Porter, A. L., Youtie, J., Shapira, P., Schoeneck, D. J., Tang, L. and Mehta, P. (2007). "Profiling Nano R&D," *Nano-Giga Challenges*, Phoenix, AZ.

President's Council of Advisors on Science and Technology (2005). The National Nanotechnology Initiative at Five Years: Assessment and Recommendations of the National Nanotechnology Advisory Panel, U.S. Government Printing Office.

Quillen, C. D., Webster, O. H. and Eichmann, R. (2002). "Continuing Patent Applications and Performance of the U.S. Patent and Trademark Office - Extended," *The Federal Circuit Bar Journal*, 12, 35-55.

Redner, S. (1998). "How Popular is Your Paper? An Empirical Study of the Citation Distribution," *European Physical Journal B*, 4, 131-134.

Robertson, G. G., Mackinlay, J. D. and Card, S. K. (1991). "Cone Trees: Animated 3D Visualizations of Hierarchical Information," *Proceedings of ACM SIGCHI'91,* 189-194.

Robertson, G. G., Card, S. K. and Mackinlay, J. D. (1993). "Information Visualization Using 3D Interactive Animation," *Communications of the ACM,* 36(4), 56-71.

Roco, M. C. (2001). "International Strategy for Nanotechnology Research and Development," *Journal of Nanoparticle Research*, 3(5-6), 353-360.

Roco, M. C. and Bainbridge, W. S. (2001). *Societal Implication of Nanoscience and Nanotechnology*, Kluwer Academic.

Roco, M. C. (2003). "Broader Societal Issues of Nanotechnology," *Journal of Nanoparticle Research*, 5(3-4), 181-189.

Roco, M. C. (2005a). "International Perspective on Government Nanotechnology Funding in 2005," *Journal of Nanoparticle Research,* 7, 707-712.

Roco, M. C. (2005b). "Environmentally Responsible Development of Nanotechnology: How the U.S. Government is Dealing with the Intermediate and Long-term Issues of this New Technology," *Environmental Science and Technology,* ACS, 106-112.

Roco, M. C. (2007). "National Nanotechnology Initiative - Past, Present, Future,"in *Handbook on Nanoscience, Engineering and Technology, 2nd ed.,* Taylor and Francis, 3.1-3.26.

Roco, M. C. (2008). "Possibilities for Global Governance of Converging Technologies," *Journal of Nanoparticle Research*, 10(1), 11-29 (online at Springer: DOI 10.1007/s11051-007-9269-8).

Roco, M. C., Williams, R. S. and Alivisatos, P. (Eds.) (2000). *Nanotechnology Research Directions*, Kluwer Academic Publishers, Boston.

Roussinov, D. G. and Chen, H. (1999). "Document Clustering for Electronic Meetings: An Experimental Comparison of Two Techniques," *Decision Support Systems*, 27(1), 67-79.

Salton, G. (1989). *Automatic Text Processing*, Addison-Wesley Publishing Company, Reading, MA.

Salton, G., Allan, J., Buckley, C. and Singhal, A. (1995). "Automatic Analysis, Theme Generation, and Summarization of Machine-readable Text," *Science*, 264(3), 1421-1426.

Schmoch, U. (1993). "Tracing the Knowledge Transfer from Science to Technology as Reflected in Patent Indicators," *Scientometrics*, 26, 193-211.

Schummer, J. (2004). "Multidisciplinarity, Interdisciplinarity, and Patterns of Research Collaboration in Nanoscience and Nanotechnology," *Scientometrics*, 59(3), 425-465.

Scientific American (2002). *Understanding Nanotechnology*, Time Warner Book Group, New York, NY.

Shiffrin, R. M. and Börner, K. (2004). *Mapping Knowledge Domains*, Arthur M. Sackler Colloguia of the National Academy of Sciences, National Academies of Sciences.

Shneiderman, B. (1994). "Dynamic Queries for Visual Information Seeking," *IEEE Software*, 11(6), 70-77.

Shneiderman, B. (1996). "The Eyes Have It: A Task by Data Type Taxonomy for Information Visualization," *Proceedings of IEEE Workshop on Visual Languages'96*, 336-343.

Siegel, R. W., Hu, E. and Roco, M. C. (Eds.) (1999). "Nanostructure Science and Technology," National Science and Technology Council (NSTC), White House (WH), Washington, D.C., 1999; also Kluwer Academic Publishers, Boston and Dordrecht, 1999.

Singh, J. (2003). "Social Networks as Drivers of Knowledge Diffusion," available at: SSRN DOI: 10.2139/ssrn.431872.

Small, H. (1999). "Visualizing Science by Citation Mapping," *Journal of the American Society for Information Science*, 50(9), 799-812.

Sparrow, M. K. (1991). "The Application of Network Analysis to Criminal Intelligence: An Assessment of the Prospects," *Social Networks,* 13, 251-274.

Tolle, K. M. and Chen, H. (2000). "Comparing Noun Phrasing Techniques for Use with Medical Digital Library Tools," *Journal of the American Society for Information Science*, 51(4), 352-370.

Verbeek, A., Debackere, K. and Luwel, M. (2003). "Science Cited in Patents: A Geographic "Flow" Analysis of Bibliographic Citation Patterns in Patents," *Scientometrics,* 58, 241-263.

Voutilainen, A. (1997). "A Short Introduction to Nptool," available at: http://www.lingsoft.fi/doc/nttool/intro/.

Ward, J. H. (1963). "Hierarchical Grouping to Optimize an Objective Function," *Journal of the American Statistical Association*, 58(301), 236-244.

Wasserman, S. and Faust, K. (1994). *Social Network Analysis: Methods and Applications*, Cambridge University Press, Cambridge.

Watts, D. (1999). *Small Worlds: The Dynamics of Networks between Order and Randomness*, Princeton University, Princeton, NJ.

Watts, D. and Strogatz, S. H. (1998). "Collective Dynamics of 'Small-World' Networks," *Nature*, 393, 440-442.

White, H. D., Lin, X., Buzydlowski, J. W. and Chen, C. (2003). "User-controlled Mapping of Significant Literatures," *Proceedings of the National Academy of Sciences*, 101, 5297-5302.

Willett, P. (1988). "Recent Trends in Hierarchical Document Clustering: A Critical Review," *Information Processing and Management*, 24(5), 577-597.

Wise, J. A., Thomas, J. J., Pennock, K., Lantrip, D., Pottier, M., Schur, A. and Crow, V. (1995). "Visualizing the Non-visual: Spatial Analysis and Interaction with Information from Text Documents," *Proceedings of InfoVis'95, IEEE Symposium on Information Visualization*, New York, 51-58.

Wood, J. L. (1966). "The Parameters of Document Acquisition at Chemical Abstracts Service," in *Eighth Annual Institute of Information Storage and Retrieval*, Washington, D.C.

Xu, J. and Chen, H. (2005a). "CrimeNet Explorer: A Framework for Criminal Network Knowledge Discovery," *ACM Transactions on Information Systems*, 23(2), 201-226.

Xu, J. and Chen, H. (2005b). "Criminal Network Analysis and Visualization," *Communications of the ACM*, 48(6), 101-107.

Zhu, B. and Chen, H. (2001). "Social Visualization for Computer-Mediated Communication: A Knowledge Management Perspective," *Proceedings of Workshop on Information Technology and Systems (WITS 2001) Conference*, New Orleans, LA.

Zhu, B. and Chen, H. (2005). "Information Visualization," *Annual Review of Information Science and Technology (ARIST),* 39, 139-178.

Zucker, L. G. and Darby, M. R. (2005). Socio-economic Impact of Nanoscale Science: Initial Results and Nanobank. National Bureau of Economic Research, NBER Working Paper No. 11181.

SUBJECT INDEX

Printed in the United States of America